248

Topics in Current Chemistry

Topics in Current Chemistry

Recently Published and Forthcoming Volumes

Anion Sensing
Volume Editor: Stibor, I.
Vol. 255, 2005

Organic Solid State Reactions
Volume Editor: Toda, F.
Vol. 254, 2005

DNA Binders and Related Subjects
Volume Editors: Waring, M.J., Chaires, J.B.
Vol. 253, 2005

Contrast Agents III
Volume Editor: Werner Krause
Vol. 252, 2005

Chalcogenocarboxylic Acid Derivatives
Volume Editor: Kato, S.
Vol. 251, 2005

New Aspects in Phosphorus Chemistry V
Volume Editor: Majoral, J.-P.
Vol. 250, 2005

Templates in Chemistry II
Volume Editors: Schalley, C.A., Vögtle, F., Dötz, K.H.
Vol. 249, 2005

Templates in Chemistry I
Volume Editors: Schalley, C.A., Vögtle, F., Dötz, K.H.
Vol. 248, 2005

Collagen
Volume Editors: Brinckmann, J., Notbohm, H., Müller, P.K.
Vol. 247, 2005

New Techniques in Solid-State NMR
Volume Editor: Klinowski, J.
Vol. 246, 2005

Functional Molecular Nanostructures
Volume Editor: Schlüter, A.D.
Vol. 245, 2005

Natural Product Synthesis II
Volume Editor: Mulzer, J.
Vol. 244, 2005

Natural Product Synthesis I
Volume Editor: Mulzer, J.
Vol. 243, 2005

Immobilized Catalysts
Volume Editor: Kirschning, A.
Vol. 242, 2004

Transition Metal and Rare Earth Compounds III
Volume Editor: Yersin, H.
Vol. 241, 2004

The Chemistry of Pheromones and Other Semiochemicals II
Volume Editor: Schulz, S.
Vol. 240, 2005

The Chemistry of Pheromones and Other Semiochemicals I
Volume Editor: Schulz, S.
Vol. 239, 2004

Orotidine Monophosphate Decarboxylase
Volume Editors: Lee, J.K., Tantillo, D.J.
Vol. 238, 2004

Long-Range Charge Transfer in DNA II
Volume Editor: Schuster, G.B.
Vol. 237, 2004

Long-Range Charge Transfer in DNA I
Volume Editor: Schuster, G.B.
Vol. 236, 2004

Spin Crossover in Transition Metal Compounds III
Volume Editors: Gütlich, P., Goodwin, H.A.
Vol. 235, 2004

Spin Crossover in Transition Metal Compounds II
Volume Editors: Gütlich, P., Goodwin, H.A.
Vol. 234, 2004

Spin Crossover in Transition Metal Compounds I
Volume Editors: Gütlich, P., Goodwin, H.A.
Vol. 233, 2004

Templates in Chemistry I

Volume Editors:
Christoph A. Schalley · Fritz Vögtle · Karl Heinz Dötz

With contributions by
M. Albrecht · J. Brüggemann · F. Diederich · K. H. Dötz · T. Friščić ·
T. D. Hamilton · H. C. Jahr · J. H. Jung · L. R. MacGillivray ·
G. S. Papaefstathiou · C. A. Schalley · S. Sergeyev · S. Shinkai ·
C. Thilgen · D. B. Varshney · F. Vögtle · T. Weilandt · B. Wenzel

Springer

The series *Topics in Current Chemistry* presents critical reviews of the present and future trends in modern chemical research. The scope of coverage includes all areas of chemical science including the interfaces with related disciplines such as biology, medicine and materials science. The goal of each thematic volume is to give the nonspecialist reader, whether at the university or in industry, a comprehensive overview of an area where new insights are emerging that are of interest to a larger scientific audience.

As a rule, contributions are specially commissioned. The editors and publishers will, however, always be pleased to receive suggestions and supplementary information. Papers are accepted for *Topics in Current Chemistry* in English.

In references *Topics in Current Chemistry* is abbreviated Top Curr Chem and is cited as a journal.

Visit the TCC content www.springerlink.com

Library of Congress Control Number: 2004108949

ISSN 0340-1022
ISBN 978-3-642-06142-4 ISBN 978-3-540-44505-0 (eBook)
DOI 10.1007/978-3-540-44505-0

springeronline.com

Originally published by Springer-Verlag Berlin Heidelberg New York in 2004
Softcover reprint of the hardcover 1st edition 2004

Cover design: KünkelLopka, Heidelberg/design & production GmbH, Heidelberg
Typesetting: Fotosatz-Service Köhler GmbH, Würzburg

Printed on acid-free paper 02/3020/xv – 5 4 3 2 1 0

Volume Editors

Priv.-Doz. Dr. Christoph A. Schalley
c.schalley@uni-bonn.de

Prof. Dr. Fritz Vögtle
voegtle@uni-bonn.de

Prof. Dr. Karl H. Dötz
doetz@uni-bonn.de

Kekulé-Institut für Organische Chemie und Biochemie
Gerhard-Domagk-Str. 1
53121 Bonn, Germany

Editorial Board

Topics in Current Chemistry
Also Available Electronically

For all customers who have a standing order to Topics in Current Chemistry, we offer the electronic version via SpringerLink free of charge. Please contact your librarian who can receive a password for free access to the full articles by registration at:

springerlink.com

If you do not have a subscription, you can still view the tables of contents of the volumes and the abstract of each article by going to the SpringerLink Homepage, clicking on "Browse by Online Libraries", then "Chemical Sciences", and finally choose Topics in Current Chemistry.

You will find information about the

- Editorial Board
- Aims and Scope
- Instructions for Authors
- Sample Contribution

at springeronline.com using the search function.

Preface

It is at the same time intriguing from an intellectual point of view and most fruitful with respect to the development of new strategies and applications to transfer concepts from a macroscopic, daily life world to the nanoscopic realm of chemistry. One example is templated synthesis, a term which has been coined by *Daryle H. Busch* in his seminal paper on the synthesis of tetradentate ligands formed around a metal ion through self-condensation of aminobenzaldehyde (*J. Am. Chem. Soc.* 1964, *86*, 4834). Since then, the template idea has spread over almost all areas of chemistry ranging from DNA replication to the synthesis of interlocked molecules and to the fabrication of inorganic materials on the basis of organic templates. If a term begins to become "fashionable", it often looses some of its conceptual power due to exaggerations and misuse. One might argue that this is also true for "template". Nevertheless, the rapidly increasing number of publications on template effects and templation appearing in any synthesis-relevant journal indicates how important the template strategy in all its facets has become.

The present monograph intends to shed light on a selection of aspects of "template chemistry" by combining chapters from areas as different as templated solid state synthesis, metal-mediated self-assembly processes, organometallic synthesis, the formation of mechanically interlocked molecules, and, last but not least, the production of inorganic materials based on organic templates such as gels. Each chapter has its own scientific focus, although some overlap may exist. We do not consider this a disadvantage, because it provides views from different angles on the same topics. It is obvious that this volume in the Topics in Current Chemistry series cannot be comprehensive at all and follow-up volumes on other aspects of templates in chemistry may be necessary – and are planned – to at least provide a rough overview covering the most important aspects of template chemistry. Based on the rapid development of this area of research, other problems become more and more urgent. One of them is a comprehensive and clear definition of what we should call a template. In view of the many different aspects described by this term, it becomes increasingly difficult to express in a simple, but straightforward way what the concept of templation means, what it does not mean, and how it is related to catalysis. We do not intend to provide a final answer here, but rather like to initiate a discussion in the scientific community about this issue. Such a discussion may sharpen the language used and, thus, may significantly help to clarify the concepts behind.

Finally, most templates developed in the past have been found by serendipity and empiricism. It would be highly desirable, if this volume contributes a bit to make templates an issue of design and careful planning and in such a way can be implemented into novel synthetic strategies of the future.

Bonn, December 2004

Christoph A. Schalley
Fritz Vögtle
Karl Heinz Dötz

Contents

Contents of Volume 249
Templates in Chemistry II

Volume Editors: Christoph A. Schalley · Fritz Vögtle · Karl Heinz Dötz
ISBN 3-540-23087-4

Topics in Current Chemistry (2004) 248: 1–61
DOI 10.1007/978-3-540-44505-0

Spacer-Controlled Multiple Functionalization of Fullerenes

Carlo Thilgen (✉) · Sergey Sergeyev · Francois Diederich

Laboratorium für Organische Chemie, ETH Zürich, Wolfgang-Pauli-Strasse 10, 8093 Zürich, Switzerland
thilgen@org.chem.ethz.ch, diederich@org.chem.ethz.ch

Dedicated to Professor Fritz Vögtle at the occasion of his 65th birthday

Abstract The chapter provides a survey of the development and applications of the tether- and template-directed regio- and, in the occurrence, stereoselective multifunctionalization of fullerenes over the past ten years. After a presentation of the first tether-directed remote functionalization of C_{60}, a broad spectrum of applications is reviewed according to the involved reaction types. The most frequently used chemistry consists of additions of tethered 2-halomalonates (double Bingel reactions) and 1,3-dienes (double Diels-Alder reactions). The former, in particular, were used for the only known tether-directed functionalization of a higher fullerene (C_{70}) and also for most of the rare examples of three- and fourfold one-pot tether-directed derivatizations of C_{60}. Other, commonly used reactions are [3+2] cycloadditions, notably of azides and, in a few cases, of vinylcarbenes, and azomethine ylides. Some interesting examples of intramolecular [2+2] cycloadditions between fullerene moieties are also included as they are in fact spacer-controlled dimerizations of the carbon spheres. Throughout the account, particular emphasis is put on the diastereoselective generation of chiral fullerene functionalization patterns by use of enantiomerically pure tethers. As compared to tether-directed multiple additions to fullerenes, regioselective functionalizations with non-covalent templates have remained rather rare. Two important examples are reported, one involving a reversible reaction with 9,10-dimethylanthracene in solution, the other one a topochemical anthracene transfer.

Keywords Fullerenes · Multiple functionalization · Tether · Spacer · Template · Regioselectivity · Stereoselectivity

List of Abbreviations

ap	Antiperiplanar
BET	Back electron transfer
CD	Circular dichroism
CIP	Cahn-Ingold-Prelog
CTV	Cyclotriveratrylene
CV	Cyclovoltammetry
DABCO	1,4-Diazabicyclo[2.2.2]octane
DBU	1,8-Diazabicyclo[5.4.0]undec-7-ene
DCC	*N,N'*-Dicyclohexylcarbodiimide
DIBAL-H	Diisobutylaluminium hydride
DMA	9,10-Dimethylanthracene
DMAP	4-(Dimethylamino)pyridine
DPV	Differential pulse voltammetry
ESI-MS	Electrospray ionization mass spectrometry
HOBT	1-Hydroxy-1*H*-benzotriazole

HOMO	Highest occupied molecular orbital
HSVM	High-speed vibration milling
IUPAC	International Union of Pure and Applied Chemistry
LUMO	Lowest unoccupied molecular orbital
MALDI-TOF	Matrix-assisted laser-desorption ionization time-of-flight
NMR	Nuclear magnetic resonance
ONIOM	Our own *n*-layered integrated molecular orbital and molecular mechanics method
PET	Photoelectron transfer
PM3	Parametrization method 3
SCF-CI-DV MO	Self-consistent-field configuration-interaction dipole-velocity molecular orbital
SET	Single electron transfer
UV/vis	Ultraviolet/visible

1 Introduction

As the chemistry of fullerenes was explored [1–4], it soon became clear that regio- and, in the event, stereoselective multiple functionalization of the spheroidal carbon polyenes was a key issue to be addressed in order to make pure multiadducts with well-defined addition patterns available on a reasonable scale and without having to resort to tedious purification protocols. Also, a considerable number of interesting addition patterns is not available by simple, consecutive additions due to the intrinsic reactivity of fullerene derivatives [5]. The availability of specific multiadduct regio- and stereoisomers is, however, a conditio sine qua non for the full exploitation of the unique three-dimensional fullerene scaffolds and the associated π-chromophores, for example, in advanced materials chemistry [6–11]. In the search for a rational approach to the regioselective formation of multiadducts of C_{60} and possibly other fullerenes, Diederich and co-workers successfully tried the tether-directed remote functionalization in 1994 [12]. This method had been introduced by Breslow and co-workers to control chemical selectivity in a biomimetic way and it allowed them to carry out reactions at specific positions in steroids and in long alkyl chains [13]. The first step of this approach consists in the attachment to the molecular backbone of an anchor carrying a reactive group at the end of a tether. In the ideal case, conformational preferences of the spacer and steric constraints allow the reactive group to reach and attack only a specific position within the molecular conjugate. Thanks to the remarkable directing effects of many reported tethers in relatively unselective reactions or to their effective competition against the intrinsic reactivity of fullerene derivatives, the spacer-controlled remote functionalization has become the method of choice for selective multiple additions to fullerenes [14–17]. If a tether is solely used as an auxiliary to attain the desired selectivity in a synthesis, its persistence in the final product may be unwanted and it should be removable, thus acting as a template [18]. Template-type easy-to-remove tethers have admittedly remained scarce in fullerene chemistry, and further efforts need to be devoted to their development. On the other hand, some non-spacer-based template-directed syntheses have been used with great success in the generation of certain multiaddition patterns of buckminsterfullerene. In combination with other resources such as the exploitation of the intrinsic reactivity of fullerene adducts in further sequential functionalizations or the complete removal of certain addends at a given stage of a synthesis, the templated and spacer-controlled multiple functionalizations have allowed access to a large number of well-defined, multiply functionalized fullerene derivatives over the past decade [14–17].

2
Bisadducts of C_{60}: Addition Patterns and Chirality

Multiple addition to the fullerene sphere can result in the formation of numerous isomeric products. Thus, double, triple, and fourfold addition of symmetrical and identical addends to C_{60} can afford, in theory, 8, 46, and 262 regioisomers, respectively. The IUPAC has recently presented recommendations for systematic names of fullerenes and fullerene derivatives [19]. However, trivial descriptors, first introduced by Hirsch and co-workers [20], have been widely used for the discussion of the structure of C_{60}-adducts, since they provide a simple description of regioisomeric bis- and, to some extent, of trisadducts, comparable to *ortho*, *meta*, and *para* in benzene chemistry. The following consideration will be limited to products of double addition of C_{2v}-symmetric addends across 6-6 bonds (common edge of two six-membered rings) of C_{60}. According to the system of Hirsch and co-workers, the C_{60} sphere is divided into three sections with regard to the position of the second addend in reference to the first: the former can lie in the same hemisphere (*cis*), at the borderline between hemispheres (*equatorial* or *e*), or in the opposite hemisphere (*trans*). Within the same hemisphere, there are three sets of four double bonds each (*cis*-1, *cis*-2, and *cis*-3), and there are four different bond types in the opposite hemisphere (*trans*-1 (one bond), *trans*-2, *trans*-3, and *trans*-4 (three sets of four bonds each)) (Fig. 1). If both C_{2v}-symmetrical addends are identical, the resulting eight possible relative arrangements would give eight possible regioisomeric bisadducts. In case of two different C_{2v}-symmetrical addends, two different constitutional isomers with *equatorial* addition pattern can arise. Looking from one of the functionalized *e* type bonds to the other, one can see either the edge or the face of a cycle fused to the distal *e* bond. Accordingly, the position of the viewer is described as e_{edge} and e_{face}, respectively, with regard to the distal *e* bond. The total number of possible regioisomers increases, therefore, to nine.

An interesting aspect of *cis*-3, *trans*-3, and *trans*-2 addition patterns of C_{60} is their inherent chirality, that is, even addition of two identical addends without chirality elements of their own gives a chiral molecule [21, 22]. Many higher

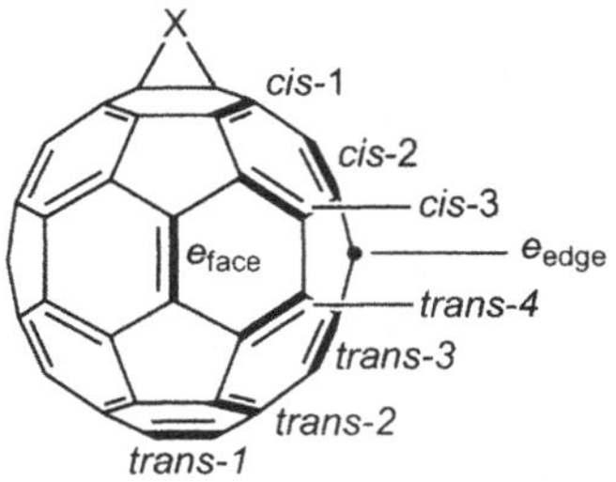

cis-1	C_s
cis-2	C_s
cis-3	C_2
e	C_s
trans-4	C_s
trans-3	C_2
trans-2	C_2
trans-1	D_{2h}

Fig. 1 Possible regioisomers for bisadducts of C_{60}, their trivial designations, and their symmetries in case of identical, C_{2v}-symmetrical addends

adducts of C_{60} have also an inherently chiral functionalization pattern, for example, C_3-symmetrical *e,e,e* or D_3-symmetrical *trans*-3,*trans*-3,*trans*-3 trisadducts. There are also fullerene adducts with noninherently chiral addition patterns – their chirality is due to the presence of structurally different addends [21,22]. Thus, in case of two different addends, *cis*-1, *cis*-2, and *trans*-4 addition patterns become noninherently chiral. The analogy of this situation to that of a center of chirality with tetrahedral coordination should be noted.

The configuration of chiral fullerene addition patterns may in principle be described by indicating the absolute configuration of each stereogenic center on the fullerene surface in terms of the Cahn-Ingold-Prelog (CIP) descriptors (*R*) or (*S*). However, application of this system may be very lengthy and not straightforward due to the highly branched carbon skeleton of fullerene derivatives. Furthermore, as the configuration of individual stereogenic centers in fullerene derivatives cannot generally be inverted independently of the others, the consideration of the fullerene sphere as a single stereogenic unit is advantageous [21, 22], and the configurational descriptors ($^{f,s}C$) and ($^{f,s}A$) (f=fullerene, s=systematic (numbering), *C*=clockwise, *A*=anticlockwise) were introduced in this context [19, 22]. They relate to the direction of the numbering commencement, i.e., a path traced from C(1) to C(2) to C(3) in a fullerene derivative numbered according to the IUPAC rules [19], taking into account the lowest set of locants for all addends, in particular.

3 The First Tether-Directed Remote Functionalization of C_{60}

In the search for a rational approach toward the regioselective synthesis of multiple adducts of C_{60} and possibly other fullerenes, Diederich and co-workers initially targeted *e* bisadducts of buckminsterfullerene by the tether-directed remote functionalization [12, 23]. A methanofullerene carboxylic acid group, readily attached to the fullerene core by the Bingel reaction [24], was chosen as an anchor (Scheme 1) and a 1,3-butadiene moiety, known to undergo irreversible Diels-Alder addition to 6-6 bonds of C_{60} [25, 26], served as reactive

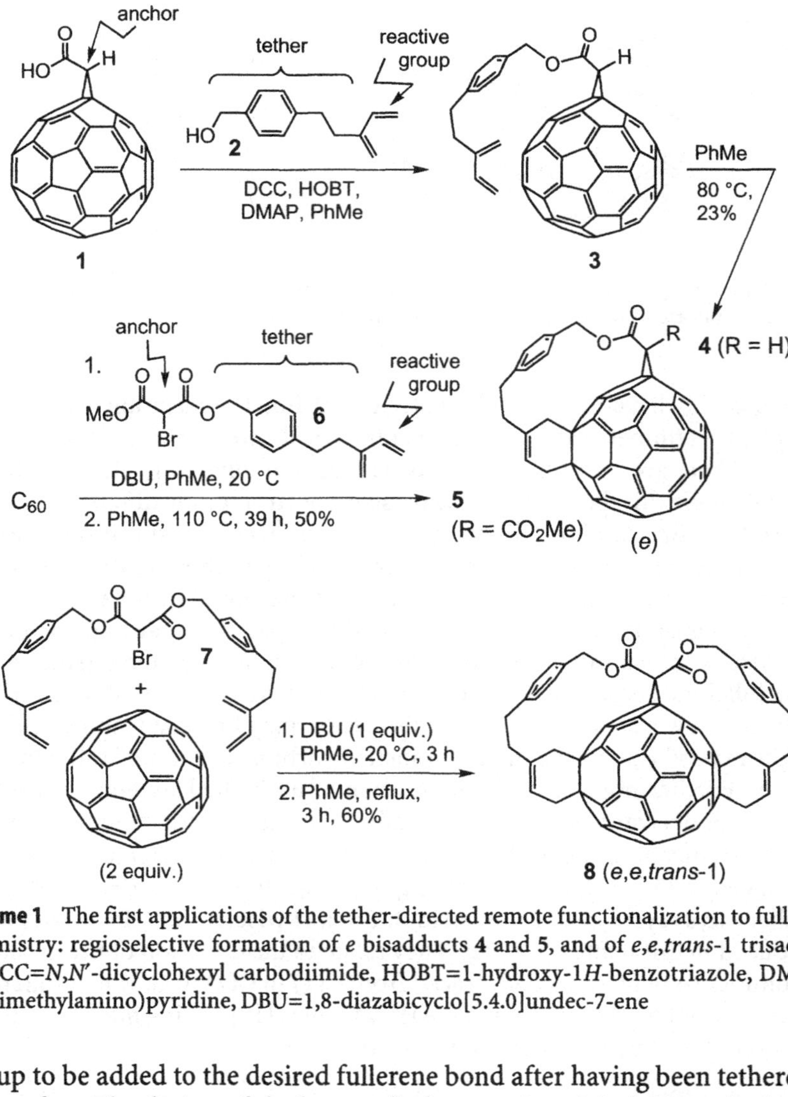

Scheme 1 The first applications of the tether-directed remote functionalization to fullerene chemistry: regioselective formation of *e* bisadducts **4** and **5**, and of *e,e,trans*-1 trisadduct **8**. DCC=*N,N'*-dicyclohexyl carbodiimide, HOBT=1-hydroxy-1*H*-benzotriazole, DMAP=4-(dimethylamino)pyridine, DBU=1,8-diazabicyclo[5.4.0]undec-7-ene

group to be added to the desired fullerene bond after having been tethered to the anchor. The design of the latter relied on semiempirical PM3 calculations using the relative heats of formation of possible regioisomeric bisadducts as a selection criterion. It was expected that a tether incorporating a (4-methylphenyl)ethane unit (see Scheme 1) should show a high selectivity for the targeted e_{face} over the neighboring *cis*-3 and *trans*-4, or the alternative e_{edge} bonds (cf. Fig. 1) [12]. This prediction was supported by more elaborate modeling taking into account the relative populations of potentially reactive conformations of the tether-reactive group conjugate [27].

The tether-reactive group conjugate was attached to the pre-fixed anchor by esterification of wine-red methanofullerenecarboxylic acid **1** with alcohol **2** (Scheme 1) [12]. Subsequent intramolecular reaction of the tethered 1,3-butadiene group of **3** with the fullerene core had to be done under high dilution (ca. 10^{-4} mol/l) to prevent polymerization. ^{13}C NMR spectroscopy of the brown product provided clear proof for the exclusive formation of the desired C_s-symmetrical *e* bisadduct **4**. A more soluble analog, **5**, was prepared starting from C_{60} and the preformed anchor-tether-reactive group conjugate **6** by sequential Bingel and Diels-Alder additions [23, 28]. Later electrochemical investigations on "mixed" bisadduct **5** suggested that a bis(ethoxycarbonyl)methano bridge could be electrochemically removed from the fullerene (electrochemical retro-Bingel reaction [17, 29, 30]) in a chemoselective way, i.e., in the presence of a different addend which is left unaffected [31]. This possibility opened up new synthetic avenues in terms of reversible functionalization of carbon cages and template strategies [17].

Application of the above strategy to an anchor carrying two tethered butadiene moieties (**7**) regioselectively provided the orange-brown, C_{2v}-symmetrical trisadduct **8** (Scheme 1) with two cyclohexene rings fused to the e_{face} positions, after heating of the primarily formed wine-red Bingel-type monoadduct in dilute toluene solution [23]. Its overall *e,e,trans*-1 addition pattern cannot be obtained by stepwise addition of independent (untethered) reagents.

Bingel reaction of the readily available **8** with 5 equiv. of diethyl 2-bromomalonate and 1,8-diazabicyclo[5.4.0]undec-7-ene (DBU) in chlorobenzene afforded the bright red C_s-symmetrical tetrakisadduct **9** (Scheme 2, top). By treating **8** with a tenfold excess of the reagents, the bright yellow, C_{2v}-symmetrical hexakisadduct **10**, displaying a pseudo-octahedral arrangement of addends (cf. Sect. 5.3), was obtained in 73% yield [12, 23]. Its structure, confirmed by X-ray crystallography [32, 33], includes a cyclophane subunit with eight benzenoid rings placed at the corners of a cube and interconnected by biphenyl type bonds radiating from positions 1, 3, and 5 (Fig. 2a). With respect to the I_h-C_{60} skeleton, the bond length alternation in the benzenoid rings of **10** is reduced from ca. 6 to 4 pm. The same trend was observed in the crystal structures of analogous hexakisadducts [25, 34].

Further treatment of tetrakisadduct **9** with diethyl 2-bromomalonate and DBU afforded pentakisadducts **11** (C_{2v}-symmetrical) and **12** (C_s-symmetrical) (Scheme 2, bottom) [23, 32]. In order to achieve chromatographic separation which was not possible with **11** and **12**, analogs **13** and **14** with more polar diester residues were prepared in a 72:28 ratio from tetrakisadduct **15** [23, 28, 32]. At first view, the observed selectivity may seem surprising, knowing that in each case the last reacting bond has three neighbors in *e* positions which are supposed to confer similar activating effects. However, the geometric arrangement of the neighboring *e* bonds is different for the two considered reactive positions: whereas in the more abundantly formed **11** and **13** the last reacting bond is positioned doubly e_{face} and singly e_{edge} relative to the neighboring addends, the reverse is true for **12** and **14**. It can be inferred that double bonds in e_{face} position relative to an existing addend are more reactive than e_{edge} bonds (cf. also Sect. 5.3), a conclusion that had also been reached computationally [35].

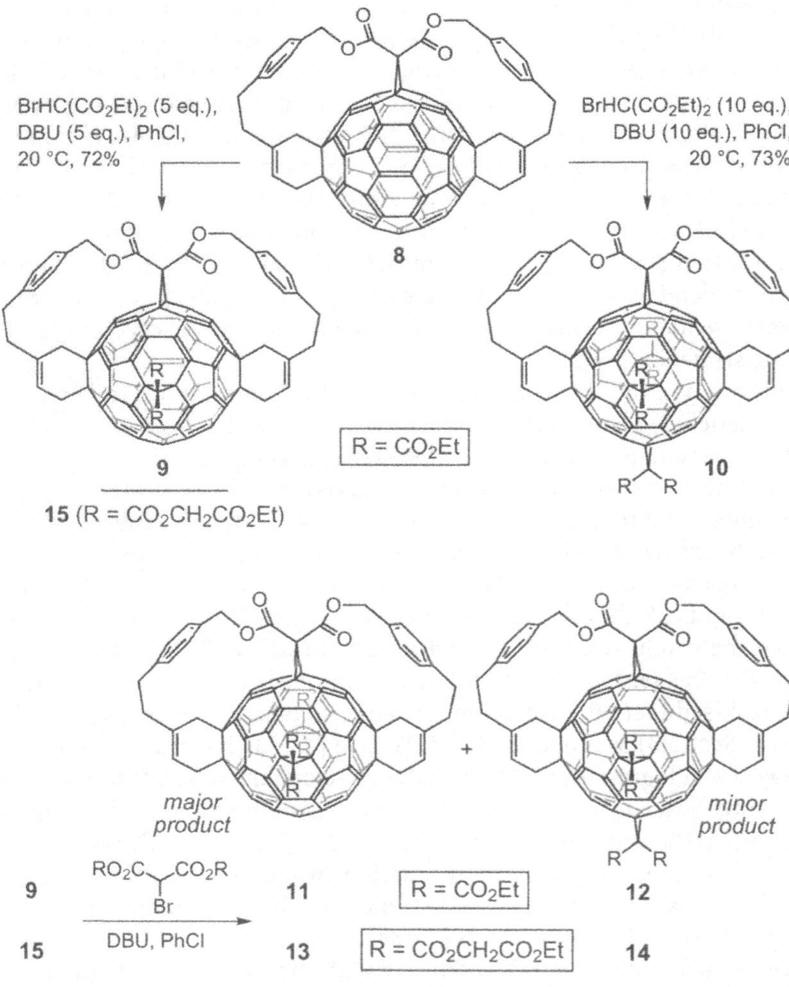

Scheme 2 *Bingel* type cyclopropanations starting from trisadduct **8** and leading to tetrakis-, pentakis-, and hexakisadducts of C_{60} with the addend fusion sites belonging to a group of six pseudo-octahedrally arranged fullerene bonds

Endohedral ^{3}He complexes of multiadducts such as **5** (Scheme 1), **8**, **10**, and **13–15** (Scheme 2) were used to study π-electron ring current effects as a function of the degree and pattern of fullerene functionalization [36]. Considerable upfield shifts relative to i^3HeC_{60} were observed up to the bisadduct stage whereas the resonances of tetrakis- to hexakisadducts were shifted only slightly further upfield. This was rationalized in terms of the compensation of deshielding due to the functionalization-induced decrease in diamagnetic π-electron ring currents that extend around the fullerene sphere by the shielding that results from the weaker paramagnetic ring currents of the pentagons and the increased number of localized benzenoid substructures in the higher adducts.

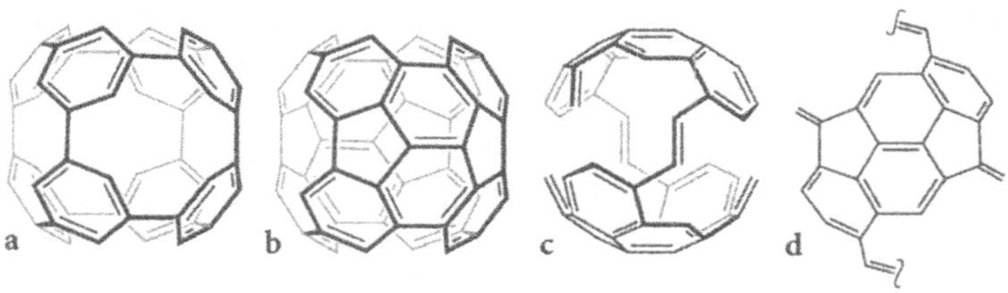

Fig. 2a–d **a** The "cubic cyclophane" substructure with eight benzenoid rings, resulting from functionalization of six pseudo-octahedrally arranged [60]fullerene double bonds. **b** The residual conjugated π-chromophore of tetrakisadduct **24** (Schemes 4 and 22), composed of two fourfold-interconnected tetrabenzopyracylene substructures. **c** The conjugated π-chromophore of hexakisadduct (±)-**49** (Scheme 8) viewed as a whole. **d** The conjugated π-chromophore of hexakisadduct (±)-**49** (Scheme 8) viewed as a single moiety

Cyclopropanation of the remaining *e* type double bond of **13** by 3-bromo-1,5-bis(trimethylsilyl)penta-1,4-diyne in Me_2SO in the presence of DBU afforded a hexakisadduct which, after deprotection of the acetylene groups, was subjected to Eglinton-Glaser conditions [23, 32]. Cyclooligomerization yielded trimeric D_{3h}-**16** and tetrameric D_{4h}-**17** (Fig. 3) as solubilized derivatives of C_{195} and C_{260}, respectively, two representatives of a new class of fullerene-acetylene hybrid carbon allotropes with the general formula $C_{n(60+5)}$. Matrix-assisted laser-desorption ionization time-of-flight (MALDI-TOF) mass spectrometry

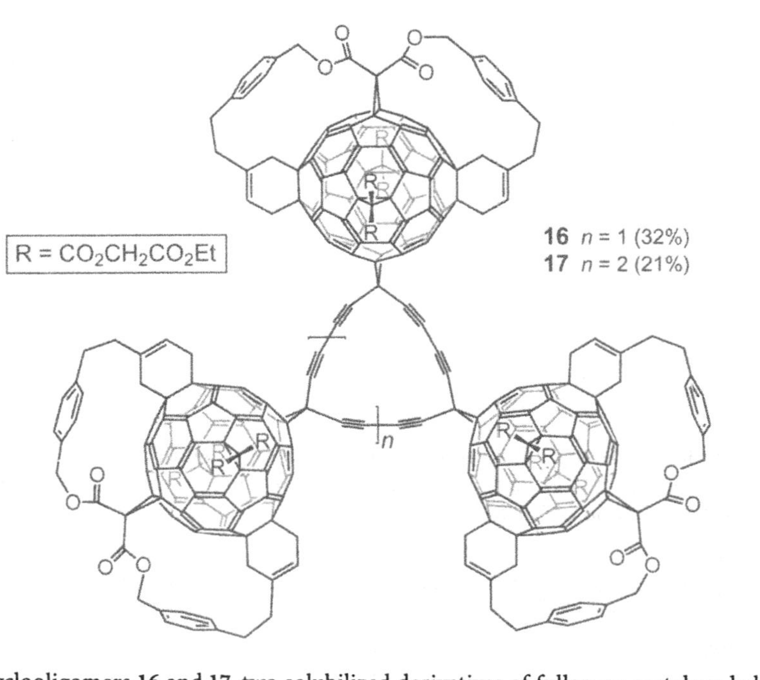

Fig. 3 Cyclooligomers **16** and **17**, two solubilized derivatives of fullerene-acetylene hybrid carbon allotropes C_{195} and C_{260}, respectively

indicated a sequential loss of fullerene units, leading to ions corresponding to mono-fullerene-adducts of the cyclocarbons [37] cyclo-C_{15} and cyclo-C_{20}, respectively [23, 32].

Higher adducts of C_{60} were obtained in an interesting sequence starting with the addition of diazomethane to pentakisadduct **11** (Scheme 3) [38, 39]. Kinetically controlled thermal extrusion of dinitrogen from the primary adduct, pyrazolofullerene **18**, was proposed to pass through a Möbius-aromatic transition state in an eight-electron orbital-symmetry-controlled $[2\pi_s+2\pi_s+2\sigma_s+2\sigma_a]$ process [38–40]. It afforded hexakisadduct (±)-**19** via an intermediate norcaradiene substructure that cycloreverts to the cycloheptatriene unit present in homofullerene (±)-**19**. Further addition of diazomethane to the regenerated *e* type double bond (see (±)-**19**, Scheme 3) afforded according di- and trihomofullerene derivatives. In contrast, photolysis of **18** presumably proceeds through a diradical mechanism, leading to a 1:1 mixture of homofullerene (±)-**19** and methanofullerene **20**.

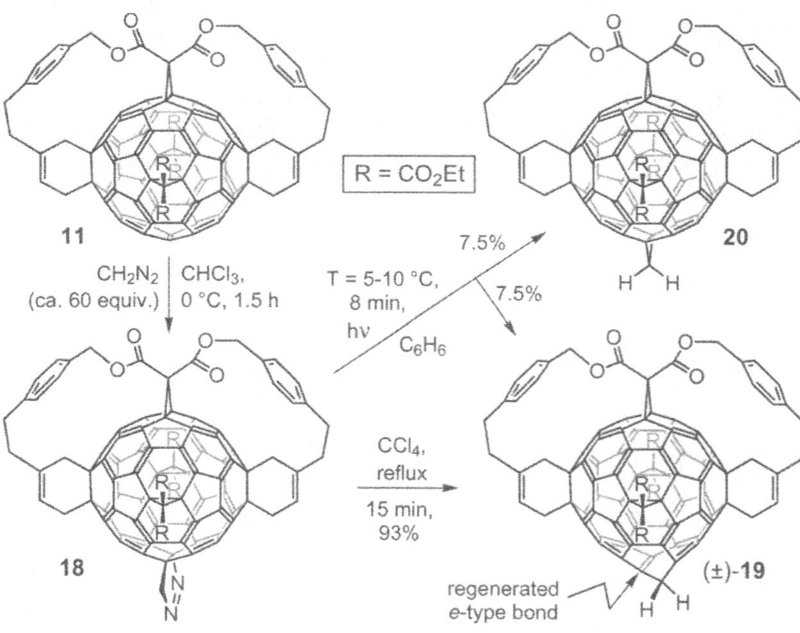

Scheme 3 Regioselective addition of diazomethane to pentakisadduct **11** and different, condition-dependent pathways for the transformation of pyrazolofullerene **18** to methanofullerene **20** and/or homofullerene (±)-**19**

To turn the tether-directed remote functionalization method described in this section into a truly templated synthesis, it was necessary to remove the tethers together with the fullerene-fused cyclohexene rings after they had been used as activating and *e*-directing groups in the synthesis of tetrakis-, pentakis-, and hexakisadducts starting from **8** (cf. Scheme 2). Due to the thermal stability of fullerene-fused cyclohexene rings [25,26], this could not be achieved by retro-Diels-Alder reaction [41] and the Diederich group resorted to an elegant procedure developed by Rubin and co-workers [42,43]. A solution of **10** containing C_{60} as sensitizer for the generation of singlet oxygen [44,45] was irradiated while a stream of dioxygen was bubbled through (Scheme 4). The $^1O_2$ ene-reaction at the two cyclohexene rings yielded the allylic hydroperoxides **21** as a mixture of stereo- and constitutional isomers with endocyclic double bonds [46,47]. The mixture **21** was reduced with triphenylphosphane to the corresponding mixture of isomeric allylic alcohols **22** which was transformed by acid-catalyzed dehydration into a fullerene-derived bis(cyclohexadiene) (not isolated). Addition of dimethyl acetylenedicarboxylate to the reaction mixture afforded, in a Diels-Alder/retro-Diels-Alder sequence, C_{2v}-symmetrical tetrakisadduct **23** in 42% overall yield starting from **10.** Transesterification of **23** with K_2CO_3 in THF/EtOH gave the D_{2h}-symmetrical octakis(ethyl ester) **24** which was characterized by X-ray crystallography [47]. Tetrakisadduct **24** was also obtained by Kräutler and co-workers by a completely different template-directed approach (see Sect. 5.8) [48]. It includes two reactive 6-6 bonds at opposite poles in a *trans*-1 arrangement which can be exploited for further molecular scaffolding [41,49,50]. Application of the above procedure to pentakisadduct **14** (Scheme 2) and sub-

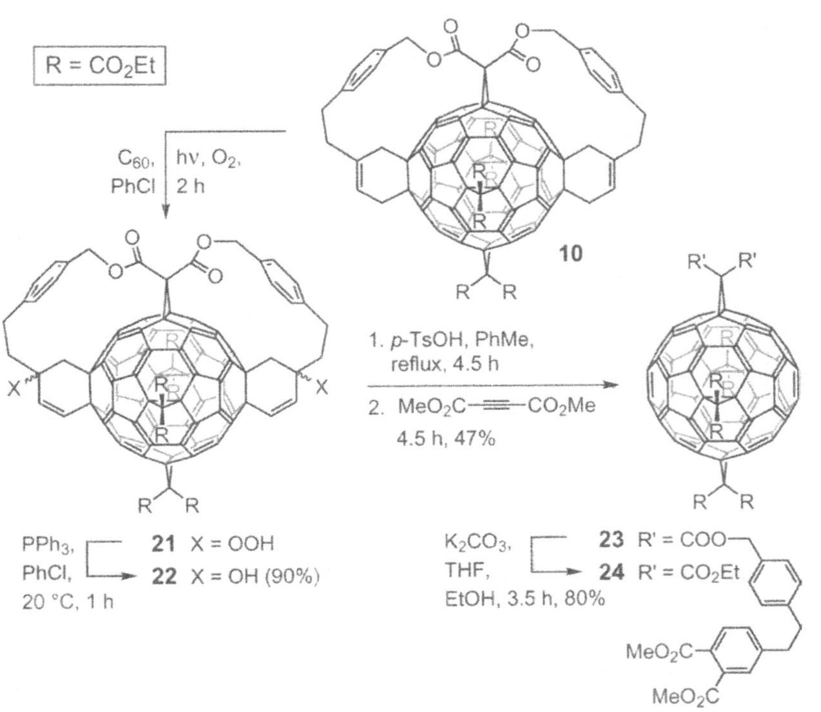

Scheme 4 Preparation of tetrakisadduct **24** by removal of the fullerene-fused cyclohexene rings and the tethers from hexakisadduct **10**, thus turning the tether-directed remote functionalization outlined in Schemes 1 and 2 into a truly templated method

sequent transesterification afforded the according *e,e,trans*-1 trisadduct (**25**, cf. Scheme 17) [47].

As opposed to the bright red tetrakisadduct **9** (Scheme 2), tetrakisadducts **23** and **24** (Scheme 4) are green-yellow in CH_2Cl_2 solution [46, 47]. Despite the four methano bridges along the equator of the carbon spheres, both molecules still have an extended conjugated π-electron system consisting of two tetrabenzopyracylene substructures interconnected by four biphenyl-type bonds (Fig. 2b). More generally, it can be said that methano bridges along an equatorial belt, as in **24**, lead to the smallest possible perturbation of the fullerene π-chromophore [46, 47, 51]. Thus, the electronic properties of **24** remain quite similar to those of pristine C_{60} whereas those of its regioisomers deviate significantly.

Systematic studies of physical properties and chemical reactivity of many of the above and other multiadducts of C_{60} revealed [28, 46, 47] that with increasing reduction of the conjugated fullerene π-chromophore (i) the optical (UV/Vis) HOMO-LUMO (Highest Occupied Molecular Orbital, Lowest Unoccupied Molecular Orbital) gap widens, (ii) the number of reversible 1-electron reductions decreases, (iii) the computed LUMO energy increases and the electron affinity decreases, (iv) the reactivity of the fullerene toward nucleophiles or as ene component in cycloadditions decreases, and (v) the efficiency in the photosensitized generation of $^1O_2$ decreases [52]. As to the nature of the addends, cyclopropa-fusion appears to induce the weakest perturbation of the fullerene chromophore, a conclusion reached also through ^{3}He NMR investigations of fullerene derivatives with endohedrally incarcerated ^{3}He [53].

The above-described first tether-directed remote functionalizations with fullerenes made use of both the Bingel [24] and the Diels-Alder reaction. Most of the subsequently reported applications of tether- or template-directed syntheses in fullerene chemistry were based on either of these reactions and will be discussed below. Also, whereas the above functionalization scheme started from reagents in which anchor and tethered reactive group(s) constitute different chemical functions that were reacted successively with the carbon sphere, many of the succeeding implementations simply involved two or several identical tethered reactive groups added to the fullerene in a one-pot procedure so that the group(s) attached first acted as anchor(s) for the remaining one(s) during the macrocyclization reaction.

4 Tether-Directed Remote Functionalizations Based on the Bingel Reaction

4.1 Tether-Directed Double Bingel Additions

Addition of malonates or similar activated methylene compounds to the formal double bonds of a fullerene sphere under formation of cyclopropane derivatives is often referred to as "Bingel reaction" [24]. A general approach for the tether-directed double Bingel addition is depicted in Scheme 5. Typically, tethered diols **26** are transformed into the corresponding bismalonates and subsequently converted to bis-2-halomalonates. Reaction of the latter with fullerenes affords dicyclopropafullerenes of type **27**. A very convenient one-pot modification consists in treating a solution of the fullerene in toluene with malonate, I_2, and a non-nucleophilic base (e.g. DBU) [54]; it involves in situ generation of a 2-iodomalonate which reacts immediately with the fullerene.

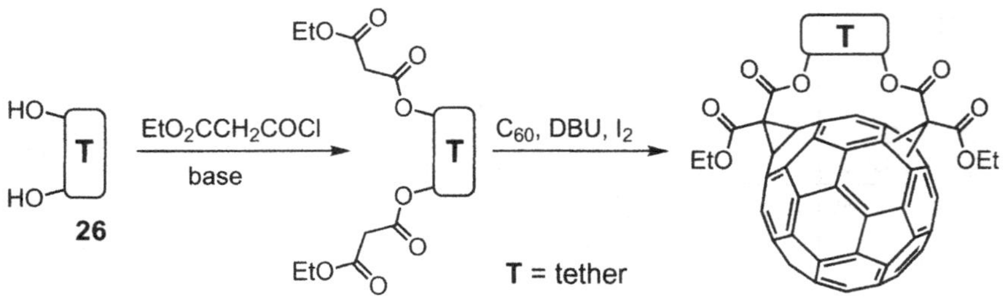

Scheme 5 General scheme of the tether-directed double Bingel reaction of bismalonates with C_{60}

4.1.1 Xylylene Tethers and Related Spacers

A variety of C_{60}-bisadducts was prepared starting from bismalonates including *ortho*-, *meta*-, or *para*-xylylene tethers (Scheme 6). As expected, the regioselectivity of the reaction depends strongly on the substitution pattern of the benzene ring. Thus, one-pot addition of in situ generated bis-iodomalonates derived from *o*- or *m*-xylene afforded exclusively *cis*-2 bisadducts **28** and **29**, respectively (Scheme 6, top), while the *para*-analog gave mainly *trans*-4 bisadduct **30** together with a minor amount of the racemic *e* isomer (±)-**31** [55, 56]. Similar tethers, derived from substituted *p*-xylenes, also afforded *trans*-4 bismethanofullerenes (cf. **30**) in yields up to 56% [57]. Interestingly, the regioselectivity attained with these spacers strongly depends on the nature of the leaving group in position 2 of the malonate: in contrast to the iodomalonate generated in situ, the according preformed chloromalonate afforded only the *e* bisadduct (cf. **31**) in 65% yield while the bromo derivative gave a ca. 1:1 mixture of *e* and *trans*-4 isomers in 53% overall yield [58]. The unexpected change in regioselectivity was explained by either a carbanionic (for chloromalonate) or a carbenoid-type (for iodomalonate) intermediate effecting the initial attack on the fullerene.

The regioselectivity of addition depends also on the nature of the activated methylene compound. Thus, addition of bis-β-ketoesters derived from *m*- (**32**) and *p*-xylene (**33**) resulted in formation of *cis*-2 and *e* bisadducts, respectively (Scheme 6, bottom), thus showing a regioselectivity similar to the geometrically

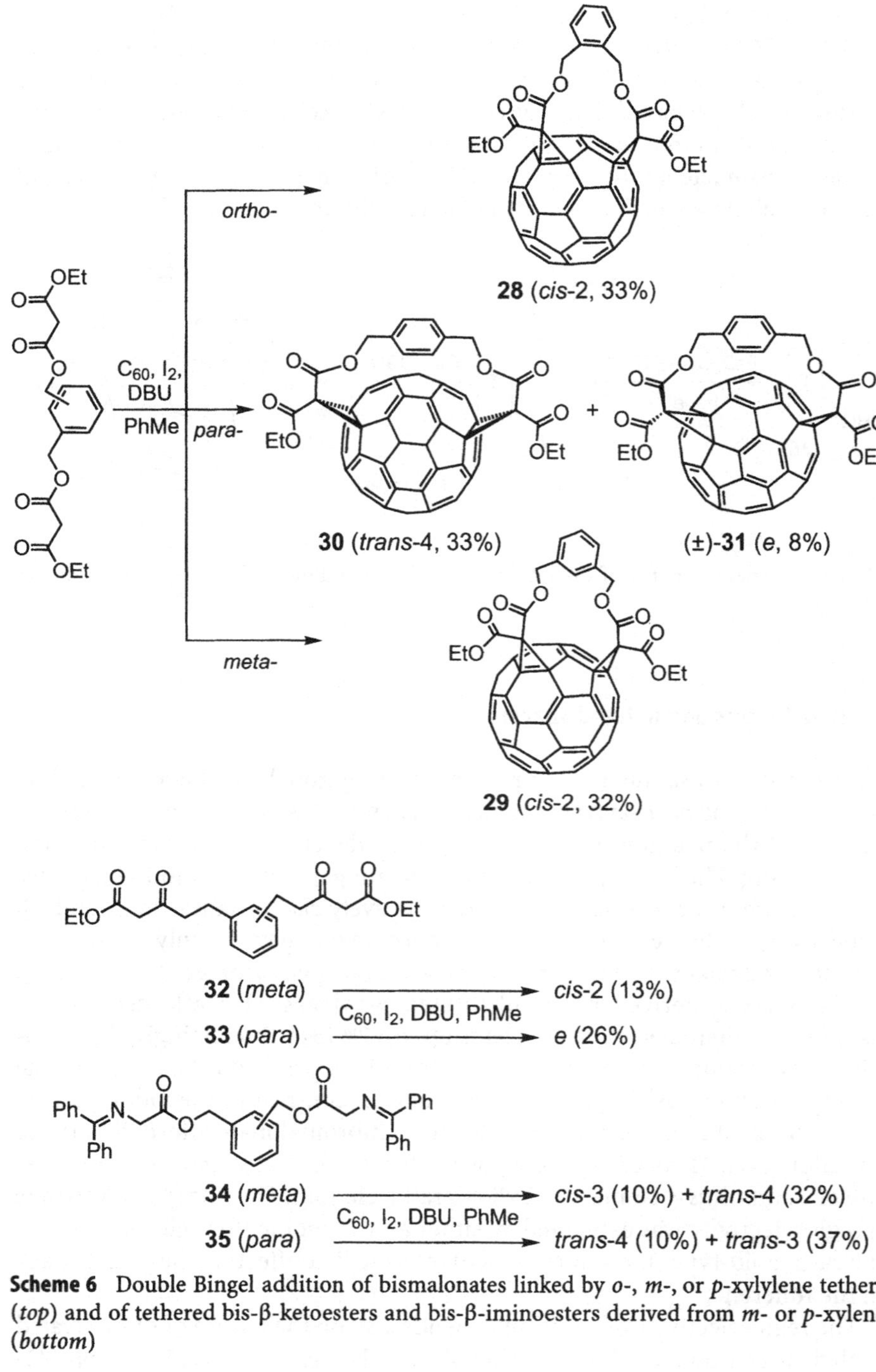

Scheme 6 Double Bingel addition of bismalonates linked by *o*-, *m*-, or *p*-xylylene tethers (*top*) and of tethered bis-β-ketoesters and bis-β-iminoesters derived from *m*- or *p*-xylene (*bottom*)

analogous bismalonates [59]. On the other hand, addition of the according bis-β-iminoesters gave a mixture of *cis*-3 and *trans*-4 bisadducts (from **34**) and *trans*-4 and *trans*-3 bisadducts (from **35**) [60, 61]. A number of other bismalonates with tethers derived from simple arenes including *o*-phenylenediamine [62] and 9,10-disubstituted anthracene [63] showed various regioselectivities in the double Bingel reaction with C_{60}.

Addition to C_{60} of a bismalonate including a *p*-xylylene-type tether which is simultaneously a part of a crown ether was a key step in the synthesis of the first fullerene-containing catenane, (±)-**36**, obtained as *trans*-4 bisadduct with complete regioselectivity (Fig. 4). Interestingly, whereas *trans*-4 bisadduct **30** (Scheme 6) produced by the addition of *p*-xylylene-tethered bismalonate has C_s symmetry and is therefore achiral, bridging of the fullerene sphere in (±)-**36** by the 2,5-disubstituted hydroquinone ring incorporated in a crown ether introduces an element of planar chirality and reduces the overall symmetry of the molecule to C_1 [64].

Fullerene-derived dendrimers have been synthesized in a large structural variety using the tether-directed remote functionalization with *m*-xylylene tethers over the past years as they appear to be particularly promising for supramolecular chemistry and materials science applications [65–67]. Thus, various amphiphilic fullero-dendrimers including oligoethyleneglycol, carbohydrate, or cholesterol subunits were demonstrated to form stable Langmuir monolayers which can also be transferred onto solid substrates yielding high-quality Langmuir-Blodgett films [68–72]. Fullerene dendrimer **37** with six 4′-cyanobiphenyl groups (Fig. 4) was shown to form a stable mesogenic liquid

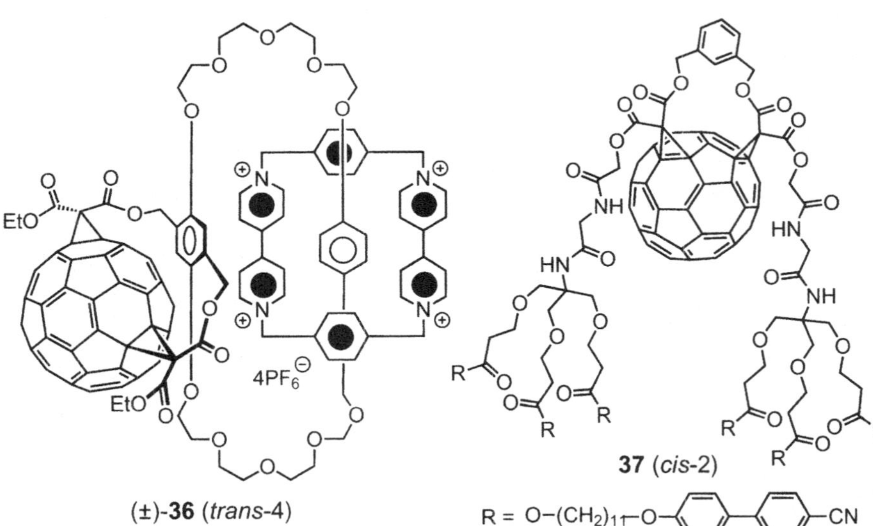

(±)-**36** (*trans*-4)

37 (*cis*-2)

Fig. 4 Examples of a catenane ((±)-**36**) and a fullerene-derived dendrimer (**37**) prepared using the tether-directed double Bingel reaction

crystalline phase [49]. Assembling phenanthroline ligands with attached fullero-dendritic wedges around a copper(I) center gives supramolecular arrays including up to 16 fullerene chromophores with the copper-phenanthroline complex "buried" inside the dendritic "black box" [73, 74]. The characteristic electrochemical oxidation of Cu(I) to Cu(II) (E=+0.60 V for $[Cu(phen)_2]^{+/2+}$) cannot be observed any more in such systems as the central electroactive site is totally blocked by the surrounding dendrimer structure.

Materials combining C_{60} with extended π-conjugated oligomers have received growing attention as promising systems for solar energy conversion. The polycondensation of carboxy-functionalized *cis*-2 bisadducts of C_{60} with tetraethyleneglycol gave access to fullerene-containing polymers that were used as π-acceptor components together with poly(*p*-phenylenevinylene) type donors in photovoltaic cells [75]. Linear copolymers with alternating C_{60} and oligo (*p*-phenylenevinylene) subunits represent another promising donor-acceptor system for photovoltaic applications [76].

Stepwise tether-directed remote functionalization with pre-anchored, *m*-xylylene-tethered bismalonates gave regioselective access to tris-, tetrakis-, and hexakis (cf. Sect. 4.1.3 and Scheme 8) adducts of C_{60} in multiple macrocyclization reactions [55]. Thus, treatment of **38** (Scheme 7), in which two tethered malonates are anchored to [60]fullerene, with DBU and I_2 in toluene under high dilution conditions yielded (±)-**39** as a single trisadduct. In accord with the typical directing effect of *m*-xylene tethers (see above), the newly formed methano bridges of (±)-**39** are both located in *cis*-2 position relative to the original cyclopropane ring of **38**. NMR experiments showed that (±)-**39** is C_2-symmetrical and has, therefore, a *cis*-2, *cis*-2, *trans*-3 overall addition pattern. In extension of this strategy, a triply bridged tetracyclopropa-homolog of (±)-**39**, again with *cis*-2 relationships between each pair of neighboring addends, was prepared from a doubly anchored reactive bismalonate [55].

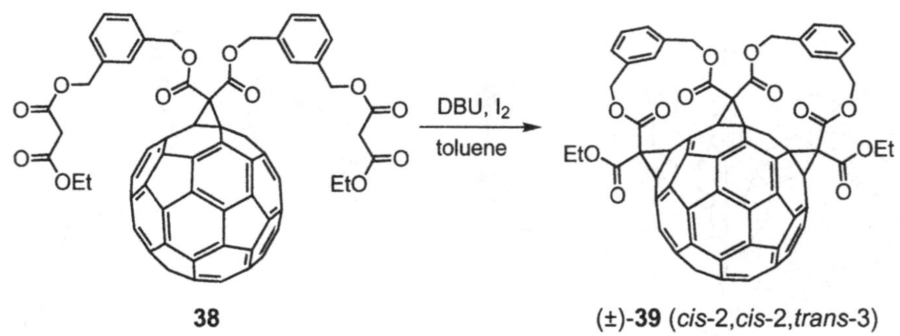

38 (±)-**39** (*cis*-2,*cis*-2,*trans*-3)

Scheme 7 Synthesis of a trisadduct of C_{60} by stepwise tether-directed Bingel addition ("clipping" reaction)

4.1.2
Porphyrin Tethers

Molecular dyads combining strongly electron-accepting C_{60} and various electron-donating subunits attracted much attention as model systems mimicking natural photosynthetic reaction centers and as optoelectronic molecular-level devices [77, 78]. In particular, a wide variety of conjugates containing porphyrin derivatives as electron donors have been constructed in recent years [79–81]. In many cases, the porphyrin is attached to the fullerene by a single linker and, consequently, distance and orientation between the two components of such dyads are often not well defined. Use of the porphyrin unit as a spacer in the tether-directed remote functionalization leads to cyclophane-type structures and allows precise control of the geometrical parameters which have a crucial influence on the photophysical and electron-transfer properties of donor-acceptor dyads. In line with this principle, modified Bingel reaction with a porphyrin-bismalonate conjugate regioselectively provided the *trans*-1 bisadduct **40** in 17% yield (Fig. 5) [82, 83]. Zinc(II) complexation by the porphyrin-bismalonate affected neither yield nor regioselectivity of the macrocyclization. Interestingly, the use of a very similar bismalonate with the longer ethylene instead of methylene linkers between the reactive ester groups and the diphenylporphyrin core afforded exclusively *trans*-2 bisadduct (±)-**41** with a shorter distance between the addends [84].

In contrast to (±)-**41** (and also to **40**) in which porphyrin and fullerene moieties are locked in a face-to-face arrangement, an average edge-to-face orientation is realized in an *e* analog regioselectively prepared in 12% yield from a Bingel reagent in which the two malonate "arms" are attached to *meso* positions 5 and 10 of the porphyrin (see Fig. 5 for locants) as opposed to the substitution of positions 5 and 15 in the precursor of (±)-**41** [85]. While the face-to-face alignment with short interchromophoric distance (ca. 3 Å) in (±)-**41** leads to appreciable π-π stacking, the edge-to-face alignment in the *e* congener hampers

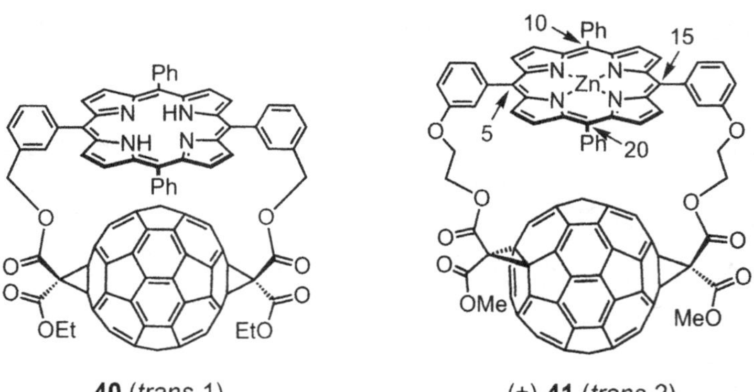

40 (*trans*-1) (±)-**41** (*trans*-2)

Fig. 5 Porphyrin-fullerene dyads prepared regioselectively by tether-directed double Bingel reaction

such interactions. Different donor-acceptor orientations have a strong impact on the back electron transfer (BET) dynamics: the life-time of the charge-separated state of the face-to-face dyad (±)-**41** is 4 orders of magnitude lower than that of the *e* type edge-to-face isomer [86, 87]. Interestingly, cyclovoltammetric (CV) studies showed that the electronic interactions between fullerene and porphyrin in **40** or its complex with zinc are relatively small despite the close proximity between donor and acceptor units [82]. The formation of supramolecular associates by complexation of (±)-**41** with nitrogen-containing ligands such as DABCO (1,4-diazabicyclo[2.2.2]octane) or pyridine derivatives and the photoinduced electron transfer in such associates were also investigated [88].

4.1.3 Crown Ether Tethers

Dibenzo crown ethers have been employed as templates to access various *trans*-type bisadducts of C_{60} with a rather large distance between addends. Thus, addition of a bismalonate with an *anti*-disubstituted dibenzo[18]crown-6 spacer (**42**, substitution at positions 2 and 13 of the dibenzo-fused heterocycle, see Scheme 9) afforded *trans*-1 bismethanofullerene (±)-**43** (Fig. 6) in 30% yield together with a minor amount (3%) of *trans*-2 isomer [89]. Both regioselectivity and yield could be further increased by addition of KPF_6 to the reaction mixture: under these conditions, *trans*-1 bisadduct (±)-**43** was formed as a unique isomer in 50% yield [90]. This dramatic improvement is undoubtedly due to complexation of K^+ ions by the crown ether which makes the tether conformationally more rigid and, thereby, enhances the selectivity of the second cyclopropanation. The high yield is particularly remarkable taking into account that formation of the *trans*-1 bisadduct is strongly disfavored on both statistic and electronic grounds in the double addition of diethyl 2-bromomalonate to C_{60} [20, 35]. Reaction of an isomeric tether derived from *syn*-disubstituted dibenzo[18]crown-6 (**44**, substitution at positions 2 and 14 of the dibenzo-fused heterocycle, see Scheme 9) regioselectively afforded *trans*-3 bisadduct (±)-**45** (Fig. 6). According to variable temperature NMR measurements, this

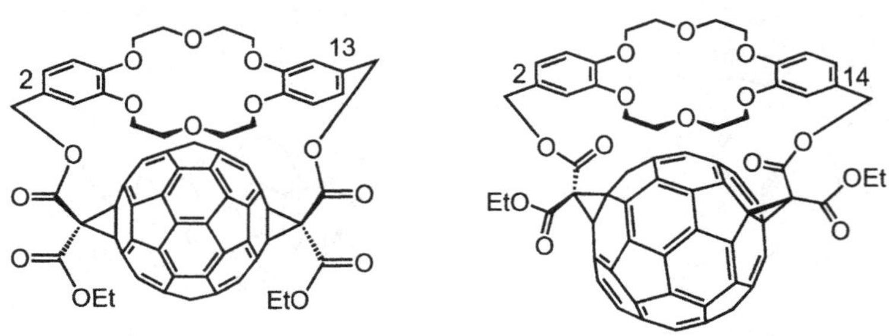

(±)-**43** (*trans*-1) (±)-**45** (*trans*-3, *in-out*)

Fig. 6 Crown ether-fullerene conjugates prepared by double Bingel reaction

bisadduct represents a rare example of an *in-out* bridged dicyclopropafullerene (cf. also Sect. 4.1.5 and Fig. 8, compound (±)-**69**).

CV investigations on the C_{60}-derived crown ether (±)-**43** allowed the first detection of a cation-complexation-induced effect on the redox properties of the fullerene. Thus, addition of KPF_6 to a solution of (±)-**43** leads to a 90 mV anodic shift of the first reduction potential of the fullerene moiety [90]. This behavior is due to an electrostatic effect of the K^+ ion fixed in close proximity to the carbon sphere and represents a sensoric signal associated with the molecular recognition event. In addition, (±)-**43** was very stable against chemical reductive retro-cyclopropanation [91] by magnesium under various conditions, although such transformation was smoothly performed [17, 92] electrochemically [29].

The crown ether-fullerene conjugate (±)-**46** (Scheme 8) with *trans*-1 addition pattern was employed as an important intermediate in the synthesis of a hexakisadduct of C_{60} involving another, fourfold tether-directed remote functionalization (cf. also Sect. 4.1.1 and Scheme 7) [50, 93]. Acid-catalyzed cleavage of the crown ether template and the *tert*-butyl ester groups afforded the corresponding tetraacid which was transformed over several steps into tetrol **47** and, finally, into tetramalonate **48** (Scheme 8). Fourfold intramolecular Bingel addition

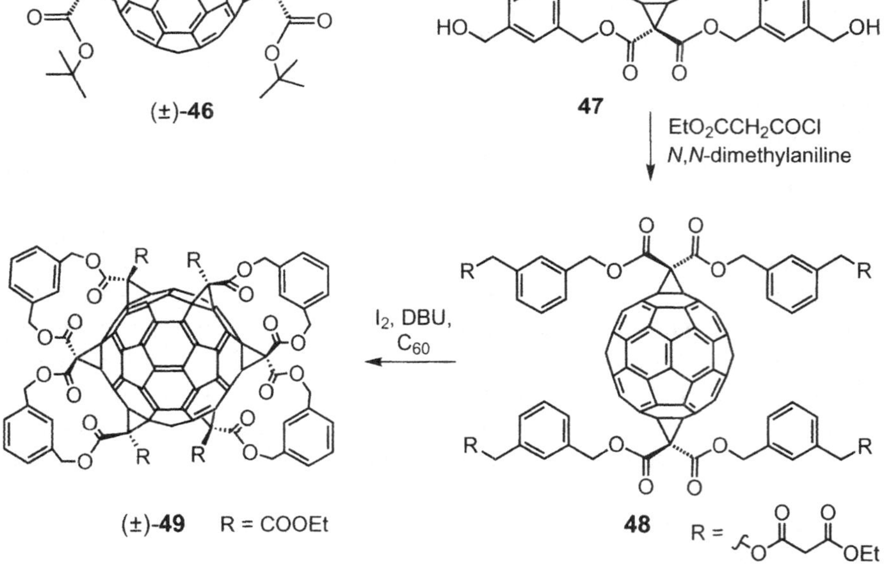

Scheme 8 Transformation of *trans*-1 bisadduct (±)-**46** into hexakisadduct (±)-**49** with all cyclopropane rings arranged in a double-helical fashion along an equatorial belt of C_{60}

afforded the shiny red hexakisadduct (±)-**49** as sole product in 10% yield. As was confirmed by X-ray crystallography, all six cyclopropane rings in (±)-**49** are arranged in a double-helical fashion along an equatorial belt of C_{60}. In (±)-**49**, π-electron conjugation between the two unfunctionalized poles of the carbon sphere is maintained via two (*E*)-stilbene-like bridges (Fig. 2c,d). As a result, (±)-**49** features very different chemical reactivity and physical properties when compared to hexakisadducts with a pseudo-octahedral addition pattern (cf. Sect. 3 and 5.3, Fig. 2a, Fig. 10, and Scheme 2 (compound **10**)). Its reduction under CV conditions is facilitated by 570 mV and it readily undergoes further, electronically favored Bingel additions at the two sterically well accessible central polar 6-6 bonds under formation of heptakis- and octakisadducts [50, 93].

Bismalonates tethered by dibenzo[18]crown-6 ionophores have also been successfully employed in Bingel additions to C_{70} [94–96]. To the best of our knowledge, this is the only known example of tether-directed remote functionalization of a higher fullerene. Generally, monoaddition to C_{70} occurs preferentially at the C(1)-C(2) bond (cf. (±)-**50**, Scheme 9), and the second addition takes place at one of the five bonds radiating from the polar pentagon of the opposite, unfunctionalized hemisphere [97–100]. In the case of achiral C_{2v}-symmetrical addends, three constitutional isomers, two of which represent pairs of enantiomers, are formed. According to a Newman-type projection looking down the C_5 axis of C_{70}, they were named "twelve o'clock", "two o'clock", and "five o'clock" isomers (Scheme 9, top) [97].

Macrocyclization of C_{70} with bismalonate **42** afforded two diastereoisomeric pairs of enantiomers, (±)-**50** and (±)-**51**, in a ca. 1:1 ratio, with complete regioselectivity [94, 95]. Similarly to (±)-**45** (Fig. 6), the chirality of these compounds is associated with an inherently chiral addition pattern [21, 22, 98, 100, 101] in combination with a planar chiral unit resulting from the restricted conformational mobility of the bridging crown ether. Interestingly, the "five o'clock" addition pattern of (±)-**50** and (±)-**51** is the most disfavored in the sequential addition of diethyl malonate to C_{70}. A notable template effect was observed when the synthesis of (±)-**50** and (±)-**51** was carried out in the presence of KPF_6: whereas the regioselectivity remained unaffected, the diastereoselectivity increased and the yield went up from 41% to 68%.

When bisadduct (±)-**50** was allowed to react with a second equivalent of the tethered bismalonate **42**, an inseparable mixture of four products was obtained. In contrast, when bismalonate **44** with a *syn*-disubstituted dibenzo[18]crown-6 moiety was reacted under the same conditions with (±)-**50** or (±)-**51**, tetrakisadducts (±)-**52** and (±)-**53**, respectively, with C_{70} sandwiched between two crown ethers were regio- and stereoselectively obtained in 31% and 26% yield, respectively [95, 96].

The addition of KPF_6 or $NaBF_4$ to a solution of (±)-**50** or (±)-**51** increases the first reduction potential of the fullerene by 70 mV, as measured by CV. An even more pronounced effect is observed for (±)-**52** and (±)-**53** when the fullerene is squeezed in between two K^+ ions, and the potential of the redox couple $0/1^-$

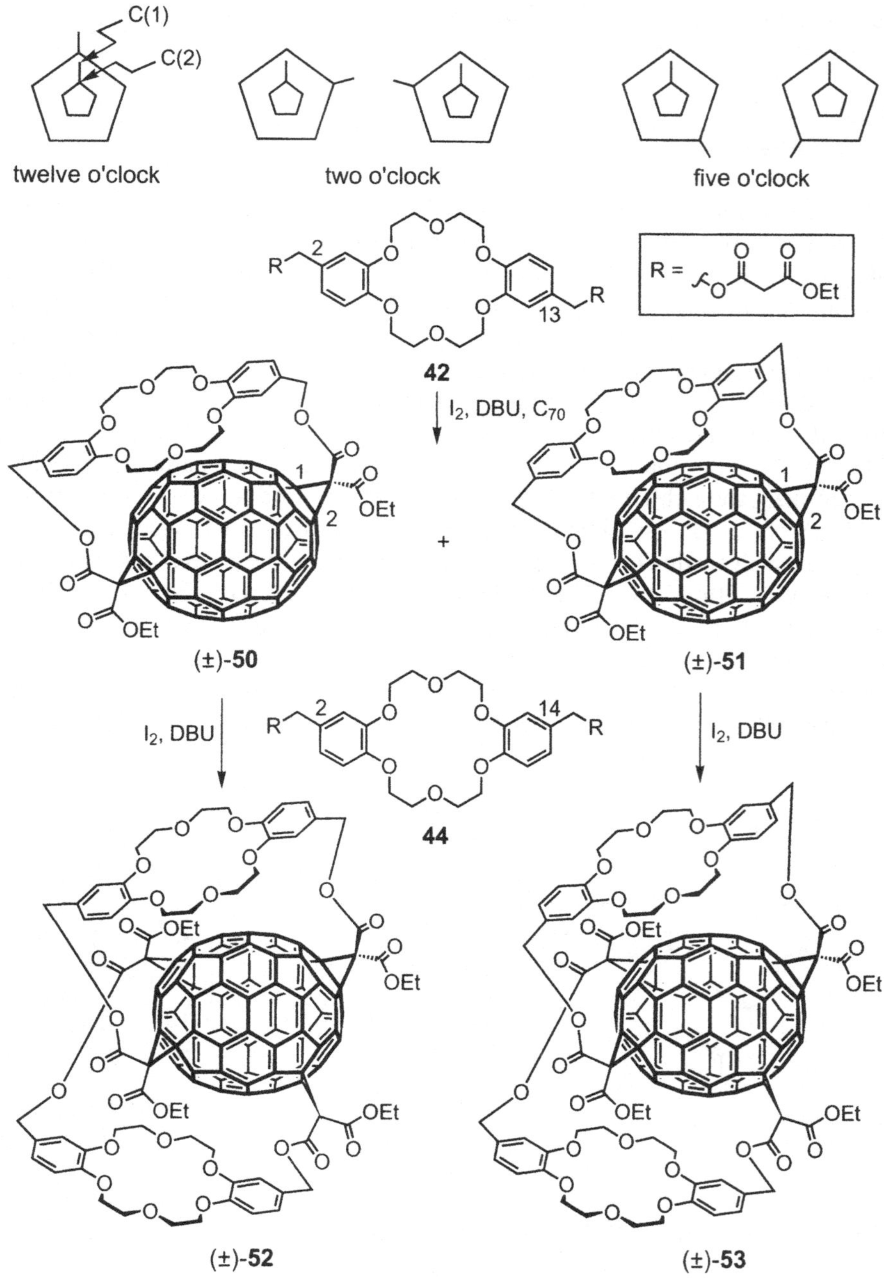

Scheme 9 Single and double tether-directed remote bisfunctionalization of C_{70} with bis-malonate-crown ether conjugates

is raised by 170 mV in comparison to the value measured in the absence of potassium salt [95, 96].

4.1.4 Chiral Tethers

A particular interest in *cis*-3, *trans*-3, and *trans*-2 bisadducts of C_{60} arises from the fact, that these addition patterns are inherently chiral (see above) [21, 22, 98, 100, 101]. The residual fullerene π-chromophores of such adducts are characterized by very strong Cotton effects over the entire UV/Vis range of the circular dichroism (CD) spectra [21, 97, 99, 102, 103], and many efforts were directed at preparing them by stereoselective synthesis. In the first asymmetric synthesis of *cis*-3 bisadducts of C_{60}, bismalonates containing tethers derived from threitol were employed (Scheme 10). Starting from optically pure (*R*,*R*)-**54** or (*S*,*S*)-**54**, the enantiomeric *cis*-3 bisadducts (*R*,*R*,^{f,s}*C*)-**55** and (*S*,*S*,^{f,s}*A*)-**55**, respectively, were obtained in 15% yield. The major products of these additions (21%) were *cis*-2 bisadducts (*R*,*R*)-**56** and (*S*,*S*)-**56**, respectively, which are chiral exclusively due to stereogenic centers in their addends [55, 104]. Separate transesterifications of (*R*,*R*,^{f,s}*C*)-**55** and (*S*,*S*,^{f,s}*A*)-**55** with ethanol yielded tetraethyl esters with (^{f,s}*C*)- and (^{f,s}*A*)-configuration, respectively, as pure enantiomers exhibiting strong CD bands between 250 and 750 nm with Δε values approaching 150 M^{-1} cm^{-1}. Their absolute configurations could be established as [CD(−)488]-(^{f,s}*C*) and [CD(+)488]-(^{f,s}*A*) by comparison of their CD spectra

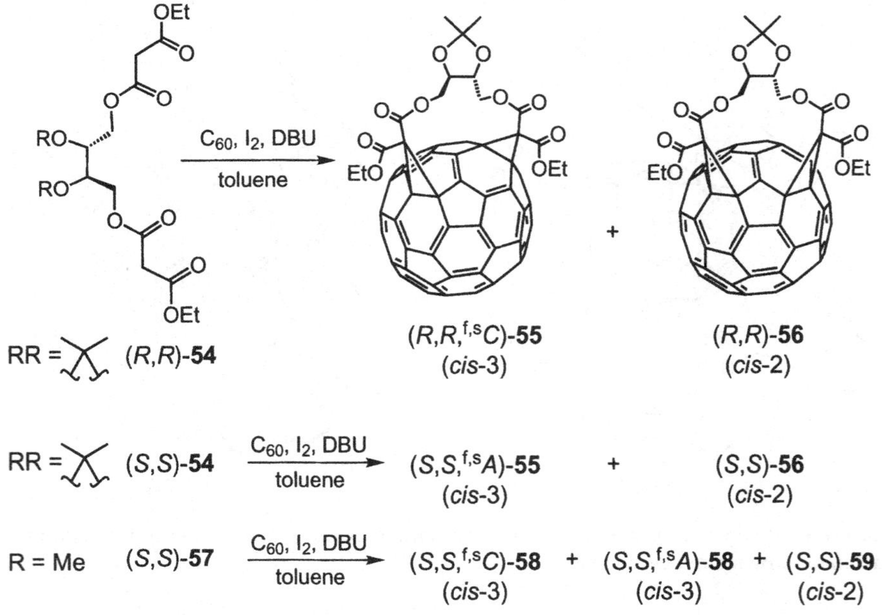

Scheme 10 Synthesis of chiral Bingel-type bisadducts of C_{60} using enantiomerically pure bismalonates derived from threitol

with those calculated using the π-electron SCF-CI-DV MO (self-consistent-field configuration-interaction dipole-velocity molecular orbital) method [105].

In contrast, addition of the open-chain threitol derivative (*S*,*S*)-**57** to C_{60} afforded only minor amounts of two diastereoisomeric *cis*-3 bisadducts (*S*,*S*,^{f,s}*A*)-**58** and (*S*,*S*,^{f,s}*C*)-**58** in a 2.5:1 ratio which reflects a rather poor diastereoselectivity. The major product of this reaction was the corresponding *cis*-2 bisadduct (*S*,*S*)-**59** (40%) [106].

Excellent regio- and diastereoselectivity was achieved, on the other hand, using a bismalonate derived from the very simple chiral (*R*,*R*)-butane-2,3-diol tether, which gave only *cis*-3 bisadduct **60** with (^{f,s}*A*)-configuration of the fullerene unit in 24% yield and >97% de, according to HPLC analysis (Fig. 7) [107]. Accordingly, addition of the (*S*,*S*)-configured bismalonate isomer afforded the enantiomeric (*S*,*S*,^{f,s}*C*)-**60**. The absolute configuration of the inherently chiral *cis*-3 addition pattern generated from the (*R*,*R*)-configured bismalonate was assigned on the basis of a conformational analysis combining semiempirical calculations and ^{1}H NMR spectroscopy. According to the calculations, the substituents of the glycolic fragment of (*R*,*R*,^{f,s}*A*)-**60** adopt a staggered conformation with *gauche* relationships between the methyl as well as between the ester groups and an antiperiplanar (*ap*) arrangement of the H-atoms (see Newman projection of Fig. 7). As to the unobserved (*R*,*R*,^{f,s}*C*)-**60** diastereoisomer, the computations anticipated *gauche* relationships between the ester groups as well as the H-atoms and an *ap* arrangement between the methyl residues. The coupling constant ^{3}J(H,H) of 7.9 Hz measured for the vicinal protons of the glycolic fragment clearly indicate that they adopt an *ap* arrangement corresponding to the most stable conformer of (*R*,*R*,^{f,s}*A*)-**60**. This result is in accord with the near-identity of the CD spectra of (*R*,*R*,^{f,s}*A*)-**60** and (*S*,*S*,^{f,s}*A*)-**55** (Scheme 10) [107].

According to the conformational preferences discussed above, tethers derived from *trans*-cyclohexane-1,2-diol with both oxy substituents fixed in a 1,2-di-

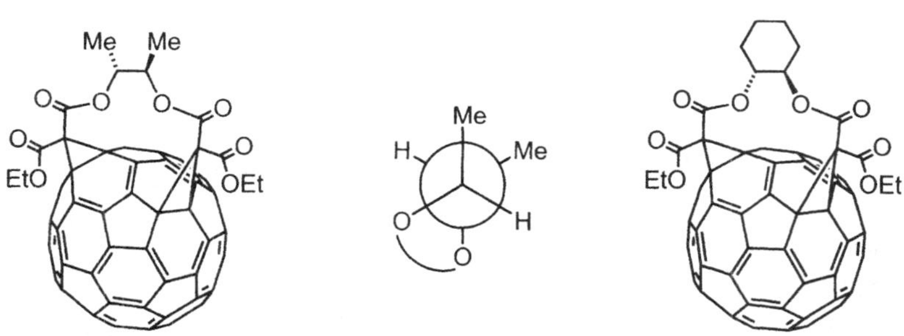

(*R*,*R*,^{f,s}*A*)-**60** (*cis*-3) (*R*,*R*,^{f,s}*A*)-**61** (*cis*-3)

Fig. 7 Chiral *cis*-3 bisadducts of C_{60} prepared diastereoselectively from bismalonates derived from enantiomerically pure (*R*,*R*)-2,3-butanediol and (*R*,*R*)-*trans*-cyclohexane-1,2-diol (*left and right*) and Newman projection of the most favorable conformation of their glycolic fragment (*center*)

equatorial *gauche* arrangement should be particularly suitable for the diastereoselective formation of a *cis*-3 bisadduct. Indeed, addition of the bismalonate derived from (*R*,*R*)-cyclohexane-1,2-diol to C_{60} afforded exclusively *cis*-3 bismethanofullerene (*R*,*R*,^{f,s}*A*)-**61** in 32% yield and with complete diastereoselectivity. If the geometry of the alkyl groups at the glycolic C-atoms of the tether component deviates from a *gauche* relationship as in the case of tethers derived from *exo*, *cis*- and *trans*-norbornane-2,3-diol or from *trans*-cyclopentane-1,2-diol, hardly any macrocyclic product was formed [107].

Asymmetric synthesis of chiral C_2-symmetrical *trans*-2 or *trans*-3 bisadducts with the addends located in opposite hemispheres of C_{60} remained unknown until recently. The main challenge consisted in finding a large but conformationally constrained chiral tether. Very recently, derivatives of Tröger's base were used in the regio- and diastereoselective synthesis of bisadducts of C_{60} with the inherently chiral *trans*-2 functionalization pattern [108]. Tröger's base is a chiral amine with two bridgehead nitrogen atoms as stereogenic centers. Rigidity, C_2-symmetry, and a folded geometry with nearly perpendicular planes of aromatic rings makes it a suitable candidate for targeting inherently chiral addition patterns. Indeed, double Bingel addition of enantiomerically pure

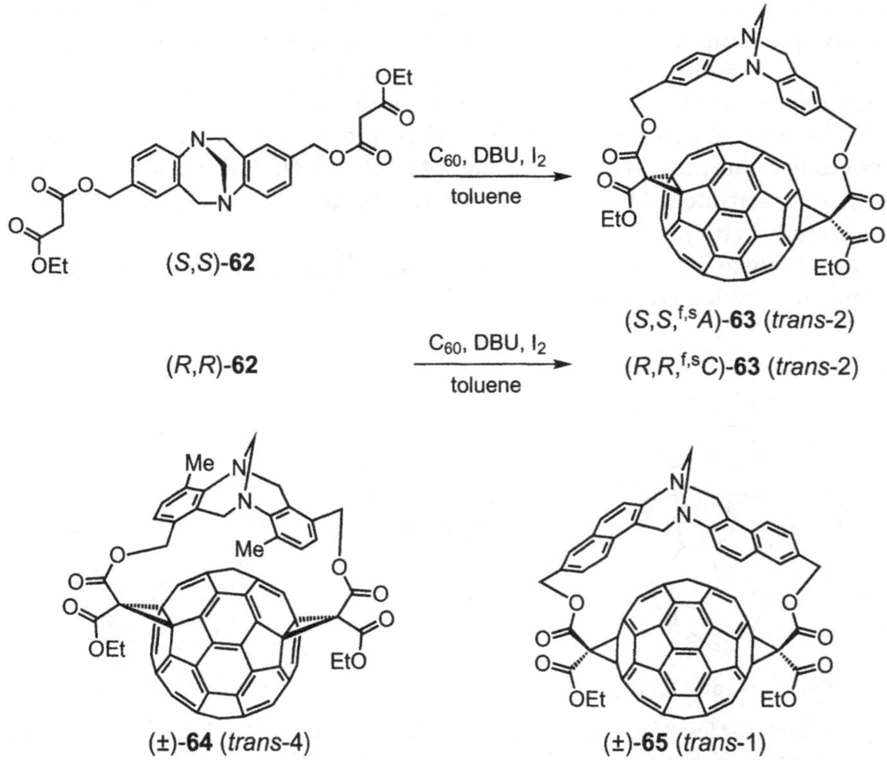

Scheme 11 Tröger's base in the tether-directed remote functionalization of C_{60} to give chiral *trans*-1, *trans*-2, and *trans*-4 bisadducts

(*S*,*S*)-**62** and (*R*,*R*)-**62** afforded two enantiomeric adducts (*S*,*S*,^{f,s}*A*)-**63** and (*R*,*R*,^{f,s}*C*)-**63**, respectively, with perfect diastereoselectivity (Scheme 11). The high asymmetric induction in the double addition of (*S*,*S*)-**62** and (*R*,*R*)-**62** to C_{60} is particularly remarkable given the very large distance between the two reacting fullerene bonds spanned by the tether. In addition, other derivatives of Tröger's base were successfully used as racemates in the regioselective tether-directed remote functionalization of C_{60} to give remarkably high yields of *trans*-4 and *trans*-1 bisadducts (±)-**64** (38%) and (±)-**65** (58%), respectively, with achiral fullerene functionalization patterns.

4.1.5 Further Tethers

In some cases, very simple aliphatic molecules were quite efficient in the regioselective tether-directed Bingel macrocyclization. Thus, a bismalonate featuring a plain ethane-1,2-diyl tether afforded the according *cis*-3 bisadduct of C_{60} (12%, analog of **60**, Fig. 7) along with a considerable amount (13%) of the corresponding monoadduct [107]. An interesting aspect of *cis*-2 bisadduct (±)-**66** (Fig. 8) which was regioselectively synthesized using a but-2-ene-1,4-diol-derived tether is the planar chiral unit represented by the *trans*-substituted ethene moiety [104, 109]. Double Bingel reaction of a bismalonate tethered by the 1,10-phenanthroline-2,9-dimethylene moiety yielded bismethano fullerene derivative (±)-**67** (13%) with the *e* addition pattern together with the corresponding *trans*-4 isomer (9%), while the 9,9′-spirobifluorene-2,2′-dimethylene tether afforded exclusively the *e* bisadduct (±)-**68** (44%) [55]. Macrocyclization with a rather large tether derived from 2,9-diphenyl-1,10-phenanthroline gave two diastereoisomeric bisadducts, (±)-**69** and (±)-**70** in 20 and 5% yield, respectively. The absorption spectra of both (±)-**69** and (±)-**70** are nearly identical and both consistent with the *trans*-3 addition pattern. The NMR spectra, however, are very different and reveal C_1 symmetry for (±)-**69** while (±)-**70** is C_2-symmetrical. Based on these considerations, (±)-**69** and (±)-**70** are identified as *in-out* (cf. also Sect. 4.1.3 and Fig. 6, compound (±)-**45**) and *in-in* diastereoisomers, respectively [55].

Addition of bis-β-ketoester **71** to C_{60} afforded two *cis*-2 bisadducts, (±)-**72** and **73**, the latter resulting from concomitant isomerization of the double bond (Scheme 12). In a similar addition of the saturated diethyl 3,9-dioxoundecanedioate, the *cis*-2 bisadduct was formed regioselectively in 7% yield [59].

Fullerene-calix[4]arene conjugate **74** (Fig. 8) was prepared by stepwise addition to C_{60} of the corresponding calixarene with two appended mixed amide-esters of malonic acid [110]. Its similarity in shape with the FIFA World Cup inspired the authors to dedicate this nanoscale trophy to the French football team who won the 1998 world championship. The synthesis of electron donor/acceptor dyads comprising [60]fullerene and π-extended tetrathiafulvalenes by Bingel-type macrocyclization afforded two regioisomers the structures of which could not be definitively assigned [111].

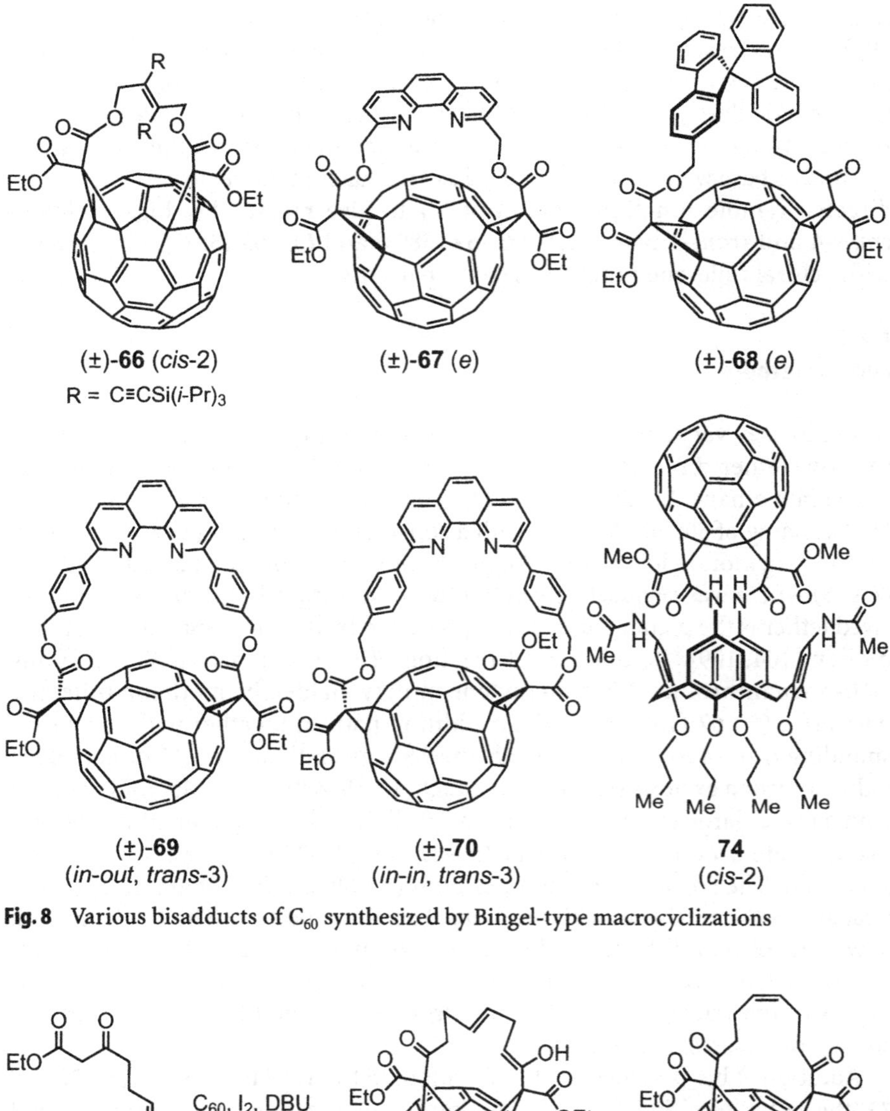

Fig. 8 Various bisadducts of C_{60} synthesized by Bingel-type macrocyclizations

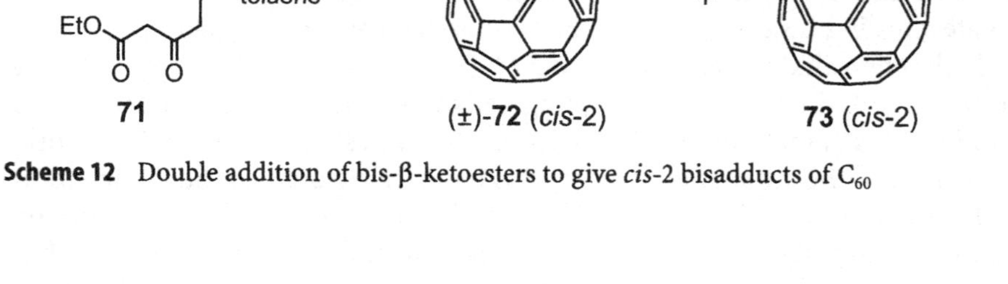

Scheme 12 Double addition of bis-β-ketoesters to give *cis*-2 bisadducts of C_{60}

4.2 Tether-Directed Multiple Functionalizations of C_{60} by Addition of Bis- to Tetrakismalonates

The reaction of racemic, C_3-symmetrical cyclotriveratrylene (CTV)-trismalonate conjugate (±)-**75** (Scheme 13) with C_{60} represents the first reported

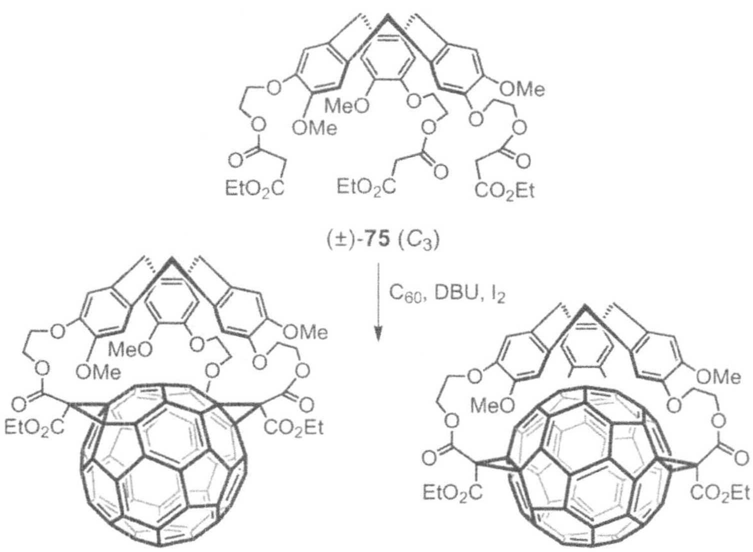

Scheme 13 Synthesis of C_3-symmetrical trisadducts of C_{60} by addition of a trismalonate derived from cyclotriveratrylene. For clarity, only one stereoisomer is shown for each addition pattern

tether-directed triple Bingel addition [112]. The two racemic products resulting from this reaction were assigned as *e,e,e* and *trans*-3,*trans*-3,*trans*-3 regioisomers (±)-**76** and (±)-**77**, respectively. Both are characterized by an interesting superposition of inherently chiral fullerene addition pattern (possible configurations ($^{f,s}C$) and ($^{f,s}A$)) and a conformationally chiral CTV unit (possible configurations (*P*) and (*M*) [113]). A particularly fascinating aspect of both trisadducts is their topological chirality which has been demonstrated in a detailed mathematical analysis [114]. Whereas each of the theoretically possible four "classical" *e,e,e* stereoisomers (($P,^{f,s}A$)-**76**, ($M,^{f,s}C$)-**76**, ($M,^{f,s}A$)-**76**, and ($P,^{f,s}C$)-**76**) corresponds to a unique topological stereoisomer, the stereoisomers of the *trans*-3,*trans*-3,*trans*-3 trisadduct are interconvertible pairwise (($P,^{f,s}A$)-**77**↔($M,^{f,s}A$)-**77**, and ($M,^{f,s}C$)-**77**↔($P,^{f,s}C$)-**77**) by a continuous deformation of the structure making the C_{60} sphere pass through the nine-membered ring of the CTV unit. In other words, there are four classical but only two topological stereoisomers for the *trans*-3,*trans*-3,*trans*-3 trisadduct.

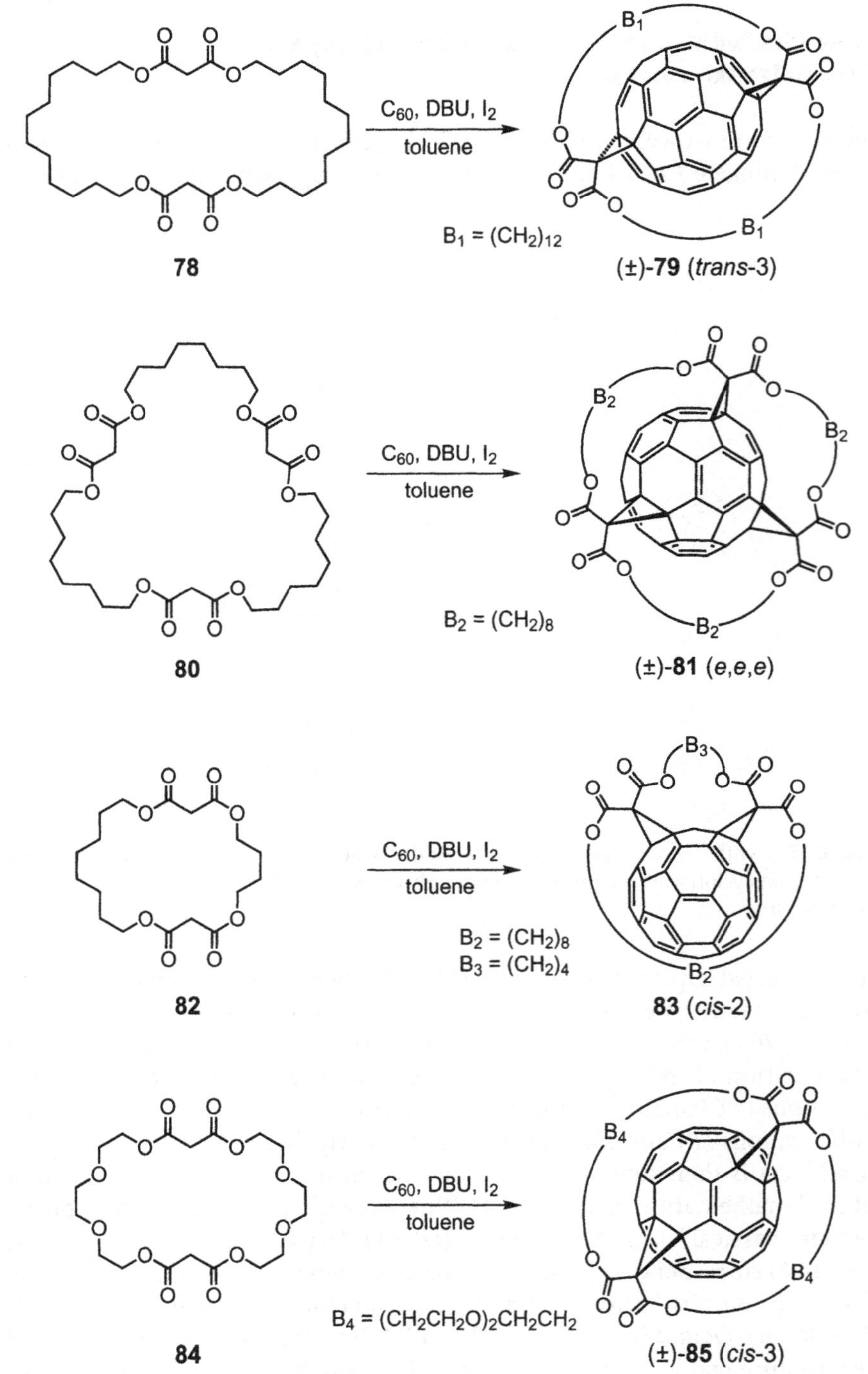

Scheme 14 Synthesis of bisadducts and a trisadduct of C_{60} starting from macrocyclic oligomalonates

To achieve a high degree of regioselectivity in additions of tethered reagents to fullerenes, the tether should be conformationally rather rigid in general. Hirsch and co-workers suggested an alternative concept using flexible cyclo-oligomalonates [115], an approach that was successfully employed for the single-stage synthesis of bis- to tetrakisadducts (Scheme 14). The regioselectivity of these reactions is governed by the distribution of strain within the macrocycle and the specific addition pattern formed depends on the size of the spacers. Cyclodi- and trimalonates containing identical alkane-α,ω-diyl spacers selectively form bis- and trisadducts of C_{60} with rotational symmetry. In contrast, when macrocycles with unequal spacers are used, other addition patterns are preferred. Thus, reaction of C_{60} with macrocyclic bismalonate **78** containing oligomethylene chains of equal length afforded C_2-symmetrical *trans*-3 bisadduct (±)-**79** and addition of cyclotrimalonate **80** gave 42% of C_3-symmetrical *e,e,e*-trisadduct (±)-**81** together with a small amount (2%) of the otherwise inaccessible *trans*-4,*trans*-4,*trans*-4 isomer (Scheme 14). On the other hand, bismalonate **82** with unequal oligomethylene chains yielded C_s-symmetrical *cis*-2 bisadduct **83**. In addition, various tetraadducts were obtained by reaction of C_{60} either with a cyclotetramalonate or with two cyclodimalonates, and macrocyclization with two cyclotrimalonates afforded an S_6-symmetrical hexakisadduct with a pseudo-octahedral addition pattern [115].

Wilson and co-workers reproduced the triple addition of macrocyclic trismalonate **80** under the conditions described by Hirsch and co-workers using i^3HeC_{60} instead of pristine C_{60} as a starting material [116]. In this case, ^3He NMR was used as an additional analytical tool to monitor the regioselectivity of triple addition and the singlet at –12.085 measured for the solely obtained trisadduct is characteristic for a trisadduct with *e,e,e* functionalization pattern [117]. Replacement of the octane-1,8-diyl residues of **80** by oligoethyleneglycol-type 3,6-dioxaoctane-1,8-diyl fragments afforded a single trisadduct the addition pattern of which was tentatively assigned as *trans*-4,*cis*-3,*cis*-3. A high regioselectivity was also observed for the addition of cyclodimalonate **84** to C_{60} which afforded *cis*-3 bisadduct (±)-**85** as single product isomer in up to 60% yield [116].

5 Tether-Directed Remote Functionalizations Based on the Diels-Alder Reaction

5.1 Bis-Functionalization of C_{60} with *o*-Quinodimethanes Connected by Oligomethylene-, Podand-, and Crown Ether-Type Tethers

Nishimura and co-workers investigated the regioselectivity of the Diels-Alder addition to C_{60} of two *o*-quinodimethane entities generated from bis(bromomethyl) precursors **86–89** (Scheme 15, top) by reductive 1,4-elimination (cf. [118, 119]) and attached via ether linkages to oligomethylene tethers $(CH_2)_n$

of different lengths (n=2–5) [16]. The identification of product isomers was complicated by the dissymmetry of the individual addends in such a way that even if the *cis*-1 as well as all *trans* bisadducts (cf. Fig. 1) were excluded for steric reasons, the macrocyclization could lead to eight constitutional isomers [120, 121]. Assignments were based on a combination of NMR spectroscopy (symmetry), MM2 calculations (steric energies), UV/vis spectroscopy (comparison of typical fingerprints to those of known bisadduct isomers with the same residual fullerene π-chromophore), and chromatographic separability on a chiral stationary phase (proof of chirality). Di- and trimethylene tethers afforded *cis*-2 (**90** and **91**) and *cis*-3 ((±)-**92** and (±)-**93**) bisadducts (Scheme 15,

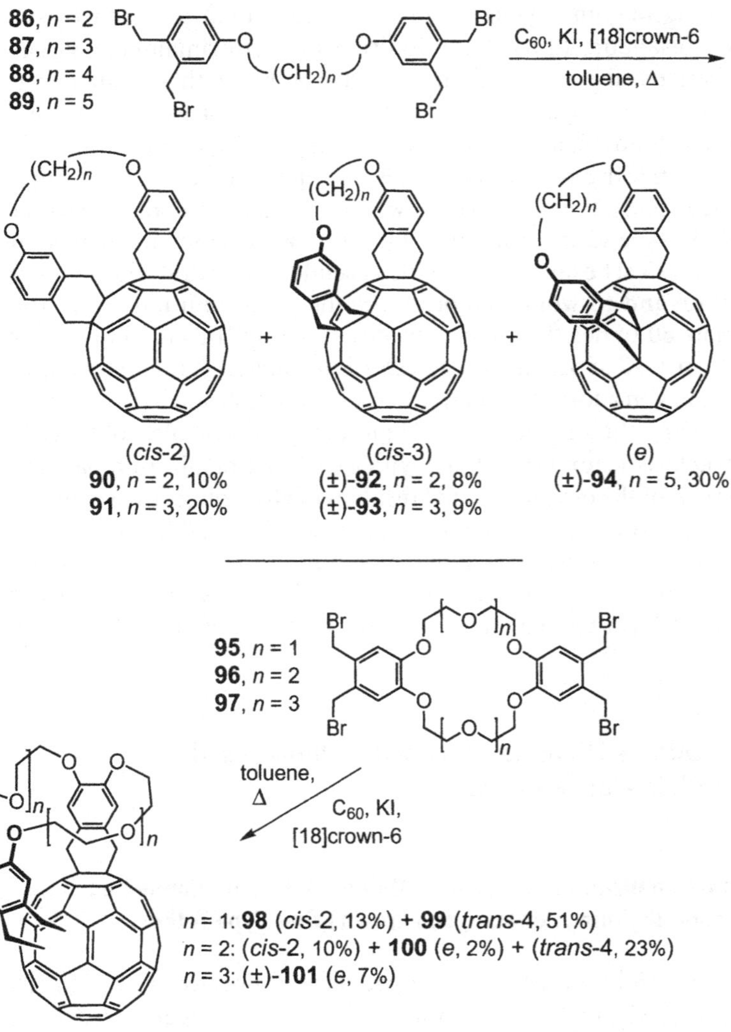

Scheme 15 Regioselective addition of singly (*top*) and doubly (*bottom*) tethered *o*-quinodimethane moieties to C_{60}

top), the latter two featuring an inherently chiral addition pattern [21, 22, 100, 101]. For n=5, e bisadduct (±)-**94** was obtained as sole isolable isomer, whereas a complex product mixture resulted for n=4 [120, 121]. Interestingly, the chirality of the e isomer (±)-**94** cannot be associated with stereogenic centers outside the fullerene sphere but with a noninherently chiral addition pattern [21, 22, 100, 101] of the C_{60} moiety; it is related to the asymmetrical substitution of the e_{edge} (cf. Fig. 1) addend.

Cleavage of the oligomethylene chains of **90**-(±)-**94** with BBr_3 in benzene produced the corresponding bisphenols, making the whole sequence a templated synthesis [120]. The chiral *cis*-3 ((±)-**92** and (±)-**93**) and e ((±)-**94**) bisadducts as well as the corresponding bisphenols were successfully resolved [120, 122] by HPLC on silica gel coated with cellulose tris(3,5-dimethylphenyl carbamate) [123]. In this context, the chromatographic resolution of a series of variously functionalized chiral *trans*-3 and *trans*-2 bisadducts of C_{60} by Nishimura and co-workers should also be mentioned [124].

Exchange of the pentamethylene chain of tetrabromide **89** (Scheme 15, top) for a 3-oxapentane-1,5-diyl analog led to a lower selectivity in the Diels-Alder macrocyclization with C_{60} despite the fact that both tethers have the same length in terms of number of atoms. Next to the e isomer (oxa-analog of **94**, 26%), the podand-type spacer afforded also 10% of *cis*-2 bisadduct [125]. In order to enhance the regioselectivity of the addition and also to confer ionophoric properties to the fullerene adducts, the bis(bromomethyl)benzene reactive groups were interconnected by an additional chain in the crown ether-derived tetrabromides **95**–**97** and the resulting product distributions are shown in Scheme 15 (bottom) [125, 126]. Whereas both the 3-oxapentane-1,5-diyl and the [18]crown-6 type (**95**) tethers afforded minor amounts of *cis*-2 isomers (e.g. **98** from **95**), they gave e and *trans*-4 (**99**) adducts, respectively, as main products. The formation of the *cis*-2 (**98**) rather than the *cis*-3 crown ether adduct was ascribed by Nishimura and co-workers to the higher LUMO coefficients at the *cis*-2 position [35, 127, 128], with simultaneous consideration that the steric energies for the two isomers are similar despite a more parallel bond orientation in the *cis*-3 arrangement. Besides, the higher yields of e bisadducts (2% (±)-**100** and 7% (±)-**101**) obtained with increasing crown ether ring size seems to be due to the interplay between high inherent fullerene reactivity at e positions and decreasing steric constraints imposed by longer tethers.

The complexation behavior of some of the fullerene-appended crown ethers was explored by electrospray ionization mass spectrometry (ESI-MS) and the K^+/Na^+ selectivities determined with ion-selective electrodes [126]. Strongly varying complexation abilities and higher K^+/Na^+ selectivities in comparison to underivatized dibenzo[18]crown-6 may be attributed to the fixation of certain conformations in the macropolycyclic ionophores. As opposed to fullerene-crown ether conjugates (±)-**43** and (±)-**45** (Fig. 6) in which the ionophore is fixed with tangential orientation in close proximity to the carbon sphere, cyclic voltammetry (CV) did not indicate any change of potentials upon addition of K^+ ions to a solution of **99** [126].

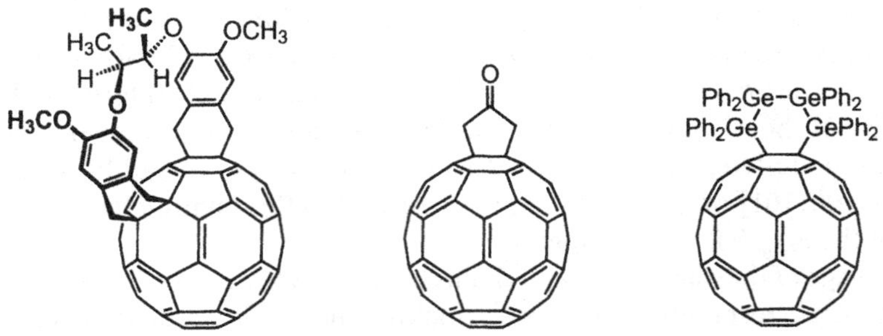

$(S,S,^{f,s}C)$-**102**
(*cis*-3)

165

166

Fig. 9 *cis*-3 Bisadduct $(S,S,^{f,s}C)$-**102** having an inherently chiral functionalization pattern generated diastereoselectively by use of an enantiomerically pure tether, and two fullerene derivatives (**165** and **166**) formed by addition of short chain-type reactive species with homofunctional ends

When using an (*S*,*S*)-configured tether synthesized stereospecifically from enantiopure commercial (*R*,*R*)-butane-2,3-diol as chiral auxiliary (cf. Sect. 4.1.4, Fig. 7 [107] and the next section [129]), Nishimura and co-workers obtained *cis*-3 bisadduct **102** (3%, Fig. 9) next to the corresponding *cis*-2 isomer (10%) [122]. Whereas the chirality of the major isomer resides in the stereogenic centers of the tether only, that of *cis*-3 bisadduct **102** is also due to the inherently chiral fullerene functionalization pattern [21, 22, 100, 101] which was formed with complete diastereoselectivity. Based on a combination of ^{1}H NMR spectroscopic analysis and computer modeling, the absolute configuration considered for **102** [122] corresponds to the IUPAC notation $(S,S,^{f,s}C)$ [19]. Due to the chiral residual fullerene π-chromophore of $(S,S,^{f,s}C)$-**102**, its circular dichroism (CD) spectrum displays very intense Cotton effects, largely exceeding those of the *cis*-2 isomer, the π-chromophore of which is only slightly perturbed by the stereogenic centers of the tether [122]. This is in full accord with similar observations made by the groups of Diederich, Hirsch, and others [21].

5.2 Regio- and Stereoselective Introduction of Boronic Acid Functions into C_{60} Using Saccharides as Imprinting Templates

The selective Diels-Alder addition of bis(*o*-quinodimethane) species to C_{60} was also investigated by Shinkai and co-workers [130–132]. The tethers, in this case, were derived from easily available saccharides bound to the *o*-quinodimethane precursor units as arylboronates (Scheme 16). This allowed not only for a regio- but also a diastereoselective double addition to C_{60} and, furthermore, for an easy removal of the tether (template) from the product. In addition, the resulting fullerene-derived bis(boronic acid)s were expected to have a high molecular recognition potential, notably for the saccharides used as templates in their formation. The entire process can, therefore, be considered as a molecular imprinting technique using C_{60} as a soluble nanosize matrix.

The most thoroughly investigated tethers in this study were L- and D-threitol which afforded 72:28 and 28:72 mixtures, respectively, of *cis*-3 bis(boronic acid)s ($^{f,s}C$)-**103** and ($^{f,s}A$)-**103** (Scheme 16) after removal of the templates and regeneration of the boronic acid functions [130, 132]. The absolute configurations of the enantiomers of **103** were identified by comparison of their CD spectra to those of other *cis*-3 bisadducts of C_{60} that had previously been assigned by the groups of Harada and Diederich [105]. Competitive complexation studies indicated that L-threitol-imprinted boronic acid ($^{f,s}C$)-**103** and D-threitol-imprinted ($^{f,s}A$)-**103** preferentially rebind their original templates with up to 48% de [131]. Further tethers tested in this study were D-mannitol-3,4-carbonate and 3-*O*-methyl-D-glucofuranose which afforded the *e* (ca. 81% diastereoselectivity) and *trans*-4 bisadducts of C_{60} as major product isomers, whereas 1-*O*-methyl-α-D-mannopyranoside was rather unselective as a spacer [132]. After removal of the sugar templates and protection of the resulting boronic acids as 5,5-dimethyl-1,3,2-dioxaborinanes (cf. Scheme 16), the purified regioisomers were assigned

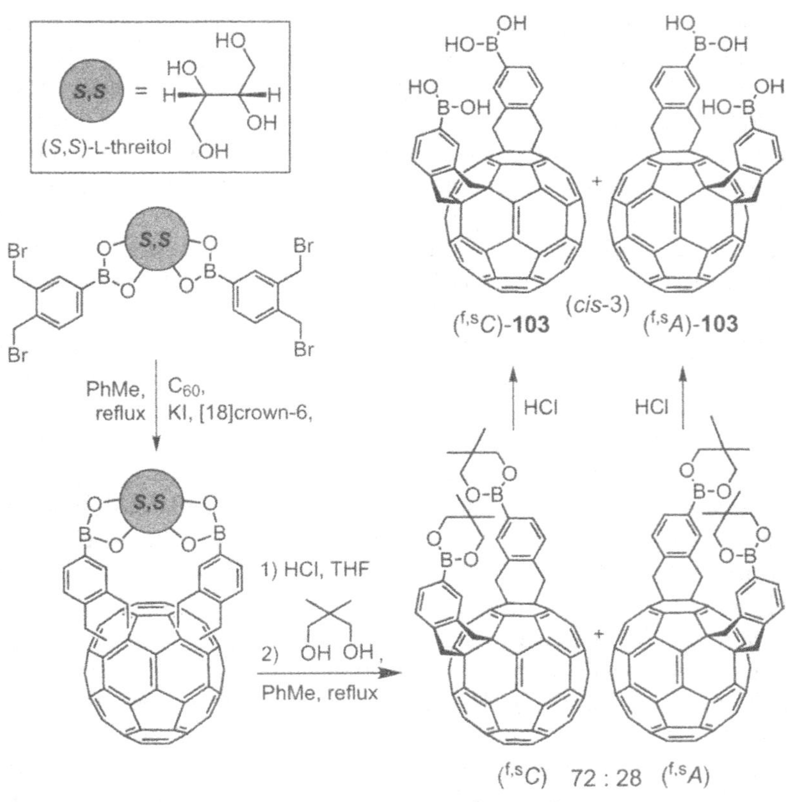

Scheme 16 Regio- and diastereoselective double Diels-Alder addition to C_{60} using a threitol derivative as imprinting template and the fullerene as a soluble nanosize matrix

mainly on the base of symmetry considerations (^1H and ^{13}C NMR) and their typical UV/Vis fingerprints. Besides, oxidation of the dioxaborinanes with H_2O_2 afforded the corresponding bis-phenols, the spectral data of which could be directly compared to those of the bis-phenols prepared by Nishimura and co-workers (cf. Sect. 5.1) [120].

The exploration of other regio- and diastereoselective double Diels-Alder reactions of C_{60} with *o*-quinodimethanes generated from bis[3,4-bis(bromomethyl)benzoates] derived from various methyl cholates and from (2*R*,3*R*)-butane-2,3-diol (cf. Sect. 4.1.4, Fig. 7 [107] and previous section, Fig. 9 [122]) showed that the magnitude of the Cotton effects induced in the π-chromophore of fullerene adducts with an achiral functionalization pattern reflect the size of the tether and the number of stereogenic centers present therein [129].

5.3 *trans*-1 Bisadduct as Key Intermediate en Route to a Sixfold Functionalization of C_{60} at Fully Addressable Octahedral Sites

A versatile tether-directed remote functionalization scheme was developed by Qian and Rubin to target *trans*-1 bisadducts of C_{60} by Diels-Alder macrocyclization [133, 134]. The selective synthesis of *trans*-1 regioisomers (cf. [50, 82, 83, 89, 90, 93, 135–137]) is particularly challenging, because they are the least likely to form on statistical and electronic grounds [20, 35, 127, 128]. A spacer based on tolane-4,4′-diyldiacetic acid (**104**, Scheme 17) was designed by computer modeling and the desired reactive 1,3-diene groups were generated in situ by cycloreversion of the cyclobutene rings of **104**. HPLC analysis of the crude mixture resulting from addition to C_{60} under high dilution conditions indicated a 1:(1.5–2) ratio of *trans*-1 (**105**) and *trans*-2 bisadducts, respectively, with no other regioisomers being detected.

The bridged *trans*-1 bisadduct **105** is a strategically protected building block with great potential in view of a sixfold functionalization of buckminsterfullerene at fully addressable positions of a pseudo-octahedral functionalization pattern (see below). Its versatility relies on the following features. (i) The remaining three double bonds at accessible *e* positions are strongly yet differentially activated toward subsequent nucleophilic attacks thanks to the additive activating and directing effects of the addends in place [12, 23, 32, 48, 133, 138, 139] (cf. also Sect. 3 and Scheme 2). They can be addressed, one at a time, in the order "doubly e_{face}" (in reference to the starting *trans*-1 bisadduct **105**), "doubly e_{edge} plus singly e_{face}" (in reference to the trisadduct resulting from the previous step), and "doubly e_{face} plus singly e_{edge}" (in reference to the tetrakisadduct resulting from the previous step). (ii) The tolane-derived tether between the fullerene-fused cyclohexene rings can be selectively removed by acid treatment, thus making a non-covalently blocked reactive *e* bond amenable to further functionalization. (iii) Elimination of the tether leaves two fullerene-fused cyclohexadiene rings in *trans*-1 arrangement that can be removed in a Diels-Alder/retro-Diels-Alder sequence [42, 43, 133, 134, 138, 139] (cf. also Sect. 3, Scheme 4, and [46, 47])

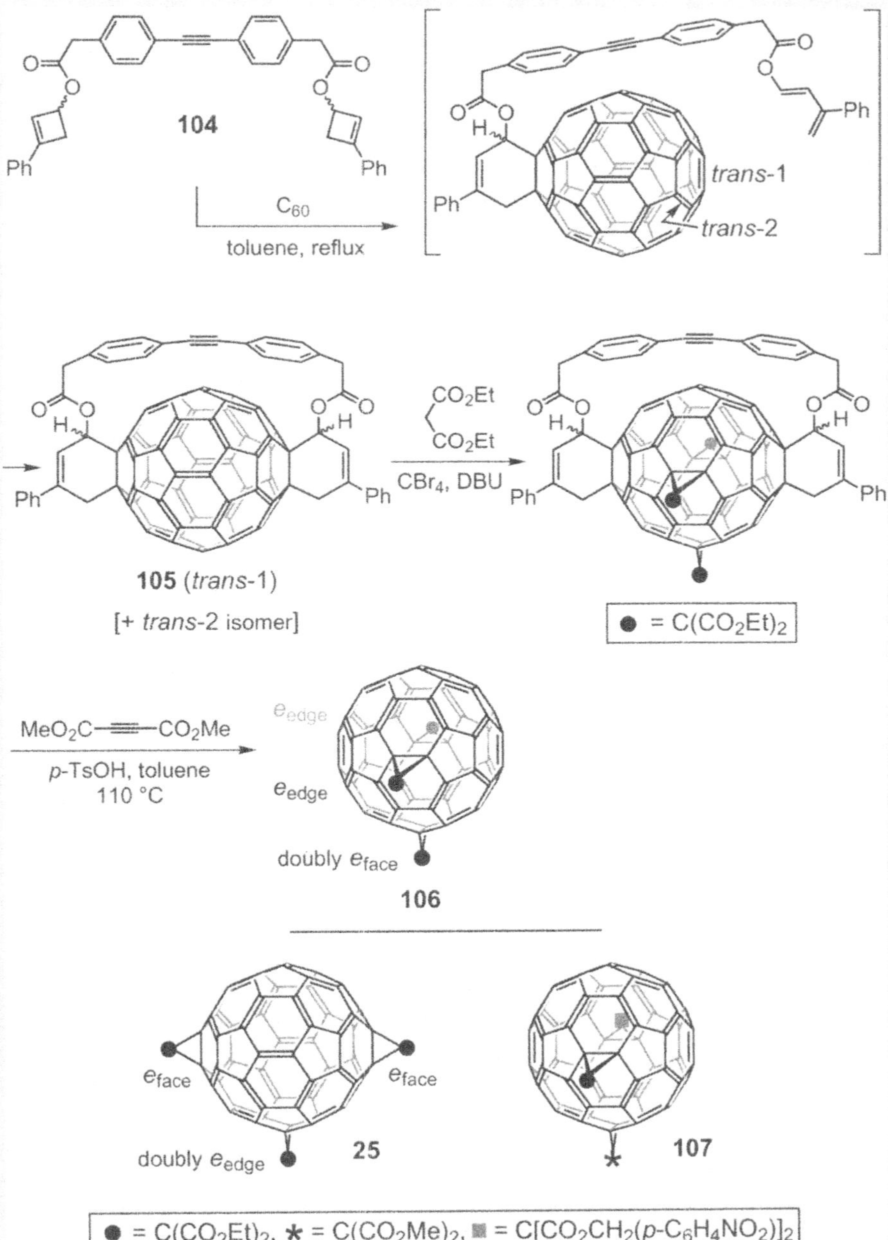

Scheme 17 Synthesis of *trans*-1 bisadduct **105** and regioselective transformation to higher adducts of C_{60} by further Bingel additions and, eventually, removal of the template

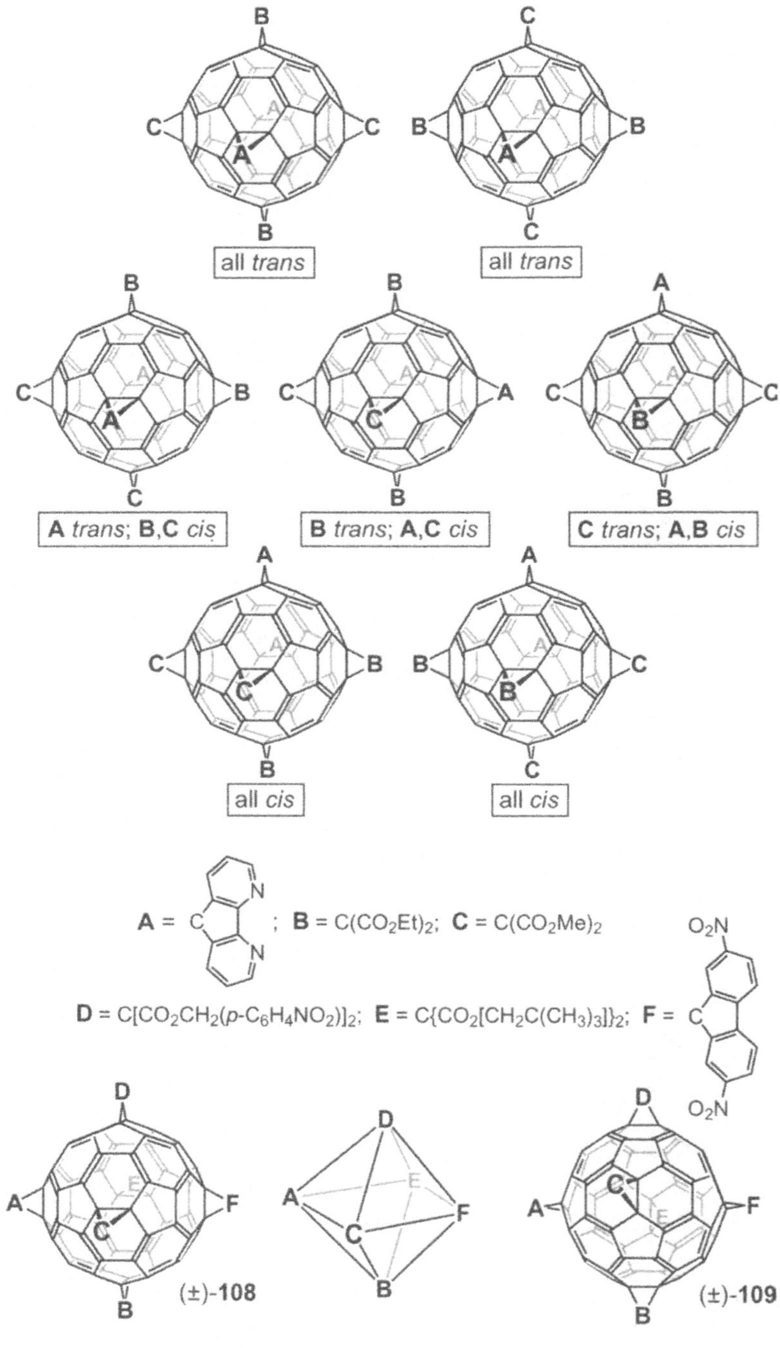

Fig. 10 The seven $A_2B_2C_2(C_{60})$ regioisomers with a pseudo-octahedral addition pattern (*top*) and two constitutional isomers $ABCDEF(C_{60})$ in which each addend has the same nearest neighbors but with inverted e_{edge} and e_{face} relationships (*bottom*)

under regeneration of the last two in a set of six pseudo-octahedrally arranged 6-6 bonds, thus allowing for a fully directed hexasubstitution if one takes into account that the last two additions can be done sequentially with different reagents. The flexibility of this templated functionalization method is illustrated by the examples described below and depicted in Scheme 17 and Fig. 10, using the potentialities (i)–(iii) to varying extents.

Three sequential Bingel type additions to **105**, with identical or different malonates, followed by removal of the tether as well as the fullerene-fused six-membered rings from the resulting pentakisadduct, afforded trisadducts **106** and **107**, respectively (Scheme 17) [133]. Interestingly, **106** is a constitutional isomer of trisadduct **25** (Scheme 17) synthesized by Diederich and co-workers starting from **14** (Scheme 2), according to the route outlined in Scheme 4 [47].

Six double bonds of C_{60}-I_h, with four *e* and one *trans*-1 interrelationships each, have their centers located at the vertices of a regular octahedron (cf. Fig. 10, bottom). If the geometry of these bonds as well as their orientation in space are taken into account, the O_h symmetry of the regular octahedron is reduced to T_h and the corresponding arrangement is often termed "pseudo-octahedral". As a consequence, the addition of six pairwise identical, symmetrical addends (A, A, B, B, C, C) to C_{60} can lead to seven diastereoisomers (Fig. 10, top) as compared to only five possible diastereoisomeric complexes resulting from the octahedral arrangement of three pairs of monodentate ligands around a transition metal (in the latter case, there is only a single all-*trans* and a single all-*cis* isomer) [138]. A complete library of the seven isomers was prepared by Qian and Rubin in an elegant parallel synthesis by application of a so-called "*mer*-3+3" regiocontrol strategy [138]. Two $B_2C(C_{60})$ and two $BC_2(C_{60})$ type *e,e,trans*-1 trisadducts were obtained as key intermediates according to routines (i) and (ii) specified above (cf. Scheme 17) [133]. Partial regioselectivity – possibly determined by an interplay between electronic and steric effects – in the functionalization of the remaining three fullerene bonds within the considered pseudo-octahedrally arranged set allowed the synthesis of the seven $A_2B_2C_2(C_{60})$ diastereoisomers to be completed (Fig. 10, top).

Complete command of addend permutation at all six positions of a group of pseudo-octahedrally arranged bonds allowed separate syntheses of ABCDEF(C_{60}) regioisomers (±)-**108** and (±)-**109** (Fig. 10, bottom) [139]. The exquisite control of the regiochemistry around the T_h-symmetrical "coordination core" of C_{60} was demonstrated, for example in the synthesis of (±)-**109**, by sequential Bingel additions of three different addends (B, C, D) to the *trans*-1 intermediate **105** (cf. above routine (i); Scheme 17), elimination of the tolanediacetic acid template, thus making a non-covalently blocked reactive position within the considered pseudo-octahedral pattern accessible, addition of the fourth malonate (E, routine (ii)), removal of the fullerene-fused cyclohexadiene rings in a Diels-Alder/retro-Diels-Alder sequence and, finally, consecutive addition of addends A and F to the regenerated fullerene double bonds (routine (iii)) [139]. Specific permutation in the order of attachment of addends B-D, on the other hand, allowed the preparation of the constitutional isomer (±)-**108**.

5.4
Regioselective Formation of Tetrahomo[60]fullerenes

Acid-catalyzed removal of the tolanediacetic acid template from **110** (Scheme 18), an analog of **105** (Scheme 17) with higher solubility, under simultaneous irradiation afforded the two constitutionally isomeric tetrahomofullerenes **111** and **112** (Scheme 18) in which the number of π-electrons on the fullerene core remains the same as in C_{60}, except that four 6-5 single bonds have been replaced by homoconjugative bridges [134, 140]. This one-pot multi-step reaction proceeded through efficient photochemically promoted [4+4] and retro-[2+2+2] intramolecular tandem rearrangements of the intermediately formed fullerene-fused cyclohexadienes **113** and **114** (Scheme 18) [141] (cf. also Sect. 7 and Scheme 27). The unexpected preference for the purple **111** (*syn*, 40% yield) over the yellowish brown **112** (*anti*, 10%) isomer in this process must arise from local electronic differences in the mono-rearranged intermediate **114** despite the relative remoteness of the inducing group.

A "mono-functionalized" dihomofullerene analog of **111** or **112** forms a Co(III) complex resulting from oxidative insertion of a CpCo (Cp=cyclopentadienide) fragment into a 6-5 formal single bond of the fullerene core [140,

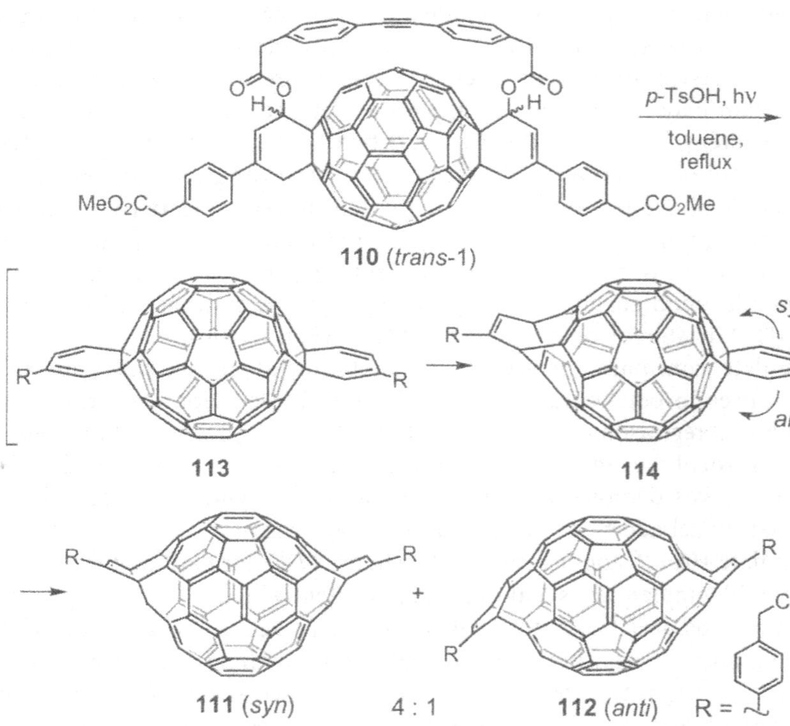

Scheme 18 Regioselective synthesis of tetrahomofullerenes **111** and **112** starting from *trans*-1 bisadduct **110**

141]. Its X-ray crystal structure shows that the cobalt atom is poised to insert into the created orifice, and such complexes constitute promising starting materials for the preparation of endohedral metallofullerene derivatives.

5.5 Regio- and Diastereoselective Formation of a *cis*-1 Bisadduct of C_{60} by Tandem Nucleophilic Addition/Diels-Alder Reaction

In a different study, Rubin and Neier investigated the addition of a series of electron-rich *N*-buta-1,3-dienyl-*O*-silyl ketene *N,O*-acetals, e.g. **115** (Scheme 19), to C_{60} – a reaction that consists of a tandem sequence of nucleophilic addition and Diels-Alder reaction [142]. Although the reactive centers of the bifunctional reagents are separated only by three bonds, the reaction cascade can be considered as a highly regioselective tether-directed "remote" functionalization of two neighboring (*cis*-1) bonds within a six-membered ring of the fullerene (Scheme 19). The interest in such compounds comes from efforts aiming at the functionalization of contiguous reactive centers in C_{60}, possibly leading to a fully saturated, planar cyclohexane substructure prone to spontaneous ring-opening in a retro-[2+2+2] fashion under creation of an orifice delimited by a 15-membered ring. Despite the fact that *cis*-1 positions in a monoadduct of C_{60} show particularly large LUMO coefficients [5, 128], *cis*-1 bisadducts are difficult to prepare by sequential addition of untethered reagents because of the mutual steric hindrance of most addends [20, 35, 127, 128].

It was demonstrated in model reactions that the tandem sequence starts with a nucleophilic Michael-type addition, presumably involving a SET (single electron transfer) followed by a radical recombination [142]. The second step consists of an intramolecularly accelerated [4+2] cycloaddition leading to diastereoisomerically pure octahydroquinolines, e.g. (±)-**116** (Scheme 19), fused to two adjacent sides of a six-membered ring of C_{60}.

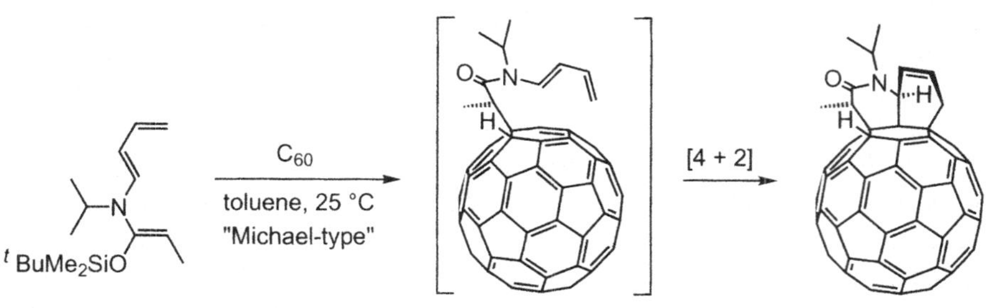

115

(±)-**116**
(*cis*-1)

Scheme 19 Tandem reaction between *N*-butadienyl *N,O*-ketene acetal **115** and C_{60}: Michael-type addition followed by intramolecular Diels-Alder reaction

5.6
Intramolecular [4+2] Cycloaddition of a [60]Fullerene-Appended Anthracene

When Irngartinger and co-workers investigated the photosensitized endoperoxide formation with C_{60}-appended anthracenes, they discovered that the refluxing of cyano-ester **117** in toluene under exclusion of dioxygen led to an intramolecular Diels-Alder addition of the arene to the fullerene moiety (Scheme 20) [143]. The conceivable *cis*-1 and *cis*-2 bisadducts are both C_1-symmetrical, and the NMR data of the product were in agreement with either structure. PM3 computational studies indicated the *cis*-1 adduct to be more stable by 3.6 kcal mol^{-1} but no definite structural assignment was made.

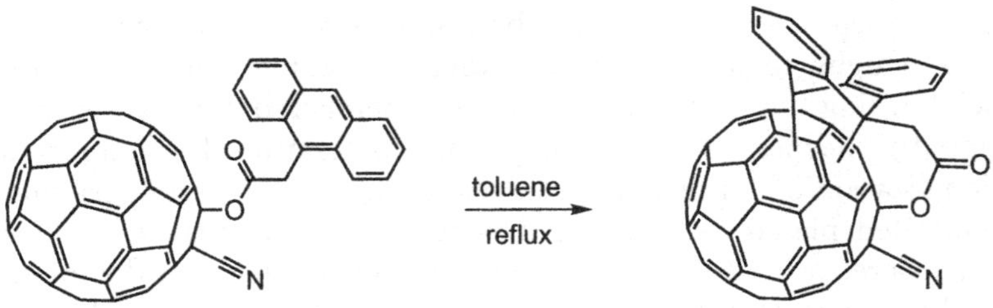

117

Scheme 20 Intramolecular [4+2] cycloaddition between the anthracene and the fullerene moiety of **117**

5.7
Activation of *e* Double Bonds in C_{60} by Reversible Addition of 9,10-Dimethylanthracene as a Template

At the end of this section on Diels-Alder-based tether-directed remote functionalizations, two procedures should be presented that do not involve the use of a tether but are nevertheless template-directed: the first is of great practical use due to its simplicity, the second is less generally applicable but all the more spectacular.

In 1994, Hirsch and coworkers isolated and structurally assigned seven out of the eight possible bisadducts resulting from stepwise addition of two diethyl bromomalonate molecules to C_{60} [20]. Whereas the *cis*-1 isomer was not detected, the *e* and *trans*-3 bisadducts (cf. Fig. 1) were formed as main products and the regioselectivity of double Bingel additions was satisfactorily correlated to a combination of enhanced frontier orbital coefficients, more reactive compressed 6-6 bonds, or the formation of thermodynamically more stable products – aspects that are all related to the characteristic cage distortion in monoadducts of C_{60} [35, 127, 128]. The high tendency for attacks at *e* bonds, expressed by enhanced frontier orbital coefficients [35, 127, 128], becomes even more pronounced as the number of existing *e* type addends increases. If the considered reaction is reversible, the thermodynamic aspect gains importance and multiadducts with an incomplete pseudo-octahedral functionalization pattern predominate in the equilibria. This is the case for the Diels-Alder addition of 9,10-dimethylanthracene (DMA), and adducts of the type $C_{60}(DMA)_n$ ($1 \leq n \leq 5$) (Scheme 21) can serve as templates for irreversible additions or addend substitutions at *e* positions. The effectiveness of the procedure is illustrated by the direct transformation of buckminsterfullerene to the "pseudo-octahedral" $C_{60}[C(CO_2Et)_2]_6$ as sole regioisomer in 23% yield (8 equiv. diethyl bromomalonate, 10 equiv. DMA; DBU) [34] or even 48% yield (10 equiv. of each DMA, diethyl malonate, and CBr_4; DBU) [144] in comparison to 14% yield and reduced regioselectivity in the absence of DMA activation [34]. As illustrated by a selection of literature references, the method has found widespread use in the preparation of mixed hexakisadducts with a pseudo-octahedral functionalization pattern by starting from a monoadduct or a lower multiadduct of C_{60} [34, 115, 145–149] or from monomeric azafullerene derivatives $RC_{59}N$ [150]. The equilibria of the Diels-Alder addition of DMA to i^3HeC_{60} and i^3HeC_{70} have been investigated in detail by Wang, Saunders, and Cross using ^3He NMR spectroscopy [151].

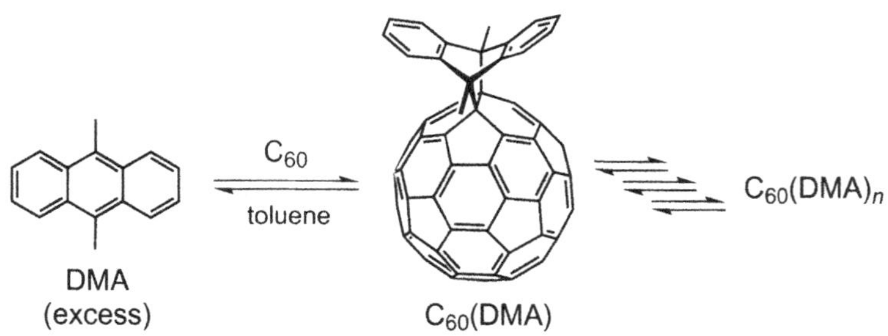

Scheme 21 Equilibria of the reversible Diels-Alder addition of an excess of 9,10-dimethylanthracene (DMA) to C_{60}. Although the multiadducts consist of mixtures of regioisomers, the addends have a preference for the bonds belonging to a pseudo-octahedrally arranged set with only *e* type nearest neighbors

5.8 Topochemically Controlled Anthracene Transfer Reaction

A highly remarkable, topochemically controlled, absolutely regioselective formation of a *trans*-1 bisadduct of C_{60} resulting from Diels-Alder addition of anthracene was discovered by Kräutler and co-workers [135]. The heating of oxygen-free crystals of monoadduct **118** (Scheme 22, top) to 180 °C for 10 min led to a solid 1:1 mixture of bisadduct **119** and C_{60} with almost complete conversion. This observation was rationalized by a model with oriented stacks of monoadduct molecules in the crystal of **118**, putting the critical *trans*-1 bond in an ideal position for anthracene transfer by satisfying not only the geometric but also the orbital-symmetry criteria of the required retro-Diels-Alder/Diels-Alder sequence. Also, the positions of atoms and molecular fragments are conserved in a remarkable way on the transition from reagents to products which constitutes an ideal premise for a topochemical transformation. As to the thermodynamics of the high-yielding reaction, the main driving force seems to be the entropy increase related to the orientational disorder [152–154] of the generated C_{60} spheres in the crystal. While a number of other crystalline monoadducts resulting from reaction of C_{60} with 1-methylanthracene, 9-methylanthracene, or 9,10-dimethylanthracene showed a similar behavior, others, such as 2-methylanthracene, or the more highly substituted 2,3,6,7-tetramethylanthracene, and 2,6-di(*tert*-butyl)anthracene simply reverted to [60]fullerene and the respective anthracene [136]. This points toward a crucial dependence of a potential topochemical reaction on steric effects of the substituents. A similar control of the reaction course by the presence and position of alkyl substituents on the arene was observed for the direct synthesis of bisadducts by heating solid C_{60} with various anthracene derivatives [137]. The diene reagents that actually provided bisadducts did so in a highly selective way under exclusive

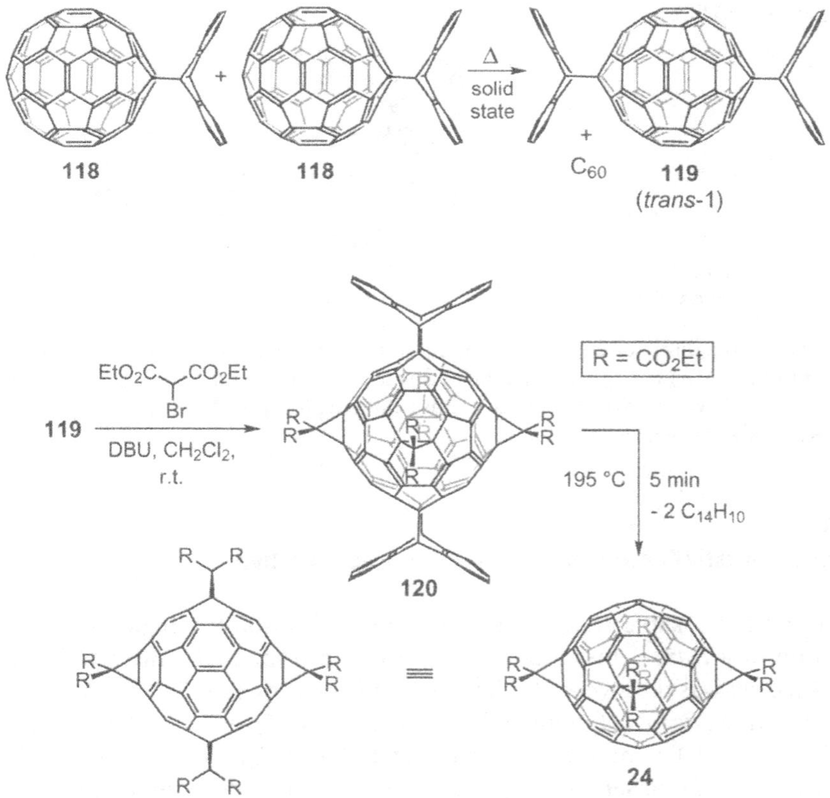

Scheme 22 Topochemical solid-state reaction of monoadduct **118** leading to a 1:1 mixture of *trans*-1 bisadduct **119** and C_{60} (*top*) and use of **119** in a templated synthesis of tetrakis-adduct **24** which is not available by direct addition of four bromomalonate molecules to C_{60} (*bottom*) (cf. Scheme 4)

formation of *trans*-1 regioisomers. This course of the solid-state cycloaddition differs strongly from the reaction of C_{60} with an excess of anthracene in CS_2 solution which leads only to minor amounts (1.6%) of the *trans*-1 isomer next to four other bisadducts and a sizeable amount of trisadducts [155].

Taking advantage of the activating and directing effects (cf. Sect. 5.3 and 5.7) of the addends in *trans*-1 bisadduct **119**, Kräutler and co-workers, by treating a suspension of the compound in CH_2Cl_2 with an excess of diethyl bromomalonate and DBU, obtained the mixed D_{2h}-symmetrical hexakisadduct **120** (Scheme 22, bottom) in 95% yield [48]. Heating of the yellow solid to 195°C for 5 min in the absence of dioxygen produced anthracene and the dark brown-green D_{2h}-symmetrical tetrakisadduct **24** with all addends located on an equatorial belt, a compound that had been obtained by Diederich and co-workers using the tether-directed remote functionalization approach described in Sect. 3 (cf. Scheme 4) [46, 47].

6 Tether-Directed Remote Functionalizations Based on [3+2] Cycloadditions

6.1 Regio- and Stereoselective Addition to C_{60} of Vinylcarbenes Generated from Tethered Bis(cyclopropenone acetal)s

Some early work on the regio- and diastereoselective tether-directed bis-functionalization of C_{60} was reported by Nakamura and co-workers [156, 157]. Their anellating reagents consisted of an oligomethylene tether of variable length bearing cyclopropenone acetal units at the ends (Fig. 11, top). Under the applied reaction conditions (150 °C, dry 1,2-dichlorobenzene), these groups are in equilibrium with minute amounts of nucleophilic vinylcarbenes which undergo [3+2] cycloaddition with 6-6 bonds of the fullerene, thus affording various bisadducts depending on the length of the tether. Due to the unsymmetrical nature of the addends, the number of theoretically possible bisadduct isomers increases from eight (for symmetrical addends) to 36 [158]. The structural assignment of the isolated products was based on symmetry considerations as reflected by NMR spectroscopy and a computational analysis of the conformational strain of the tethers. The product distribution as a function of the length of the $(CH_2)_n$ tether was as follows: *cis*-1 and *cis*-2 (n=3), *cis*-1 (n=4) (cf. Fig. 1), and *cis*-3 ((±)-**121**, n=6, Fig. 11, bottom). For n=4, no bisadduct was found, probably as a result of the ring strain of a 9- or 10-membered ring which would be generated in the double addition, and a complex bisadduct mixture was formed with an octamethylene tether. As to the possible isomerism originating from the dissymmetry of the addends, a single constitutional isomer was formed in each case with the tether-bearing C-atoms of the fullerene-fused pentagons roughly pointing toward each other (cf. Fig. 11, bottom).

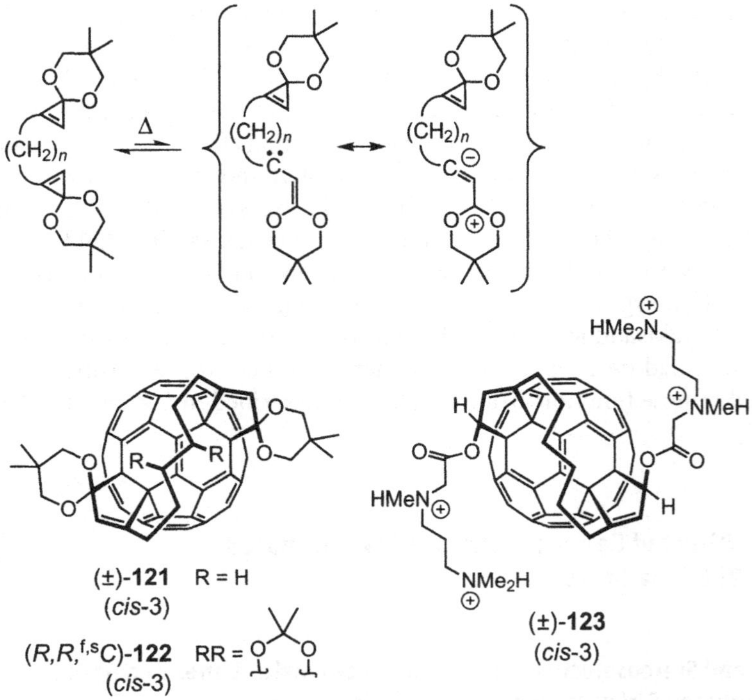

Fig. 11 Thermal equilibrium between tethered cyclopropenone acetals and the according vinylcarbenes (*top*), and *cis*-3 bisadducts of C_{60} having a [3+2] cycloaddition with suchlike reagents as a key step in their syntheses (*bottom*)

The use of a chiral, enantiomerically enriched (ee=89%) tether including an acetal of an (*R*,*R*)-configured 1,2-diol unit resulted in the diastereoisomerically pure *cis*-3 bisadduct **122** (Fig. 11, bottom) [156, 157] next to two C_1-symmetrical isomers. The formation of the latter points at some change in the directing effect of the chiral tether as compared to the achiral hexamethylene counterpart in (±)-**121** (see above). As the chiroptical contribution of side chains with stereogenic centers to the Cotton effects of fullerene derivatives with an inherently chiral π-chromophore is almost negligible [21, 97, 99, 102, 103], the absolute configuration of the *cis*-3 addition pattern of **122** could be identified as (*R*,*R*, ^{f}A) [159] (= (*R*,*R*,$^{f,s}C$) [19]) by comparison of its optical rotation $[\alpha_D]=-1875$ with the CD spectra of [CD(–)488]-($^{f,s}C$)-**55** and [CD(+)488]-($^{f,s}A$)-**55** (Sect. 4.1.4 and Scheme 10) with assigned absolute configurations [105]. Besides, reduction of the enantiomeric diones obtained from (±)-**121** by acetal cleavage and subsequent esterification of the deprotected diols with (*S*)-2-methoxy-2-phenylacetic acid allowed the separation of the diastereoisomeric esters, i.e., of bisadducts of C_{60} with enantiomeric addition patterns [159]. After separation, the diol functions could be regenerated by reduction of the esters with DIBAL-H (diisobutylaluminum hydride).

C_{60}-Derived tetraammonium ion (±)-**123** (Fig. 11, bottom), prepared in a four-step transformation from (±)-**121**, proved to be an efficient vector for gene transfection [160, 161]. It binds reversibly to duplex DNA, leads to its condensation, delivery of the complex into mammalian cells through phagocytosis, and to the transient expression of the released DNA in the target cell. The amphiphilic (±)-**123** can undergo strong ion pairing interactions with the phosphate backbone of DNA and the fullerene core provides hydrophobic binding for the DNA condensation and probably also for the delivery to the interior of the target cell. The transfection efficiencies of (±)-**123** and a related, enantiopure fullerene-based vector [162] were shown to be comparable to those of lipid-based commercial agents [160, 161].

6.2 Addition of Tethered Diazides to C_{60}

Addition to C_{60} of a number of variously tethered diazides was investigated by Luh and co-workers [163–166]. If the functional groups were connected by a simple ethylene or trimethylene chain, a single C_s-symmetrical diazadihomofullerene regioisomer (analog of **124**, Scheme 23, top) was formed in which two imino groups are inserted into the 6–5 junctions of a fluorene subunit of C_{60} [164]. The same regioisomer was obtained by Li and co-workers with the one-carbon tether of diethyl 2,2-diazidomalonate [167]. Addition of 2,2-dibenzylpropane-1,3-diyldiazide to C_{60} allowed the isolation and characterization of a *cis*-1 bis(triazolo)fullerene, presumably isomer **125** [168], as primary bisadduct (Scheme 23, top) [166]. Its thermolysis in refluxing chlorobenzene (24 h) afforded a 54:40:6 mixture of *cis*-1 bisazirinofullerene **126**, diazadihomofullerene **124**, and C_{60} in quantitative yield. When **126** was heated in chlorobenzene for 24 h, C_{60} was produced in 35% yield but no **124** was detected. There was no decomposition of **124** under the same conditions and, furthermore, no interconversion occurred between **124** and **126** [166].

Use of enantiopure, C_2-symmetrical diazides prepared from D- and L-threitol afforded diazadihomofullerenes (*R*,*R*)-**127** (Fig. 12) and (*S*,*S*)-**127**, respectively, again with two imino groups inserted into the 6-5 junctions of a fluorene substructure of C_{60} [165]. No diastereoselectivity was observed in these double additions because the fullerene functionalization pattern is C_s-symmetrical and, therefore, achiral.

In a similar way, reaction of C_{60} with diazide (*S*,*S*)-**128** (Scheme 23, bottom), prepared from 2,3-*O*-isopropylidene-L-threitol, afforded C_1-symmetrical bisadduct (*S*,*S*)-**129** which is stable for weeks in the dark [163]. Under the influence of ambient light, however, one of its imino groups rearranges regioselectively to an epimino structure in (*S*,*S*)-**130**, as demonstrated among others by analysis of ^{13}C NMR C-N couplings of ^{15}N-labeled material. Thermolysis of (*S*,*S*)-**130** cleanly converts the molecule back to the isomeric (*S*,*S*)-**129**, thus providing the first example of interconvertible azahomo-/azirinofullerene isomers.

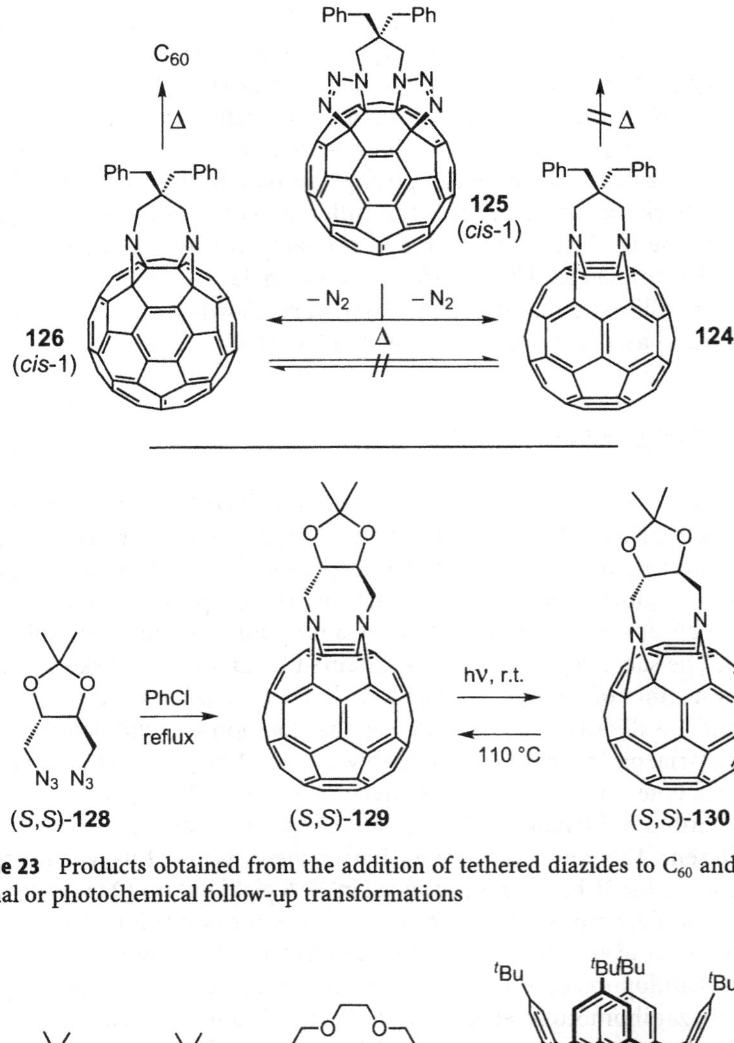

Scheme 23 Products obtained from the addition of tethered diazides to C_{60} and possible thermal or photochemical follow-up transformations

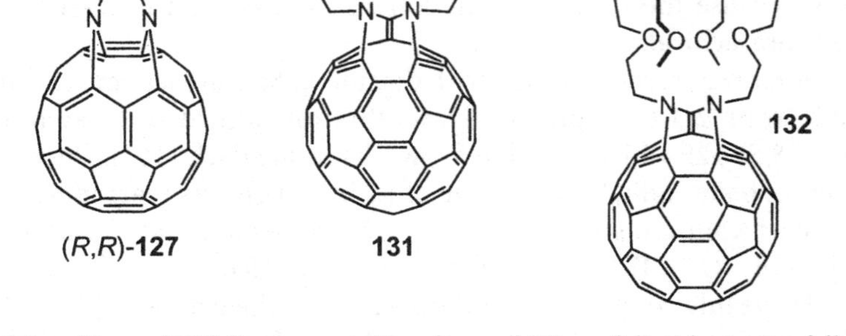

Fig. 12 Diazadihomo[60]fullerenes resulting from addition of diazides to C_{60}, followed by extrusion of dinitrogen

A number of "[60]fullerenocrowns" and "[60]fullerenocalixcrowns" with several carbon atoms of the fullerene integrated in a crown ether macrocycle, e.g. **131** and **132** (Fig. 12), were reported by the groups of Shinkai [169, 170], and Li and Chen [171, 172]. They were synthesized by the [3+2] cycloaddition/N_2-extrusion route using diazides tethered by various oligoethyleneglycol-derived chains, in some cases with an embedded or a spiro-linked calixarene. Interestingly and in contrast to the previous examples of this section, all isolated and characterized products were diazadihomo[60]fullerenes with the imino groups inserted in adjacent 6-5 bonds of the same pentagon and, thus, connected to a common carbon atom of the fullerene core. This functionalization pattern is the same as that observed for sequential additions of two independent monoazides, followed by dinitrogen extrusion [5, 173, 174]. The complexation of various salts containing alkali and alkaline earth metal ions as well as Zn^{2+}, Ag^+, and $[MeNH_3]^+$ by **131** was studied by ^1H NMR or UV/vis spectroscopy [169]. Silver(I) ions which are known to interact particularly well with the nitrogen atoms of azacrown ethers induced the largest optical absorption shifts. The evaluated binding constants are comparable to those of regular crown ethers. Some structurally related "[60]fullerenocrowns" with embedded thiophene-containing bi- and triarene moieties were used to study the photoinduced electron transfer (PET) in doped conjugated polymers [175].

6.3 Cage Opening of C_{60} Starting with Tandem [3+2]- and [4+2] Cycloadditions of a Trifunctional Diene-Diazide Conjugate

In their efforts to open up an orifice in the fullerene shell large enough to introduce atoms or small molecules (cf. Sect. 5.4), Rubin and co-workers targeted a *cis*-1,*cis*-1,*cis*-1 trisadduct by reaction of C_{60} with conjugate **133** (Scheme 24) in which two reactive azide functions are connected through *o*-phenylene spacers to a central buta-1,3-diene unit [176]. Ingeniously, the latter constitutes both a part of the tether and a reactive group for a [4+2] cycloaddition. Reaction with the three 6-6 bonds of a six-membered ring of the fullerene should afford a fully saturated planar cyclohexane moiety prone to undergo a retro-[2+2+2] ring opening reaction. Three major products, **134**, **135**, and **136** (Scheme 24) were isolated after column chromatography of the product mixture resulting from the reaction of C_{60} with **133**, and the following reaction pathway was proposed: the azide groups add first under formation of a *cis*-1 bis(triazolo)-fullerene and subsequent intramolecular Diels-Alder reaction with the third double bond of that hexagon affords a polyanellated fullerene derivative. Either intermediate can rearrange, under twofold N_2 extrusion, from a bis(triazolo)- to a diazadihomofullerene with the imino groups inserted in the two 6-5 bonds radiating from positions 1 and 4 of the targeted six-membered ring of C_{60}. The combination of these steps affords a mixture of *endo*- and *exo*-cycloadducts **137** and **138** which could not be observed directly because the electron-rich 1,4-diaminobutadiene structure within the fullerene core is very easily attacked by singlet oxygen under formation of endoperoxides **139** and **140**. Subsequent retro-[2+2+2] ring opening within the 5,6-dioxabicyclo[2.2.2]oct-2-ene subunit afforded bislactams **134** and **135** which were finally dehydrogenated to **136** under aromatization of the cyclohexa-1,4-diene ring [176].

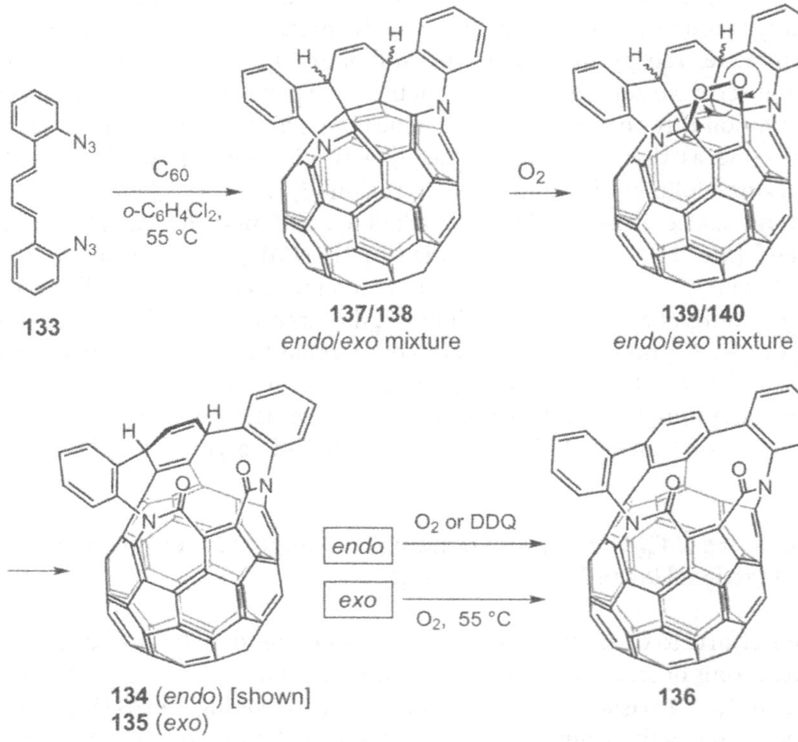

Scheme 24 Synthesis of "open-cage fullerene" **136** by double azide addition, a combination of Diels-Alder reaction and N_2-extrusion/rearrangement (→**137/138**), endoperoxide formation (→**139/140**), [2+2+2] ring opening (→**134/135**), and dehydrogenation

Bis-lactam **136** has a sizeable orifice in its carbon shell and a more soluble di(*tert*-butyl)-substituted derivative allowed reversible, direct insertion of ^3He and molecular hydrogen [177]. Hydrogen storage inside open-cage C_{60} has recently become a topic of growing interest due to potential technological applications [178, 179]. The shielding inside the cavity of **136** is illustrated by the chemical shifts $\delta=-10.10$ (^3He NMR, *o*-dichlorobenzene-d_4, dissolved ^3He as standard) and $\delta=-5.43$ (^1H NMR, $\delta[H_2^{solution}]=4.53$) measured for endohedral ^3He and H_2, respectively. The barrier to insertion was calculated to be much higher (41.4 kcal mol^{-1}) for dihydrogen with its larger surface area than for helium (24.5 kcal mol^{-1}); the value for helium was determined experimentally to 24.6±0.8 kcal mol^{-1}. Under thermodynamic aspects, the encapsulation was estimated to be thermoneutral or slightly exothermic [177]. The creation of an orifice in the carbon sphere by retro-[2+2+2] ring fragmentation and the insertion of metal ions into **136** and other "open cage" fullerene derivatives was also investigated computationally using the ONIOM method [180].

6.4 Tether-Directed Remote Functionalizations by Azomethine Ylide Addition to C_{60}

Although the [3+2] cycloaddition with azomethine ylides is one of the most popular reactions in fullerene chemistry [181, 182], it has been used only in a few cases for the tether-directed remote functionalization of C_{60} [183–185]. In a systematic study, Prato and co-workers examined the regioselectivity of the intramolecular attack of azomethine ylides tethered through oligomethylene chains $(CH_2)_n$ (n=3–5) to a C_{60}-fused isoxazoline anchor (Scheme 25) [183, 184]. The starting isoxazolofullerenes were obtained by [3+2] cycloaddition of buckminsterfullerene and isonitrile oxides carrying a protected glycine at the end of a pendant alkyl chain. The azomethine ylide dipole was generated from the attached amino acid after deprotection and treatment with formaldehyde in refluxing toluene. Due to the dissymmetry of the isoxazoline ring, the number of conceivable regioisomeric *cis*-type bisadducts is doubled and structural assignments were further complicated by the C_1-symmetry of the products. They were based on a combination of UV/vis spectroscopy, electrospray ionization mass spectrometry (ESI-MS), elemental analysis, and NMR spectroscopy in combination with semiempirical PM3 computations.

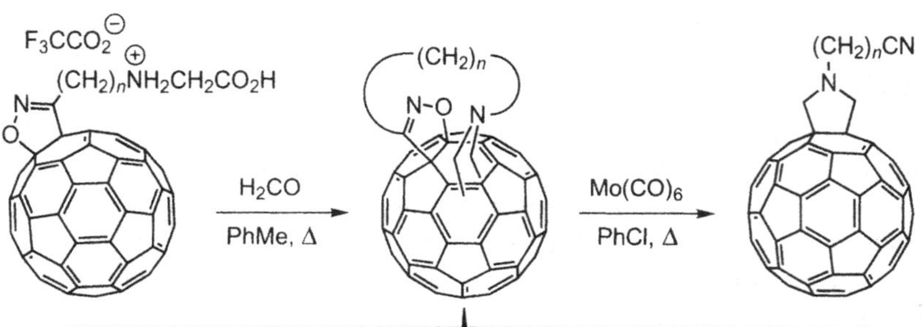

n = 3: (±)-**141** (*cis*-1, C=N-sided)
n = 4: (±)-**142** (*cis*-1, C=N-sided)
n = 5: (±)-**143** (*cis*-1, N–O-sided)
n = 5: (±)-**144** (*cis*-3, C=N-sided)
n = 5: (±)-**145** (*cis*-2, C=N-sided)

Scheme 25 Regioselective second functionalization of C_{60} by intramolecular [3+2] cycloaddition of a tethered azomethine ylide function. The fullerene-fused isoxazoline anchor can be removed by treatment with molybdenum hexacarbonyl

With the tri- and tetramethylene tethers, the macrocyclization reaction was highly selective affording *cis*-1 bisadducts (±)-**141** (65%) and (±)-**142** (67%) (pyrrolidine ring adjacent to isoxazoline-C), respectively, as sole isolated products (Scheme 25). The longer pentamethylene spacer, on the other hand, afforded a mixture of three regioisomeric bisadducts: the main product ((±)-**143**,

54% yield) was identified as the second *cis*-1 regioisomer (pyrrolidine ring adjacent to isoxazoline-O); in the second case ((±)-**144**, 31% yield), the structural analysis pointed toward a *cis*-3 isomer (pyrrolidine ring near isoxazoline-C) and the minor product ((±)-**145**, 2%) was tentatively assigned a *cis*-2 structure (pyrrolidine ring near isoxazoline-C).

Interestingly, treatment of the bisadducts with $Mo(CO)_6$ in refluxing chlorobenzene led to the removal of the fullerene-fused isoxazoline heterocycle, affording *N*-cyanoalkyl-substituted pyrrolo[60]fullerenes (Scheme 25), except in the case of (±)-**141** which essentially decomposed [183, 184].

[3+2] Cycloaddition between C_{60} and a bis(azomethine ylide) generated from sarcosine (*N*-methylglycine) and *anti*-(dibenzo[18]crown-6)-dicarbaldehyde (analog of **42**, Scheme 9) afforded a bisadduct mixture containing at least three isomers [185]. The number of possible product isomers is very high taking into account the unsymmetrical nature of the generated pyrrolidine rings (cf. Sect. 5.1). In addition, stereoisomerism can originate from the newly generated stereogenic center in each fullerene-fused pyrrolidine ring as well as from a possible restricted conformational mobility of the crown ether making it a planar chiral unit [89, 90].

7 Cyclobutadi[60]fullerene Derivatives as Products of Intramolecular Cage "Dimerization" by [2+2] Cycloaddition

The long-standing proposals of solid-state photo- and pressure-induced polymerization of fullerenes by [2+2] cycloaddition [186] obtained recent support by the first example of a controlled intramolecular photodimerization of two methanofullerene moieties, reported by Knol and Hummelen [187]. The required close proximity of the carbon spheres in precursor **146** (Scheme 26) was brought about by a *m*-phenylene spacer introduced together with the methano bridge carbon atoms by the well-established Bamford-Stevens protocol, starting from the required bis(tosylhydrazone). Refluxing of the primary adduct mixture including bis(homofullerene) **147** in *o*-dichlorobenzene led to a clean conversion to bis(methanofullerene) **146**. When the latter was irradiated with a 150-W sodium flood lamp, it afforded a clean steady-state mixture of **146** (ca. 40%) and **148** (ca. 60%) with cyclobuta-fused fullerene moieties [187]. The photodimerization process is significantly retarded by dioxygen which implies a mechanism involving the triplet excited state of the methanofullerene cages of **146**. Cycloadduct **148** was isolated in pure form and shown to be a double *cis*-3 regioisomer. Interestingly, the steric requirements for the photodimerization enforce opposite configurations of the inherently chiral *cis*-3 addition patterns of the interconnected fullerenes, thus leading to achiral **148**. Refluxing of **148** in *o*-dichlorobenzene results in complete cycloreversion to **146** within 15 min [187], a behavior similar to that reported for cyclobutadi[60]fullerene (C_{120}) which reverts to buckminsterfullerene upon heating [188, 189].

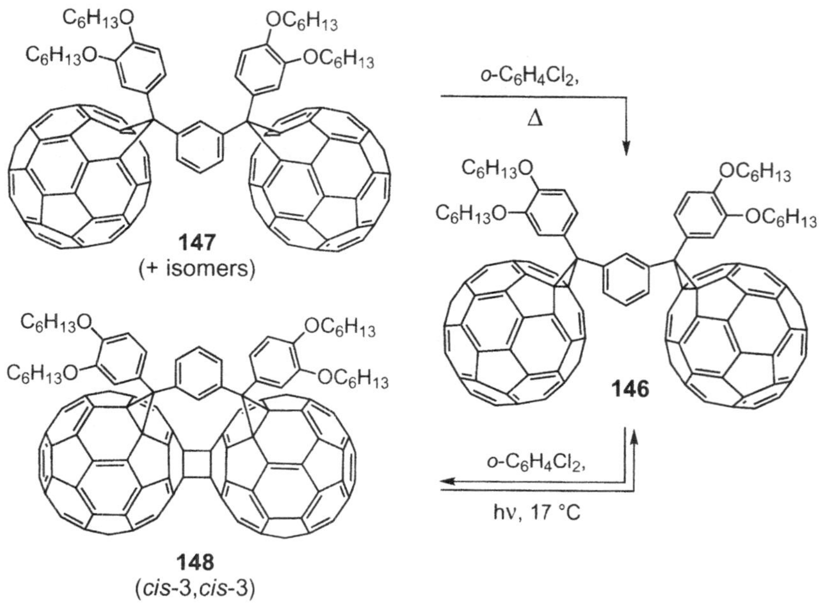

Scheme 26 Transformation of bis(homofullerene) **147** to bis(methanofullerene) **146** and photodimerization equilibrium between the latter and **148** which results from [2+2] cycloaddition between the fullerene moieties

Two related [2+2] dimerizations which may be best termed as "tether-directed *near* functionalizations" were reported by Komatsu and co-workers [190, 191]. In one of these studies [190], the authors discovered that the reaction between C_{60} and phthalazine (**149**, Scheme 27) takes a very different course in solution as compared to the solid state. When heated in 1-chloronaphthalene at 255°C, it yielded dihomofullerene derivative **150** which is an analog of compounds first reported by Rubin and co-workers (cf. also Sect. 5.4 and Scheme 18) [134, 141] and also used by other groups, often in the context of cage-opening reactions [179, 192–196]. Its formation was explained by a sequence including a [4+2] cycloaddition between heteroarene **149** and [60]fullerene, dinitrogen extrusion from the primary adduct **151** under formation of naphthofullerene intermediate **152**, an intramolecular formal [4+4] cycloaddition between its *o*-quinodimethane subunit and the fullerene core, and a succeeding retro-[2+2+2] ring opening of *cis*-1 dicyclopropafullerene **153** [190]. In contrast, when C_{60} was allowed to react with 4 equiv. of phthalazine (**149**) in the solid state under mechanochemical conditions using high-speed vibration milling (HSVM), **150** was completely absent from the reaction mixture and, instead, a difullerene analog of triptycene (**154**) was formed, probably by Diels-Alder addition of *o*-quinodimethane intermediate **152** to a second C_{60} molecule. Although this compound with two fullerene cages rigidly fixed in close proximity is stable in the solid state up to 200 °C, irradiation (275-W incandescent lamp) of a $CHCl_3$ solution or heating in *o*-dichlorobenzene gradually transforms it into **155** by intramolecular [2+2] cycloaddition. In sharp contrast to the [2+2]-dimer of buckminsterfullerene, C_{120}, and **148** (Scheme 26), heating of **155** to 200 °C leads to no appreciable cycloreversion. On the other hand and similarly to C_{120}, the cyclobuta-fusion in **155** is readily cleaved upon transfer of one electron as shown by cyclovoltammetry (CV) and differential pulse voltammetry (DPV) [190].

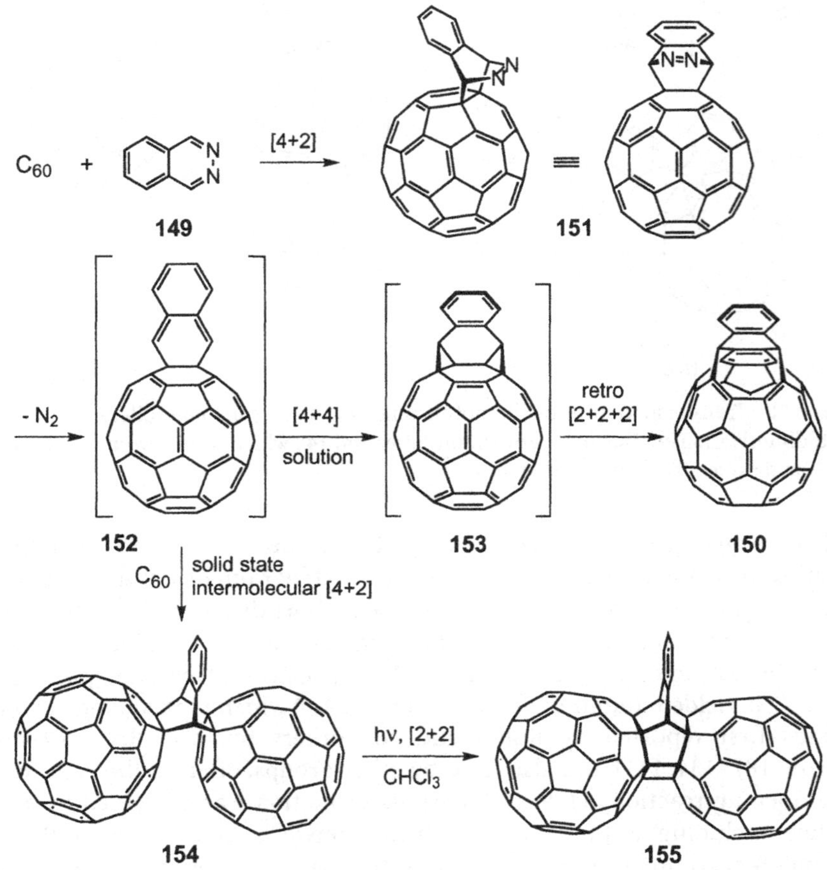

Scheme 27 Different courses of the reaction between C_{60} and phthalazine (**149**) in the solid state (high-speed vibration milling, HSVM) as compared to solution phase chemistry (1-chloronaphthalene, 255 °C)

In a similar manner, solid-state HSVM of C_{60} together with 0.5 equiv. of 3,6-di(pyridin-2-yl)-1,2,4,5-tetrazine followed by heating of the solid mixture to 150 °C afforded **156** (Fig. 13) by a Diels-Alder/N_2-extrusion/Diels-Alder sequence [191]. When a solution of **156** in *o*-dichlorobenzene was irradiated with room light, an azo-bridged "photodimer" (analog of **155**, Scheme 27) was formed by intramolecular [2+2] cycloaddition. Its cyclobuta-fusion was found to be thermally (100 °C) and electrochemically (six reduction steps) stable.

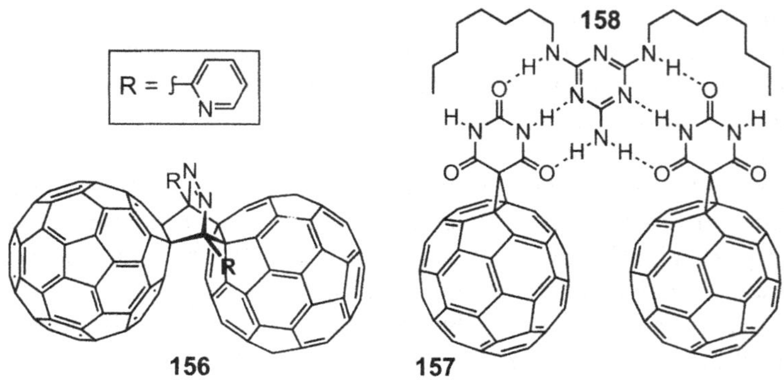

Fig. 13 Product **156** resulting from high-speed vibration milling of C_{60} with 3,6-di(pyridin-2-yl)-1,2,4,5-tetrazine, followed by heating of the solid mixture, and a supramolecular associate (**157·158·157**) prone to [2+2] photodimerization of the fullerene moieties upon irradiation

Bassani and co-workers finally reported the photodimerization of a fullerene derivative in a hydrogen-bonded supramolecular associate (Fig. 13) [197]. Thus, when a dilute *o*-dichlorobenzene solution of fullerene-barbituric acid conjugate **157** and *N,N'*-dioctylmelamine (**158**) (1:1 mixture) was irradiated, MALDI-TOF mass spectrometric analysis revealed a signal for the dimer of **157** which was not observed in the absence of the melamine derivative. This reaction represents a remote functionalization mediated by a non-covalent template.

8 Miscellaneous Tether-Directed Remote Functionalizations

The last section will group some particular examples of tether-directed remote functionalizations that are not directly related structurally or synthetically to other chapters of this review.

A complex, multiply bridged *cis*-1 type adduct of C_{60} (**159**) (Scheme 28) with a noninherently chiral addition pattern was obtained in a tandem reaction between the alkaloid scandine (**160**) and C_{60} [198, 199]. The sequence includes

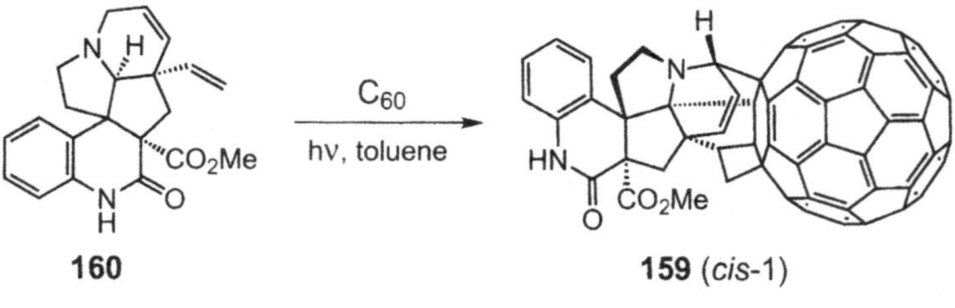

Scheme 28 Tandem reaction between the alkaloid scandine (**160**) and C_{60} under diastereoselective formation of a *cis*-1 bisadduct (**159**)

a photoinduced addition of the tertiary amine, presumably followed by [2+2] cycloaddition of the appended vinyl group at an adjacent formal double bond of the fullerene. No stereoisomer of **159** was reported and its formation can be assumed to be highly diastereoselective.

One of the most spectacular fullerene derivatives obtained by tether-directed remote functionalization is probably the peanut-shaped **161** (Scheme 29, top) with two fullerene cages interconnected by four propane-1,3-diamine chains, reported by Nakamura and co-workers [200]. The synthesis is akin to the well known oxidative addition of four piperidine molecules to a [5]radialene substructure of C_{60}, followed by epoxidation of the exocyclic double bond of the resulting local fulvene entity [201]. Accordingly, stirring of a solution of buckminsterfullerene with 16 equiv. of *N,N′*-dimethylpropane-1,3-diamine in air-saturated chlorobenzene at 21 °C under irradiation with visible light afforded the quadruply tethered bisfullerene **161** (12%) next to monomeric 1,4-adduct **162** (19%) as sole products. The structure of a near-C_{2v}-symmetrical molecular container **161** with fullerene end caps and an opening between the epoxy bridges was undoubtedly established by X-ray analysis of the orange crystals. A most notable feature of **161** is the fact that its molar UV/vis absorptivity is almost the same as that of the above-mentioned monomeric tetrapiperidinyl-epoxy-fullerene [201] despite the presence of two chromophores per molecule. This rare case of extreme hypochromicity is explained by the symmetrical, rigid structure of **161** in which two isolated fullerene chromophores with high oscillator strength and parallel transition moments are located at a short distance from each other [200].

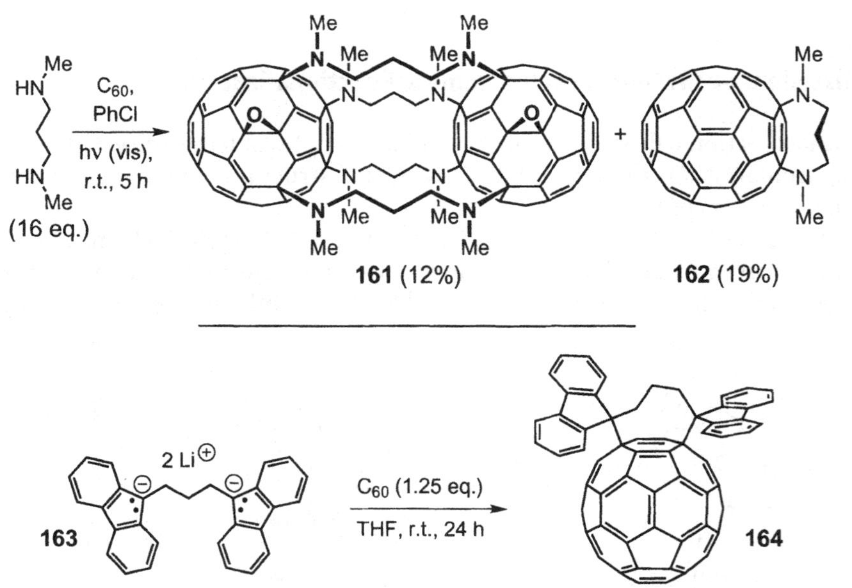

Scheme 29 Functionalization of C_{60} by addition of tethered amine (*top*) or fluorenide (*bottom*) nucleophiles, followed by oxidation

An early example of tether-directed remote functionalization was reported by Komatsu and co-workers when studying the reaction of [60]fullerene with fluorenide ions [202]: treatment of the carbon allotrope in THF under argon for 24 h with dilithium 9,9′-(propane-1,3-diyl)di(fluorenide) (**163**) afforded dispiro adduct **164** with the fluorene moieties connected to positions 1 and 4 of a six-membered ring of the fullerene (1,4-adduct) (Scheme 29, bottom). Similarly to the above addition of amines, the nucleophilic attack of the fluorenide ions on the carbon cage is followed by an oxidation.

Fullerene dianions, on the other hand, can be doubly alkylated by alkyl halides. Following this principle, Cousseau and co-workers reacted C_{60}^{2-}, generated by reduction of the neutral fullerene with a sodium alkanethiolate in acetonitrile under argon, with various halides RX, some of which carried an additional ester or keto function [203, 204]. The product was generally a mixture of 1,2- and 1,4-adduct (major product, except for R=Me). Not unsurprisingly, attack of C_{60}^{2-} by the electrophiles 1,3-diiodopropane, 1,4-diiodobutane, and 1,3-dibromopropan-2-one selectively afforded the 1,2-adduct consisting of a fullerene-fused cyclopentane, cyclohexane, and cyclopentanone (**165**, Fig. 9), respectively.

Fullerene-fused 1,2,3,4-tetrasilacyclohexane and 1,2,3,4-tetragermacyclohexane (**166**, Fig. 9) rings were each obtained as one of two products isolated from the photochemical reaction of octa(*p*-tolyl)cyclotetrasilane and octaphenylcyclotetragermane, respectively, with C_{60} [205, 206]. Stepwise addition of 1,4-diradicals, resulting from ring opening of the reagents, across a 6-6 bond of C_{60} was proposed as a possible reaction mechanism.

9
Concluding Remarks

After ten years of tether-directed remote functionalization in fullerene chemistry, a considerable number of multiadducts has become available regio- and, in some cases, stereoselectively, mainly by Bingel or Diels-Alder chemistry. All bisadduct regioisomers can in principle be prepared by use of appropriate tethers, even if the selectivity remains to be improved in a number of cases. In combination with other, sequential functionalizations and the removal of tethers or of certain addends at a given point in a synthetic sequence, an impressive number of complex functionalization patterns and, as a consequence thereof, residual fullerene π-chromophores has become accessible through spacer-controlled remote functionalization. When the used tethers include functional entities such as ionophores or porphyrins that modify existing characteristics of the fullerene or confer interesting new properties to the prepared conjugates, removal of the spacer at the end of the synthesis is not necessarily wanted. In other cases, the persistence of the tether in the final product may be tolerated but it would certainly be worthwhile to expand the spectrum of truly template-directed functionalizations of fullerenes, for example by using non-covalent templates. Also, a larger diversity of ready-made tethers for all bis- and some interesting multifunctionalization patterns would be highly desirable. One-pot multifunctionalizations are still relatively rare and some progress in this area has been made only recently. Another field that has remained almost untouched is the tether- or template-directed functionalization of higher fullerenes. In this context, the stereoselective generation of chiral fullerene functionalization patterns is of particular importance and even in the chemistry of C_{60}, the control of stereoselectivity needs to be further improved. Finally, the selective functionalization of fullerenes by methods that transcend the simple addition chemistry awaits further exploration. Promising efforts along these lines have been made in the synthesis of various homo- and azahomofullerenes, particularly in the context of cage-opening strategies.

Acknowledgements Continuing support of fullerene research by the Swiss National Science Foundation is gratefully acknowledged. We would also like to thank warmly our co-workers and collaborators whose names are mentioned in the references.

References

1. Hirsch A (1994) The chemistry of the fullerenes. Thieme, Stuttgart
2. Diederich F, Thilgen C (1996) Science (Washington DC) 271:317
3. Hirsch A (ed) (1999) Fullerenes and related structures. Top Curr Chem, vol 199
4. Wilson SR, Schuster DI, Nuber B, Meier MS, Maggini M, Prato M, Taylor R (2000) Organic chemistry of fullerenes. In: Kadish KM, Ruoff RS (eds) Fullerenes: chemistry, physics, and technology. Wiley Interscience, New York, p 91
5. Hirsch A (1999) Top Curr Chem 199:1
6. Prato M (1997) J Mater Chem 7:1097
7. Prato M (1999) Top Curr Chem 199:173
8. Prato M, Martín N (eds) (2002) Functionalised fullerene materials. J Mater Chem, vol 12(7)
9. Diederich F, Gómez-López M (1998) Chimia 52:551
10. Diederich F, Gómez-López M (1999) Chem Soc Rev 28:263
11. Guldi DM, Kamat PV (2000) Photophysical properties of pristine fullerenes, functionalized fullerenes, and fullerene-containing donor-bridge acceptor systems. In: Kadish KM, Ruoff RS (eds) Fullerenes: chemistry, physics, and technology. Wiley Interscience, New York, p 225
12. Isaacs L, Haldimann RF, Diederich F (1994) Angew Chem Int Ed 33:2339
13. Breslow R (1980) Acc Chem Res 13:170
14. Diederich F, Kessinger R (1999) Acc Chem Res 32:537
15. Diederich F, Kessinger R (2000) In: Diederich F, Stang PJ (eds) Templated organic synthesis. Wiley-VCH, Weinheim, p 189
16. Nakamura Y, O-kawa K, Nishimura J (2003) Bull Chem Soc Jpn 76:865
17. Echegoyen L, Herranz MA, Diederich F, Thilgen C (2004) Regioselective synthesis of fullerene derivatives and separation of isomers of the higher fullerenes. In: Toda F, Bishop R (eds) Separations and reactions in organic supramolecular chemistry (Perspectives in supramolecular chemistry), vol 8. Wiley, Chichester, p 137

18. Diederich F, Stang PJ (eds) (2000) Templated organic synthesis. Wiley-VCH, Weinheim
19. Powell WH, Cozzi F, Moss GP, Thilgen C, Hwu RJ-R, Yerin A (2002) Pure Appl Chem 74:629
20. Hirsch A, Lamparth I, Karfunkel HR (1994) Angew Chem Int Ed 33:437
21. Thilgen C, Gosse I, Diederich F (2003) Top Stereochem 23:1
22. Thilgen C, Herrmann A, Diederich F (1997) Helv Chim Acta 80:183
23. Isaacs L, Diederich F, Haldimann RF (1997) Helv Chim Acta 80:317
24. Bingel C (1993) Chem Ber 126:1957
25. Kräutler B, Maynollo J (1995) Angew Chem Int Ed 34:87
26. Kräutler B, Puchberger M (1993) Helv Chim Acta 76:1626
27. Friedman SH, Kenyon GL (1997) J Am Chem Soc 119:447
28. Boudon C, Gisselbrecht J-P, Gross M, Isaacs L, Anderson HL, Faust R, Diederich F (1995) Helv Chim Acta 78:1334
29. Kessinger R, Crassous J, Herrmann A, Rüttimann M, Echegoyen L, Diederich F (1998) Angew Chem Int Ed 37:1919
30. Kessinger R, Gómez-López M, Boudon C, Gisselbrecht J-P, Gross M, Echegoyen L, Diederich F (1998) J Am Chem Soc 120:8545
31. Kessinger R, Fender NS, Echegoyen LE, Thilgen C, Echegoyen L, Diederich F (2000) Chem Eur J 6:2184
32. Isaacs L, Seiler P, Diederich F (1995) Angew Chem Int Ed 34:1466
33. Seiler P, Isaacs L, Diederich F (1996) Helv Chim Acta 79:1047
34. Lamparth I, Maichle-Mössmer C, Hirsch A (1995) Angew Chem Int Ed 34:1607
35. Hirsch A, Lamparth I, Grösser T, Karfunkel HR (1994) J Am Chem Soc 116:9385
36. Rüttimann M, Haldimann RF, Isaacs L, Diederich F, Khong A, Jiménez-Vázquez H, Cross RJ, Saunders M (1997) Chem Eur J 3:1071
37. Diederich F, Rubin Y, Chapman OL, Goroff NS (1994) Helv Chim Acta 77:1441
38. Haldimann RF, Klärner F-G, Diederich F (1997) Chem Commun 237
39. Haldimann RF, Fokas C, Diederich F (2001) Helv Chim Acta 84:1635
40. Wallenborn E-U, Haldimann RF, Klärner F-G, Diederich F (1998) Chem Eur J 4:2258
41. Cardullo F (1997) ETH Dissertation No 12409, Zürich
42. An Y-Z, Viado AL, Arce M-J, Rubin Y (1995) J Org Chem 60:8330
43. An Y-Z, Ellis GA, Viado AL, Rubin Y (1995) J Org Chem 60:6353
44. Arbogast JW, Darmanyan AP, Foote CS, Rubin Y, Diederich FN, Alvarez MM, Anz SJ, Whetten RL (1991) J Phys Chem 95:11
45. Foote CS (1994) Top Curr Chem 169:347
46. Cardullo F, Isaacs L, Diederich F, Gisselbrecht J-P, Boudon C, Gross M (1996) Chem Commun 797
47. Cardullo F, Seiler P, Isaacs L, Nierengarten J-F, Haldimann RF, Diederich F, Mordasini-Denti T, Thiel W, Boudon C, Gisselbrecht J-P, Gross M (1997) Helv Chim Acta 80:343
48. Schwenninger R, Müller T, Kräutler B (1997) J Am Chem Soc 119:9317
49. Tirelli N, Cardullo F, Habicher T, Suter UW, Diederich F (2000) J Chem Soc Perkin Trans 2 193
50. Bourgeois J-P, Woods CR, Cardullo F, Habicher T, Nierengarten J-F, Gehrig R, Diederich F (2001) Helv Chim Acta 84:1207
51. Guldi DM, Asmus K-D (1997) J Phys Chem A 101:1472
52. Hamano T, Okuda K, Mashino T, Hirobe M, Arakane K, Ryu A, Mashiko S, Nagano T (1997) Chem Commun 21
53. Saunders M, Cross RJ, Jiménez-Vázquez HA, Shimshi R, Khong A (1996) Science (Washington DC) 271:1693
54. Bingel C (1994) Conference: New perspectives in fullerene chemistry and physics, Rome, Italy

55. Nierengarten J-F, Habicher T, Kessinger R, Cardullo F, Diederich F, Gramlich V, Gisselbrecht J-P, Boudon C, Gross M (1997) Helv Chim Acta 80:2238
56. Rapenne G, Diederich F (1999) New J Chem 23:1125
57. Zheng J-Y, Noguchi S, Miyauchi K, Hamada M, Kinbara K, Saigo K (2001) Fullerene Sci Technol 9:467
58. Hino T, Hamada M, Kinbara K, Saigo K (2002) Chem Lett 728
59. Nierengarten J-F, Felder D, Nicoud J-F (1998) Tetrahedron Lett 39:2747
60. Burley GA, Keller PA, Pyne SG, Ball GE (2002) J Org Chem 67:8316
61. Burley GA, Keller PA, Pyne SG, Ball GE (2000) Chem Commun 1717
62. Diekers M, Hirsch A, Luo C, Guldi DM, Bauer K, Nickel U (2000) Org Lett 2:2741
63. Diekers M, Luo C, Guldi DM, Hirsch A (2002) Chem Eur J 8:979
64. Ashton PR, Diederich F, Gómez-López M, Nierengarten J-F, Preece JA, Raymo FM, Stoddart JF (1997) Angew Chem Int Ed 36:1448
65. Nierengarten J-F (2000) Chem Eur J 6:3667
66. Nierengarten J-F, Armaroli N, Accorsi G, Rio Y, Eckert J-F (2003) Chem Eur J 9:37
67. Nierengarten J-F (2003) Top Curr Chem 228:87
68. Felder D, Gutiérrez Nava M, Carreón MdP, Eckert J-F, Luccisano M, Schall C, Masson P, Gallani J-L, Heinrich B, Guillon D, Nierengarten J-F (2002) Helv Chim Acta 85:288
69. Felder D, Gallani J-L, Guillon D, Heinrich B, Nicoud J-F, Nierengarten J-F (2000) Angew Chem Int Ed 39:201
70. Felder D, Carreón MdP, Gallani J-L, Guillon D, Nierengarten J-F, Thierry C, Deschenaux R (2001) Helv Chim Acta 84:1119
71. Zhang S, Rio Y, Cardinali F, Bourgogne C, Gallani J-L, Nierengarten J-F (2003) J Org Chem 68:9787
72. Cardullo F, Diederich F, Echegoyen L, Habicher T, Jayaraman N, Leblanc RM, Stoddart JF, Wang S (1998) Langmuir 14:1955
73. Nierengarten J-F, Felder D, Nicoud J-F (1999) Tetrahedron Lett 40:273
74. Armaroli N, Boudon C, Felder D, Gisselbrecht J-P, Gross M, Marconi G, Nicoud J-F, Nierengarten J-F, Vicinelli V (1999) Angew Chem Int Ed 38:3730
75. Gutiérrez Nava M, Setayesh S, Rameau A, Masson P, Nierengarten J-F (2002) New J Chem 26:1584
76. Gutiérrez-Nava M, Masson P, Nierengarten J-F (2003) Tetrahedron Lett 44:4487
77. Guldi DM (2000) Chem Commun 321
78. Martín N, Sánchez L, Illescas B, Pérez I (1998) Chem Rev 98:2527
79. Gust D, Moore TA, Moore AL (2001) Acc Chem Res 34:40
80. Guldi DM (2002) Chem Soc Rev 31:22
81. Imahori H, Sakata Y (1999) Eur J Org Chem 2445
82. Bourgeois J-P, Diederich F, Echegoyen L, Nierengarten J-F (1998) Helv Chim Acta 81:1835
83. Armaroli N, Marconi G, Echegoyen L, Bourgeois J-P, Diederich F (2000) Chem Eur J 6:1629
84. Dietel E, Hirsch A, Eichhorn E, Rieker A, Hackbarth S, Röder B (1998) Chem Commun 1981
85. Guldi DM, Luo C, Prato M, Troisi A, Zerbetto F, Scheloske M, Dietel E, Bauer W, Hirsch A (2001) J Am Chem Soc 123:9166
86. Guldi DM, Hirsch A, Scheloske M, Dietel E, Troisi A, Zerbetto F, Prato M (2003) Chem Eur J 9:4968
87. Guldi DM, Luo C, Prato M, Dietel E, Hirsch A (2000) Chem Commun 373
88. Guldi DM, Luo C, Da Ros T, Prato M, Dietel E, Hirsch A (2000) Chem Commun 375
89. Bourgeois J-P, Echegoyen L, Fibbioli M, Pretsch E, Diederich F (1998) Angew Chem Int Ed 37:2118

90. Bourgeois J-P, Seiler P, Fibbioli M, Pretsch E, Diederich F, Echegoyen L (1999) Helv Chim Acta 82:1572
91. Moonen NNP, Thilgen C, Echegoyen L, Diederich F (2000) Chem Commun 335
92. Herranz MA, Rivera JA, Alvarado RJ, Martín N, Thilgen C, Diederich F, Echegoyen L (2001) J Supramol Chem 1:299
93. Woods CR, Bourgeois J-P, Seiler P, Diederich F (2000) Angew Chem Int Ed 39:3813
94. van Eis MJ, Alvarado RJ, Echegoyen L, Seiler P, Diederich F (2000) Chem Commun 1859
95. van Eis MJ, Seiler P, Muslinkina LA, Badertscher M, Pretsch E, Diederich F, Alvarado RJ, Echegoyen L, Pérez Núñez I (2002) Helv Chim Acta 85:2009
96. van Eis MJ, Pérez Núñez I, Muslinkina LA, Alvarado RJ, Pretsch E, Echegoyen L, Diederich F (2001) J Chem Soc Perkin Trans 2 1890
97. Herrmann A, Rüttimann M, Thilgen C, Diederich F (1995) Helv Chim Acta 78:1673
98. Thilgen C, Herrmann A, Diederich F (1997) Angew Chem Int Ed 36:2269
99. Herrmann A, Rüttimann MW, Gibtner T, Thilgen C, Diederich F, Mordasini T, Thiel W (1999) Helv Chim Acta 82:261
100. Thilgen C, Diederich F (1999) Top Curr Chem 199:135
101. Diederich F, Thilgen C, Herrmann A (1996) Nachr Chem Tech Lab 44:9
102. Djojo F, Hirsch A, Grimme S (1999) Eur J Org Chem 3027
103. Djojo F, Hirsch A (1998) Chem Eur J 4:344
104. Nierengarten J-F, Gramlich V, Cardullo F, Diederich F (1996) Angew Chem Int Ed 35:2101
105. Goto H, Harada N, Crassous J, Diederich F (1998) J Chem Soc Perkin Trans 2 1719
106. Yoshida K, Osawa S, Monde K, Watanabe M, Harada N (2002) Enantiomer 7:23
107. Kessinger R, Thilgen C, Mordasini T, Diederich F (2000) Helv Chim Acta 83:3069
108. Sergeyev S, Diederich F (2004) Angew Chem Int Ed 43:1738
109. Nierengarten J-F, Herrmann A, Tykwinski RR, Rüttimann M, Diederich F, Boudon C, Gisselbrecht J-P, Gross M (1997) Helv Chim Acta 80:293
110. Soi A, Hirsch A (1998) New J Chem 22:1337
111. González S, Martín N, Guldi DM (2003) J Org Chem 68:779
112. Rapenne G, Crassous J, Collet A, Echegoyen L, Diederich F (1999) Chem Commun 1121
113. Collet A, Gabard J, Jacques J, Cesario M, Guilhem J, Pascard C (1981) J Chem Soc Perkin Trans 1 1630
114. Rapenne G, Crassous J, Echegoyen LE, Echegoyen L, Flapan E, Diederich F (2000) Helv Chim Acta 83:1209
115. Reuther U, Brandmüller T, Donaubauer W, Hampel F, Hirsch A (2002) Chem Eur J 8:2261
116. Zhou Z, Schuster DI, Wilson SR (2003) J Org Chem 68:7612
117. Cross RJ, Jiménez-Vázquez HA, Lu Q, Saunders M, Schuster DI, Wilson SR, Zhao H (1996) J Am Chem Soc 118:11454
118. Diederich F, Jonas U, Gramlich V, Herrmann A, Ringsdorf H, Thilgen C (1993) Helv Chim Acta 76:2445
119. Belik P, Gügel A, Spickermann J, Müllen K (1993) Angew Chem Int Ed 32:78
120. Taki M, Sugita S, Nakamura Y, Kasashima E, Yashima E, Okamoto Y, Nishimura J (1997) J Am Chem Soc 119:926
121. Nakamura Y, Taki M, Tobita S, Shizuka H, Yokoi H, Ishiguro K, Sawaki Y, Nishimura J (1999) J Chem Soc Perkin Trans 2 127
122. Taki M, Nakamura Y, Uehara H, Sato M, Nishimura J (1998) Enantiomer 3:231
123. Okamoto Y, Yashima E (1998) Angew Chem Int Ed 37:1021
124. Nakamura Y, O-kawa K, Nishimura T, Yashima E, Nishimura J (2003) J Org Chem 68:3251
125. Nakamura Y, Asami A, Inokuma S, Ogawa T, Kikuyama M, Nishimura J (2000) Tetrahedron Lett 41:2193

126. Nakamura Y, Asami A, Ogawa T, Inokuma S, Nishimura J (2002) J Am Chem Soc 124:4329
127. Hirsch A, Lamparth I, Schick G (1996) Liebigs Ann 1725
128. Djojo F, Herzog A, Lamparth I, Hampel F, Hirsch A (1996) Chem Eur J 2:1537
129. Ishi-i T, Shinkai S (1999) Tetrahedron 55:12515
130. Ishi-i T, Nakashima K, Shinkai S (1998) Chem Commun 1047
131. Ishi-i T, Iguchi R, Shinkai S (1999) Tetrahedron 55:3883
132. Ishi-i T, Nakashima K, Shinkai S, Ikeda A (1999) J Org Chem 64:984
133. Qian W, Rubin Y (1999) Angew Chem Int Ed 38:2356
134. Qian W, Rubin Y (2002) J Org Chem 67:7683
135. Kräutler B, Müller T, Maynollo J, Gruber K, Kratky C, Ochsenbein P, Schwarzenbach D, Bürgi H-B (1996) Angew Chem Int Ed 35:1204
136. Kräutler B, Müller T, Duarte-Ruiz A (2001) Chem Eur J 7:3223
137. Duarte-Ruiz A, Wurst K, Kräutler B (2001) Helv Chim Acta 84:2167
138. Qian W, Rubin Y (2000) Angew Chem Int Ed 39:3133
139. Qian W, Rubin Y (2000) J Am Chem Soc 122:9564
140. Rubin Y (1997) Chem Eur J 3:1009
141. Arce M-J, Viado AL, An Y-Z, Khan SI, Rubin Y (1996) J Am Chem Soc 118:3775
142. Rubin Y, Ganapathi PS, Franz A, An Y-Z, Qian W, Neier R (1999) Chem Eur J 5:3162
143. Irngartinger H, Weber A, Escher T (2000) Eur J Org Chem 1647
144. Camps X, Hirsch A (1997) J Chem Soc Perkin Trans 1 1595
145. Habicher T, Nierengarten J-F, Gramlich V, Diederich F (1998) Angew Chem Int Ed 37:1916
146. Timmerman P, Witschel LE, Diederich F, Boudon C, Gisselbrecht J-P, Gross M (1996) Helv Chim Acta 79:6
147. Fujiwara K, Komatsu K (2001) Chem Commun 1986
148. Hirsch A, Vostrowsky O (2001) Eur J Org Chem 829
149. Hirsch A, Vostrowsky O (2001) Top Curr Chem 217:51
150. Hauke F, Hirsch A (2001) Chem Commun 1316
151. Wang GW, Saunders M, Cross RJ (2001) J Am Chem Soc 123:256
152. Bürgi H-B, Blanc E, Schwarzenbach D, Liu S, Lu Y-j, Kappes MM, Ibers JA (1992) Angew Chem Int Ed 31:640
153. Bürgi H-B, Restori R, Schwarzenbach D (1993) Acta Crystallogr Sect B Struct Sci 49:832
154. Balch AL, Lee JW, Noll BC, Olmstead MM (1993) J Chem Soc Chem Commun 56
155. Duarte-Ruiz A, Müller T, Wurst K, Kräutler B (2001) Tetrahedron 57:3709
156. Nakamura E, Isobe H, Tokuyama H, Sawamura M (1996) Chem Commun 1747
157. Isobe H, Tokuyama H, Sawamura M, Nakamura E (1997) J Org Chem 62:5034
158. A similar increase in the number of isomers (stereoisomers, in that case) is possible with unsymmetrical tethered Bingel type addends – the according stereoisomers were designated as *in-in*, *in-out*, and *out-out* isomers; cf. [55]
159. Isobe H, Sawamura M, Nakamura E (1999) Fullerene Sci Technol 7:519
160. Nakamura E, Isobe H, Tomita N, Sawamura M, Jinno S, Okayama H (2000) Angew Chem Int Ed 39:4254
161. Isobe H, Sugiyama S, Fukui K-i, Iwasawa Y, Nakamura E (2001) Angew Chem Int Ed 40:3364
162. Isobe H, Tomita N, Jinno S, Okayama H, Nakamura E (2001) Chem Lett 1214
163. Kanakamma PP, Huang S-L, Juo C-G, Her G-R, Luh T-Y (1998) Chem Eur J 4:2037
164. Shiu L-L, Chien K-M, Liu T-Y, Lin T-I, Her G-R, Luh T-Y (1995) J Chem Soc Chem Commun 1159
165. Shen CK-F, Chien K-M, Juo C-G, Her G-R, Luh T-Y (1996) J Org Chem 61:9242
166. Shen CK-F, Yu H-H, Juo C-G, Chien K-M, Her G-R, Luh T-Y (1997) Chem Eur J 3:744

167. Dong G-X, Li J-S, Chan T-H (1995) J Chem Soc Chem Commun 1725
168. In the alternative structure, the other pair of fullerene-attached N-atoms is bridged by the tether and the double bonds of the heterocycles are shifted accordingly (cf. Sect. 6.1 and [158])
169. Ikeda A, Fukuhara C, Shinkai S (1997) Chem Lett 407
170. Kawaguchi M, Ikeda A, Shinkai S (1998) J Chem Soc Perkin Trans 1 179
171. Chen C, Li J, Ji G, Zheng Q, Zhu D (1998) Tetrahedron Lett 39:7377
172. Chen C, Li J, Ji G, Zhu D, Zheng Q (1998) Synth Commun 28:3097
173. Grösser T, Prato M, Lucchini V, Hirsch A, Wudl F (1995) Angew Chem Int Ed 34:1343
174. Hirsch A, Nuber B (1999) Acc Chem Res 32:795
175. Hasharoni K, Keshavarz-K M, Sastre A, González R, Bellavia-Lund C, Greenwald Y, Swager T, Wudl F, Heeger AJ (1997) J Chem Phys 107:2308
176. Schick G, Jarrosson T, Rubin Y (1999) Angew Chem Int Ed 38:2360
177. Rubin Y, Jarrosson T, Wang G-W, Bartberger MD, Houk KN, Schick G, Saunders M, Cross RJ (2001) Angew Chem Int Ed 40:1543
178. Murata Y, Murata M, Komatsu K (2003) J Am Chem Soc 125:7152
179. Murata Y, Murata M, Komatsu K (2003) Chem Eur J 9:1600
180. Irle S, Rubin Y, Morokuma K (2002) J Phys Chem A 106:680
181. Maggini M, Scorrano G, Prato M (1993) J Am Chem Soc 115:9798
182. Prato M, Maggini M (1998) Acc Chem Res 31:519
183. Da Ros T, Prato M, Novello F, Maggini M, De Amici M, De Micheli C (1997) Chem Commun 59
184. Da Ros T, Prato M, Lucchini V (2000) J Org Chem 65:4289
185. Smith PM, McCarty AL, Nguyen NY, Zandler ME, D'Souza F (2003) Chem Commun 1754
186. Rao AM, Zhou P, Wang KA, Hager GT, Holden JM, Wang Y, Lee WT, Bi XX, Eklund PC, Cornett DS, Duncan MA, Amster IJ (1993) Science (Washington DC) 259:955
187. Knol J, Hummelen JC (2000) J Am Chem Soc 122:3226
188. Wang G-W, Komatsu K, Murata Y, Shiro M (1997) Nature (London) 387:583
189. Komatsu K, Wang G-W, Murata Y, Tanaka T, Fujiwara K, Yamamoto K, Saunders M (1998) J Org Chem 63:9358
190. Murata Y, Kato N, Komatsu K (2001) J Org Chem 66:7235
191. Murata Y, Suzuki M, Komatsu K (2001) Chem Commun 2338
192. Inoue H, Yamaguchi H, Suzuki T, Akasaka T, Murata S (2000) Synlett 1178
193. Hsiao T-Y, Santhosh KC, Liou K-F, Cheng C-H (1998) J Am Chem Soc 120:12232
194. Inoue H, Yamaguchi H, Iwamatsu S-i, Uozaki T, Suzuki T, Akasaka T, Nagase S, Murata S (2001) Tetrahedron Lett 42:895
195. Murata Y, Murata M, Komatsu K (2001) J Org Chem 66:8187
196. Murata Y, Komatsu K (2001) Chem Lett 896
197. McClenaghan ND, Absalon C, Bassani DM (2003) J Am Chem Soc 125:13004
198. Guo L-W, Gao X, Zhang D-W, Wu S-H, Wu H-M, Li Y-J (1999) Chem Lett 411
199. Guo L-W, Gao X, Zhang D-W, Wu S-H, Wu H-M, Li Y-J, Wilson SR, Richardson CF, Schuster DI (2000) J Org Chem 65:3804
200. Isobe H, Ohbayashi A, Sawamura M, Nakamura E (2000) J Am Chem Soc 122:2669
201. Schick G, Kampe KD, Hirsch A (1995) J Chem Soc Chem Commun 2023
202. Murata Y, Komatsu K, Wan TSM (1996) Tetrahedron Lett 37:7061
203. Allard E, Rivière L, Delaunay J, Dubois D, Cousseau J (1999) Tetrahedron Lett 40:7223
204. Allard E, Delaunay J, Cheng F, Cousseau J, Ordúna J, Garín J (2001) Org Lett 3:3503
205. Kusukawa T, Kabe Y, Ando W (1995) Organometallics 14:2142
206. Kusukawa T, Shike A, Ando W (1996) Tetrahedron 52:4995

Topics in Current Chemistry (2004) 248: 63–103
DOI 10.1007/978-3-540-44505-0

Chromium-Templated Benzannulation and Haptotropic Metal Migration

Karl H. Dötz (✉) · Benjamin Wenzel · Holger C. Jahr

Kekulé-Institut für Organische Chemie und Biochemie, Gerhard-Domagk Strasse 1, 53121 Bonn, Germany
doetz@uni-bonn.de

Abstract This review highlights the recent development in intramolecular migration of metal templates along fused arene skeletons. The focus is on the haptotropic rearrangement of a $Cr(CO)_3$-fragment along polycyclic aromatic π-systems. The straightforward and regioselective synthesis of arene-$Cr(CO)_3$ complexes via chromium-templated [3+2+1]-benzannulation is discussed in the context of mechanistic and synthetic aspects. Proper tuning of the arene substitution pattern and the template coligand sphere allows the design of the haptotropic rearrangement for an application in novel organometallic molecular switches.

Keywords Arene complexes · Chromium complexes · Chromium-templated benzannulation · Haptotropic rearrangement · Intramolecular rearrangement · Ligand substitution · Molecular switch

1 Haptotropic Rearrangement

1.1 Definition

Sigmatropic rearrangements such as the hydrogen migration in prototropic shifts are also well-known for metals. Rearrangements that change the mode of the coordinative bond of π-electron systems to metal fragments encouraged Hoffmann [1] to combine the term "sigmatropic rearrangement" with the hapto-notation previously introduced by Cotton [2] to coin the term "haptotropic rearrangement". According to his definition, haptotropic rearrangements cover those cases, where an ML_n-unit changes its connectivity (hapto number) to a ligand [3].

A definition suggested by Nagashima [4] covers σ,σ-migrations like in the mercury complexes **1** and **2** described by Nesmeyanov [5] (Scheme 1), π,π-rearrangements such as in naphthalene complexes **5** and **6** discovered by Deubzer and Öfele [6], as well as σ,π-metal-metal-exchange reactions as reported by Lotz for the thiophene complexes **3** and **4** [7]. Nagashima extended the term "haptotropic rearrangement" to the slipping of a transition metal fragment from one coordination site to the other with the total electron count remaining unchanged in a polyaromatic or (cyclic) polyene ligand.

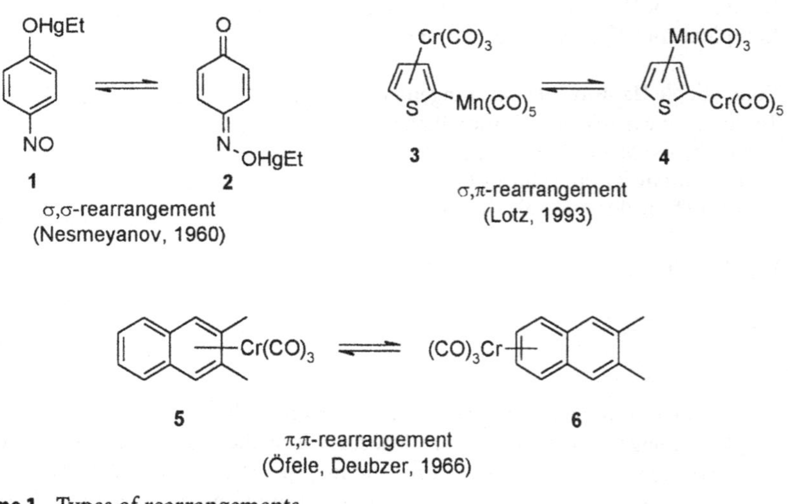

Scheme 1 Types of rearrangements

Due to the general importance of haptotropic rearrangements, which are involved in many transition metal-catalyzed processes, considerable research concentrated on the aim to improve the understanding of the relevant mechanisms and to apply these metal shifts to selective synthesis. The research in this area included a wide range of π-systems available; nevertheless most examples focused on the π,π-rearrangement reactions of metal fragments on aromatic ligands. On one hand, this was due to the easy synthetic access to this type of complexes; on the other hand, a modification of the arene substitution pattern allows one to control the kinetics and thermodynamics of the rearrangement reaction. The scope of model systems applied for this aim is discussed below.

1.2
Scope of Metal Fragments

A wide variety of metals is available for the complexation of π-ligands thus allowing the investigation of the rearrangement behavior of the complexes obtained. Due to their exceptional ability to control the hapticity of π-ligand systems, transition metals have been used in most cases, but also main group elements like magnesium [8], boron [9], silicon [10], germanium or tin [11, 12] have been studied in metal migration reactions.

Haptotropic rearrangements along π-ligands are known for complexes containing osmium [13], ruthenium [14], iron [15], copper [16] or nickel [17] complexes. Additionally, complexes in which iridium [18] or manganese [19] as well as palladium [20] or platinum fragments [21] were shifted by π,π-rearrangements, e.g., on phenalene, were reported. Nevertheless, the majority of haptotropic migration reactions refers to a variety of arene ligands coordinated to chromium or molybdenum fragments. Among them, most interest focused on arene chromium carbonyl complexes as model systems [22]. These compounds are easily available in high yields, generally by direct complexation of the arene either by heating the arene in presence of $Cr(CO)_6$ in suitable solvents or via appropriate transfer reagents like $Cr(CO)_3(NH_3)_3$ [23]; a complementary attractive and straightforward approach affording densely substituted hydroxyarenes is based on the [3+2+1]-benzannulation [24] which is discussed in detail later in this review.

1.3
Scope of π-Systems

In general, organic structures containing two or more π-bondings may be involved in haptotropic rearrangement reactions. Open-chain systems, like η^2-coordinated allenes [15] or η^4-coordinated trienes [25] have been studied most extensively as well as cyclic polyene structures; pertinent examples include the η^4,η^4-rearrangement of the $Fe(CO)_3$-fragment in cyclooctatetraene [26] or the η^6,η^6-rearrangement of $Cr(CO)_3$ in heptalene systems [27].

A problem frequently encountered in the observation of these dynamic processes is the degeneration of the rearrangement involving identical starting materials and the rearranged products (see Scheme 2); a kinetic analysis of this type of reaction, however, can be based on dynamic NMR spectroscopy. In this context, Vollhardt reported on the investigation of the degenerated η^2,η^2-isomerization of a $Ni(PR_3)_2$-fragment along the anthracene skeleton; NMR spin saturation transfer techniques were successfully applied to observe the rearrangement [28].

Scheme 2 Degenerated haptotropic rearrangement

Encouraged by the rearrangement in η^6-$Cr(CO)_3$-2,3-dimethylnaphthalene reported by Öfele and Deubzer a variety of arenes derived from fluorene, phenalene, anthracene, phenanthrene, biphenylene and other skeletons have been investigated [29]. Recent development concentrates on the application of enantiopure planar-chiral (arene)$Cr(CO)_3$ complexes in haptotropic rearrangement reactions (see later).

Besides this work on arene complexes, several research groups have studied the shift of metal fragments in non-aromatic systems. As shown by Lotz a σ,π-rearrangement occurs in a bimetallic thiophene complex (see Scheme 1); the η^1-$Mn(CO)_5$, η^5-$Cr(CO)_3$-isomer **3** rearranges irreversibly at 30 °C in acetone to the corresponding η^1-$Cr(CO)_5$,η^5-$Mn(CO)_3$-isomer **4** [7].

Reversible metal migration reactions which are of major interest in the development of organometallic molecular switches were reported by Nagashima [30]. They described a symmetric η^5,η^3-diiron complex of acenaphthylene **7** that underwent a nearly quantitative metal migration upon photoirradiation (or even in sunlight) to give the unsymmetric isomer **8**; upon storage in the dark, this isomer rearranges to form again the symmetric starting material (Scheme 3).

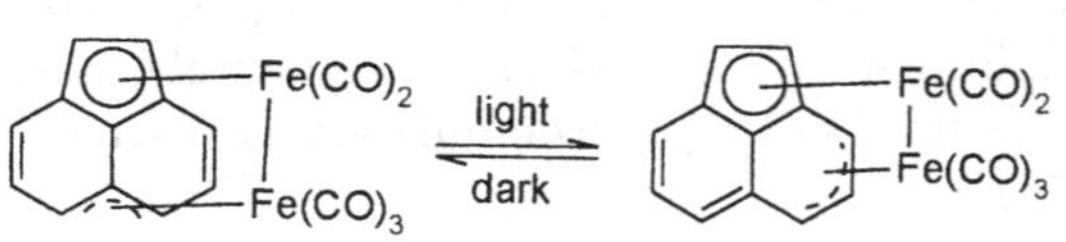

7 **8**

Scheme 3 Rearrangement of a diiron-acenaphthylene complex

Recently, the same group reported on a reversible thermo-optical rearrangement of a guiazulene diruthenium complex interconverting the two isomers **9** and **10** upon sequential thermal treatment or photoirradiation (Scheme 4) [14]. This example reveals an influence of the metal coligand sphere on the rearrange-

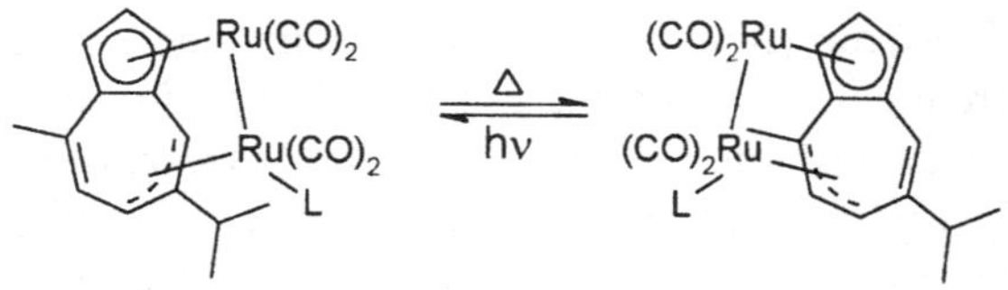

9 **10**

Scheme 4 Shift of a guiazulene-diruthenium complex with different ligands

ment: while no metal migration is observed upon substitution of CO for common phosphines and phosphites, complete rearrangement occurs with the specific cage-type phosphite complex P{$(OCH_2)_3CCH_3$}. These two studies indicate that both the metal and its coligand sphere play a crucial role in the tuning of the metal migration. This aspect is addressed in more detail later in this review.

Additional recent results in this context refer to pH-driven translocations of Ni^{2+} or Cu^{2+}-ions in organic molecules containing two distinct compartments characterized by a different binding behavior towards the metal upon pH-change. Fabrizzi designed ditopic ligands that bear two distinct tetradentate binding sites; one pattern is provided by two 2,2′-bipyridine (bpy) fragments while the other offers a combination of two secondary amine and amide functionalities (Scheme 5) [16]. As the bipyridyl unit shows a fairly good permanent binding tendency towards Cu^{2+}, the amine/amide binding site needs to be deprotonated under slightly basic conditions to become a sufficiently strong donor as required for coordination to divalent transition metal ions.

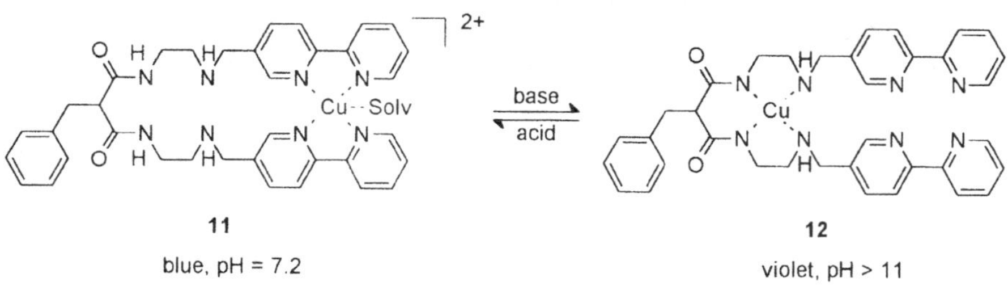

Scheme 5 pH-induced copper translocation

An equimolar solution of the oligoaza ligand and Cu^{2+}-ions shows a blue color at pH 7.2 which changes to pink-violet upon basification reaching its maximum abundance at pH≥11. The blue species is assigned to the Cu^{2+}-ion which is located in the bpy-environment and η^5-coordinated by four nitrogen atoms and one solvent oxygen atom (**11**); the violet color originates from the metal ion η^4-coordinated by the deprotonated amine/amide binding site (**12**). The direct and reverse translocation processes can be repeated at will, limited only by increasing dilution caused by subsequent addition of acid and base. The rate constants of the metal shift have been determined by stopped flow spectrophotometric experiments which – in accordance with an intramolecular rearrangement – reveal strictly first-order kinetics. Related rearrangement processes have been observed for similar quinoline-Ni^{2+} complexes [31]. Additionally, redox-driven translocations using the Fe^{2+}/Fe^{3+} couple are known [32].

The principle of coligand sphere controlled metal translocation has been elegantly demonstrated by Balzani and Sauvage in rotaxane and catenane based molecular machines [33, 34].

The latter examples are not covered by the definition given earlier, as no rearrangement between σ- or π-bonding in a polyaromatic or (cyclic) polyene ligand takes place. Since, however, a metal rearrangement between two distinct binding sites occurs analogous to the systems described before, these metal translocations may be discussed in the context of haptotropic rearrangements as well; an extension of the previous definition is suggested to cover all those cases in which the metal fragment migrates between two different distinct binding sites in an organic molecule attached to the metal fragment by either covalent or coordinative bonding.

2 Chromium-Templated Benzannulation

First reported in 1975 by Dötz [35], the chromium-mediated benzannulation of an α,β-unsaturated chromium Fischer carbene complex and an alkyne has become the most versatile reaction involving Fischer carbene complexes. Due to its octahedron configuration the chromium template assembles the carbene ligand as a C_3-synthon, a carbonyl ligand (C_1-synthon) and the alkyne (C_2-synthon) in a facial geometry suited for a stepwise [3+2+1]-cycloaddition to afford a hydroxyarene coordinated to the $Cr(CO)_3$ template (Scheme 6).

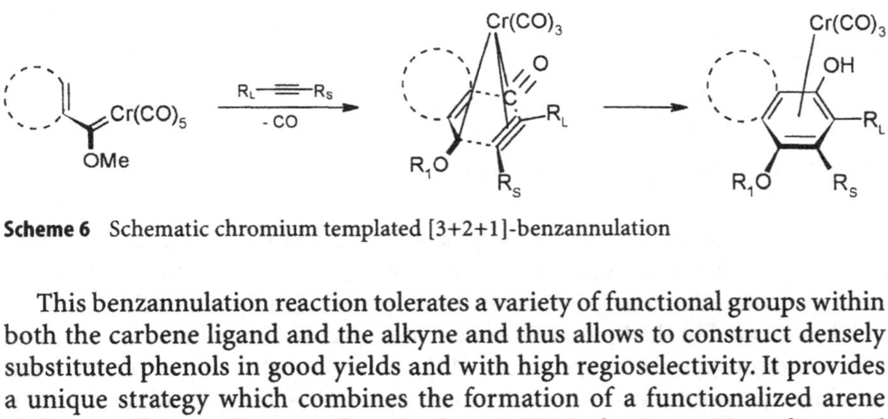

Scheme 6 Schematic chromium templated [3+2+1]-benzannulation

This benzannulation reaction tolerates a variety of functional groups within both the carbene ligand and the alkyne and thus allows to construct densely substituted phenols in good yields and with high regioselectivity. It provides a unique strategy which combines the formation of a functionalized arene skeleton with its in situ-complexation by a $Cr(CO)_3$-fragment. Since the metal fragment is known to activate the arene towards deprotonation, benzylic substitution and, in particular, towards *anti*-addition of nucleophiles, the chromium-templated benzannulation has gained considerable attention in stereoselective synthesis [24, 36]. Moreover, the mild non-acidic conditions are attractive for and compatible with application in natural product synthesis, like the formation of the allocolchicin derivative **14** from the carbene precursor **13** depicted in Scheme 7 [37, 38].

On the other hand, the fact that the benzannulation of aryl(alkoxy)carbene chromium complexes, when carried out under kinetic control (i.e., generally below 50 °C), leads to an exclusive coordination of the newly formed phenolic ring, allows for a regioselective metal labeling even in the synthesis of poly-

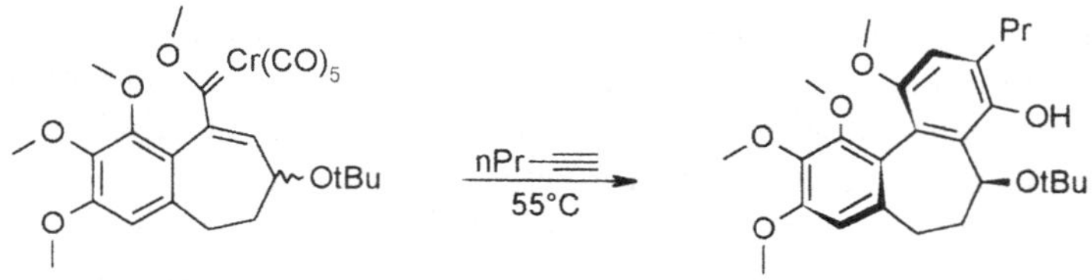

13 14

Scheme 7 Atropisomer-selective benzannulation to an allocolchicinoid derivative

cyclic arene skeletons. Thus, it is the methodology of choice for the selective complexation of fused arenes as required for studies towards haptotropic metal migration reactions (for detailed discussion, see below).

The scope of Fischer chromium carbenes and alkynes employed in the benzannulation is obvious from a broad pattern of compatible functional groups. Generally, solutions of the starting materials in a donor solvent are kept at 40–80 °C for 2–3 h; in some cases, dry-state [39], photo-irradiation [40], ultrasound [41] and, most recently, microwave-conditions [42] have been reported to give similar or, in isolated cases, even superior yields of benzannulation products. However, in general, decomplexation occurs under these conditions affording uncoordinated hydroquinones or – after oxidative work-up – quinones.

The alkynes used in the benzannulation may bear various substituents, such as aryl, amide, ketone or acetal groups. There is an obvious dependence on both their steric and electronic properties; *ortho*-substituted arylalkynes and strong electron-withdrawing substituents, if not compensated by donor functionalities in either the carbene ligand or the alkyne [43], result in moderate to poor yields [44].

The α,β-unsaturated chromium carbenes tolerate a broad substitution pattern, such as vinylic structures, *ortho*-, *meta*- and *para*-substituted aromatic as well as heteroaromatic groups. As outlined for the alkyne component, electronic properties of the aromatic or vinylic substituents have a predominant influence on the yields [45], but, in addition, steric aspects also have to be considered [46].

In some cases and under specific conditions the benzannulation pathway to give hydroxyarenes may have to compete with the formation of 4- and 5-membered ring systems such as cyclobutenones, cyclopentadienes, indenes or furans [36, 47]. The chemodiversity reflects the tendency for sequential alkyne and CO insertion into the chromium carbene bond and the stereochemistry of the intermediates resulting therefrom. The chemoselectivity depends on the nature of the metal and the substitution pattern of the carbene ligand as well as on the reaction conditions. Besides the concentration and the temperature used, the solvent has been recognized to play a major role [47, 48].

Less electron-poor chromium carbenes like aminocarbene complexes strongly favor the formation of five-membered rings irrespective of whether the amino donor is attached directly or in conjugation to the carbene carbon. While chromium alkenylaminocarbenes mainly yield aminophenols [49, 50], arylaminocarbene complexes form exclusively indenes [51]; their conjugated analogues **15** give phenols **16** and indenes **17** with the latter predominating (Scheme 8) [52]. If, however, the donor ability is reduced by *N*-BOC-protection (**18**), phenol formation prevails [53, 54]. Additionally, aminostabilized but electron-poor carbenes result in exclusive phenol formation [45].

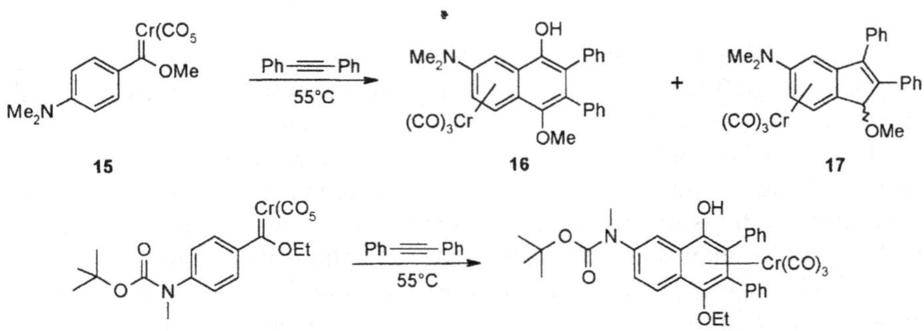

Scheme 8 Chemoselective benzannulation via amidation of the amino donor

Examples to replace the alkyne for triple bond systems bearing a heteroatom are scare and confined to isolated reports on phosphaalkynes. The kinetically stabilized *tert*-butylphosphaethyne and a 2-furylcarbene complex were reacted in a benzannulation to give a furanophosphahydroquinone [55]. The use of isonitriles in the benzannulation reaction is documented in reactions with 1-metalla-1,3,5-hexatrienes **20** to give *ortho*-alkoxyanilines **22** indicating that only the terminal carbon atom is incorporated into the arene ring [56]. The same chromium carbene, however, undergoes alkyne insertion to give cyclooctatrienones **21** in good yields (Scheme 9) [57].

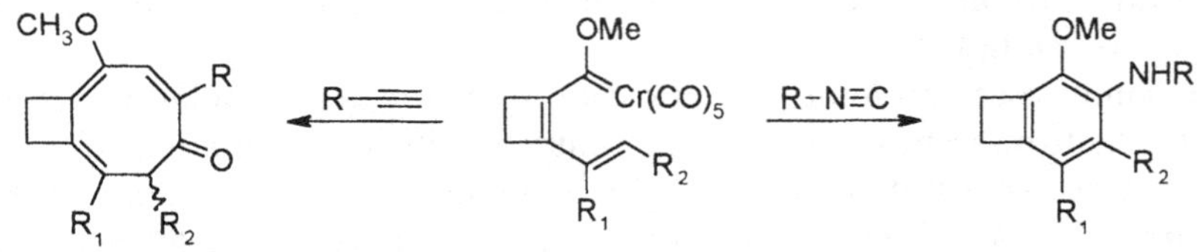

Scheme 9 Chemoselectivity with alkynes and heteroalkynes

2.1
Selective Synthesis of (Arene)Cr(CO)$_3$-Complexes

2.1.1
Regioselectivity

The regioselectivity of the [3+2+1]-benzannulation is mainly determined by the alkyne substitution pattern; reaction conditions as given by temperature and concentration play a minor role [48, 58]. Due to steric interactions in the η^1,η^3-vinylcarbene intermediate resulting from alkyne insertion (Scheme 10), the sterically more demanding substituent is preferentially located next to the phenolic group in the final product [24, 36, 59].

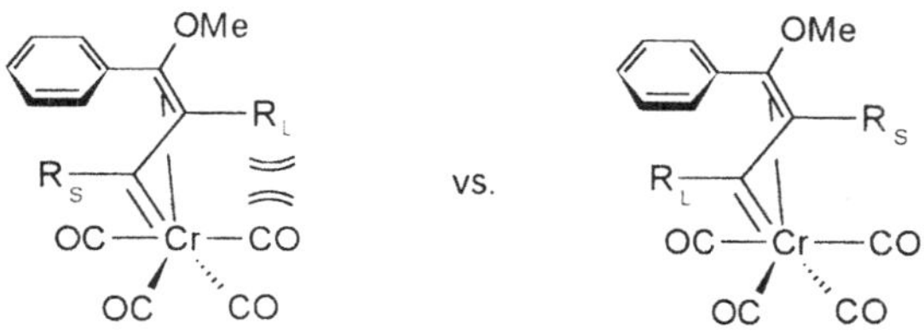

Scheme 10 Regioselectivity of alkyne insertion

Terminal alkynes applied in the reaction afford one single regioisomer, whereas internal unsymmetrical alkynes result in the formation of regioisomeric mixtures. The steric control reaches its limits if diarylalkynes bearing two electronically different *para*-substituents were incorporated [58]. The steric preference is overruled with alkynylstannanes [60] and alkynylboronates [61] which reveal a contrasteric regiopreference. This outcome was rationalized in terms of a carbonyl oxygen to tin or boron Lewis acid/base interaction in the metallahexatriene intermediate (Scheme 11). This argument, however, is not in line with the incorporation of 1-methylthio-2-phenylethyne which also is found to occur in a contrasteric manner [62].

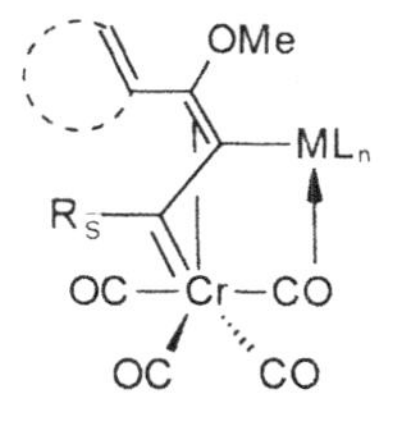

ML$_n$ = SnR$_3$, B(OR)$_3$

Scheme 11 Inverse regioselectivity

Another contrasteric example results from the intramolecular benzannulation of enyne-carbene complexes **23** which affords a *meta*-methoxyphenol skeleton as part of *para*-cyclophanes **28** (Scheme 12) [63]. Their formation, which requires a formal cleavage of the C≡C bond, was suggested to involve a Lewis-acid catalyzed rearrangement of a paddalane-intermediate bearing a benzvalenone core (**25–27**) which may be formed by a [2+2]-cycloaddition of the vinylketene intermediate **24**.

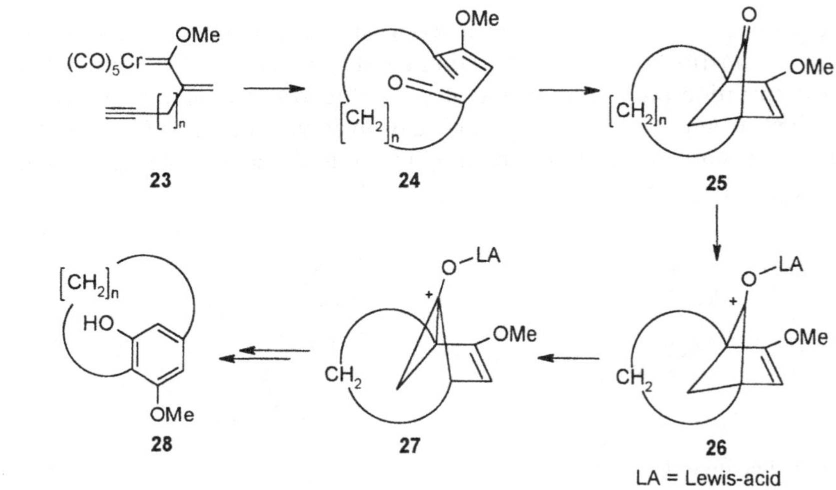

Scheme 12 *Meta*-cyclophanes by intermediate-rearrangement

2.1.2 Diastereoselectivity

As a result of the intrinsic unsymmetric arene substitution pattern, the $Cr(CO)_3$-coordinated benzannulation products bear a plane of chirality and – in the absence of any chiral information – are formed as a racemic mixture of enantiomers (Scheme 13).

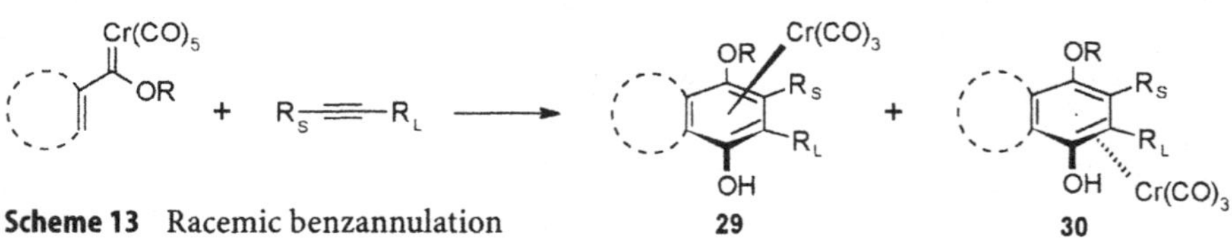

Scheme 13 Racemic benzannulation

Enantiopure as well as diastereopure (arene)$Cr(CO)_3$-complexes are of general interest in stereoselective synthesis. On the other hand, they have also been applied to haptotropic rearrangement reactions in which they may help to shed light on the reaction mechanisms [29]. Until recently, the separation of chiral $Cr(CO)_3$-complexes by chiral HPLC was a tedious process [64], and the availability of single diastereomers was very limited. Thus, considerable efforts were directed towards a diastereoselective version of the chromium-mediated benzannulation. Since all bond forming steps between the synthons of the benzannulation reaction occur within the coordination sphere of the metal, diastereoselectivity may generally be expected by placing a chiral information into the chromium template, the alkyne or the carbene ligand. Chiral information incorporated into α-alkoxyalkynes was found to allow for good to excellent diastereoselectivities depending on the steric bulk of the chiral auxiliary used [65]. An alternative strategy using a chiral carbene *C*-side chain has been pursued with a racemic mixture of the methoxycyclohexenylcarbene complex **31** reacted with pent-1-yne and resulted in a 88% de preference of the *anti*-annulation product **32** with respect to the $Cr(CO)_3$-fragment and the methoxy substituent (Scheme 14) [66].

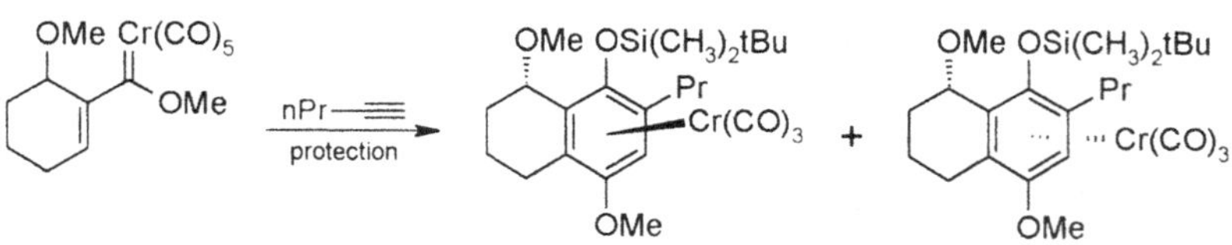

31 (rac) **32** (rac)-anti *94* **33** (rac)-syn *6*

Scheme 14 Substrate-controlled diastereoselectivity

In contrast, the benzannulation of glucal-derived chromium carbenes leads to products bearing the $Cr(CO)_3$-fragment on the sterically more congested face in good to excellent diastereoselective excesses [67]. This stereochemical outcome seems to result from a preferred approach of the alkyne occurring from the less congested side opposite to the bulky protecting silyl groups shielding one face of the glucal carbene complex.

The most general strategy aims at the incorporation of the chiral information into the heteroatom carbene side chain. It allows a straightforward protocol for chiral alcohol auxiliaries readily available from the chiral pool. They may be incorporated into the carbene complex via alcoholysis of acyloxycarbene precursors and, after the benzannulation, they may be recovered by cleavage of the aryl ethers making this sequence promising for synthetic applications.

Diastereoselectivities up to 82% de have been achieved in the annulation of menthyloxycarbene complexes with *tert*-butylacetylene [68]; a careful balance of carbene and alkyne substitution patterns is required, since other combinations of terpene alcohol auxiliaries and terminal alkynes afforded far less attractive diastereoselectivities not exceeding 20% de [69]. A favorable combination of a chiral chromium carbene and a chiral alkyne has been elaborated to a desired "matched" situation: the benzannulation of a chromium *O*-glycoside vinyl carbene by an *O*-perprotected ethynyl glucoside afforded the $Cr(CO)_3$-coordinated arenes in virtually complete diastereoselectivity [70].

Attempts to use the planar chirality produced in the benzannulation for chirality transfer have been reported for the synthesis of biaryls (Scheme 15). The

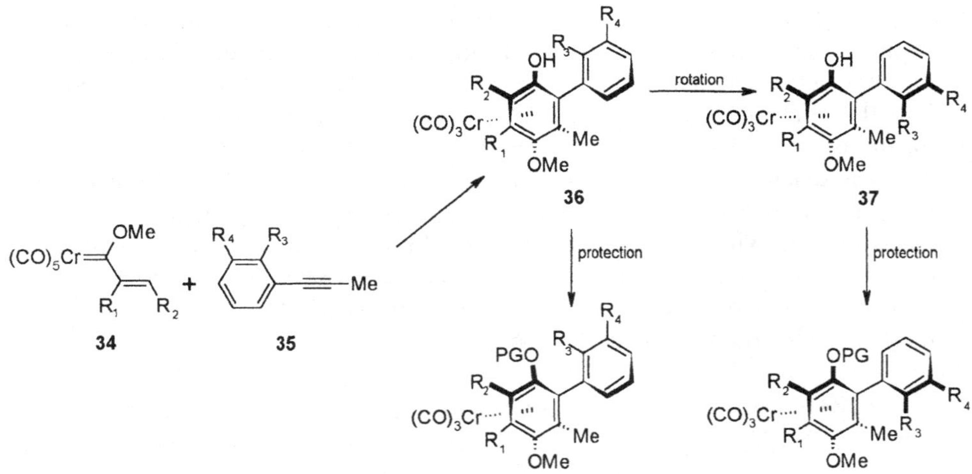

Scheme 15 Atropisomeric biphenyls via [3+2+1]-benzannulation

atropisomerism in appropriately substituted biphenyl systems imposed upon benzannulation of **34** and **35** was frozen by *O*-protection of the newly formed hydroxy group. In the unprotected phenylphenol the chromium-coordinated ring is able to rotate interconverting the primarily formed kinetic *syn*-isomer **36** to the thermodynamic isomer **37** containing the bulky aryl substituents on the adjacent ring in *anti*-position to the $Cr(CO)_3$-fragment. Variation of in-situ protection vs subsequent protection allows one to adjust the steric bulk of the *ortho*-hydroxy group and to suppress further rotation which results in either the *syn-* (**38**) or the *anti-*(**39**) biphenyl system [71].

Apart from attempts to induce chirality by chiral substrates or chiral auxiliaries, the product itself can force the benzannulation to provide a single isomer. This principle has been elucidated in the formation of [2.2]metacyclophanes

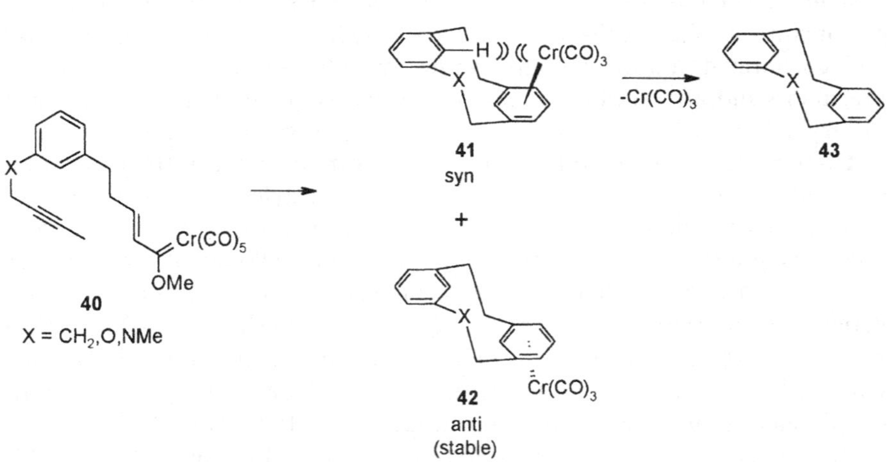

Scheme 16 Diastereomeric [2.2] metacyclophane synthesis

from **40** by intramolecular benzannulation, which due to its steric restraint leads to the *anti*-isomer **42** isolated in up to 36% yield as the only isomer observed (Scheme 16) [72,73]. Based on a comparison with homologous naphthalenophane complexes this result indicates a selective decomplexation of the *syn*-isomer **41** to **43** which reflects the close proximity of the $Cr(CO)_3$ fragment and the inner hydrogen atoms of the other benzene deck.

2.2 Angular vs Linear Benzannulation

If Fischer carbene complexes of *ortho*-unsubstituted polycyclic arenes are employed in the benzannulation reaction, a competition between angular (**44**) and linear (**45**) annulation products has to be envisaged (Scheme 17).

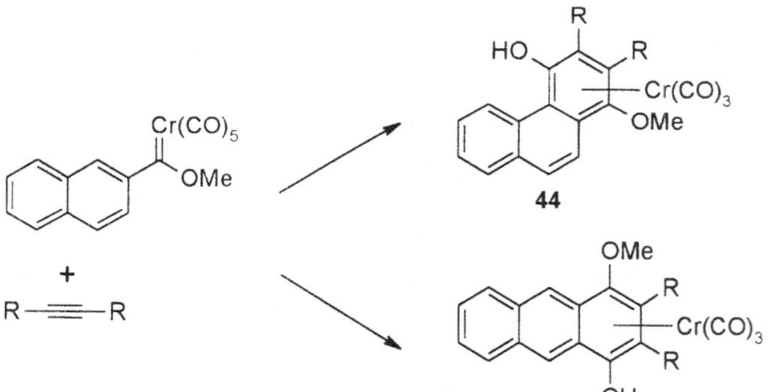

Scheme 17 Linear and angular benzannulation

Nevertheless, angular phenanthrene benzannulation products are generally observed in reactions of 2-naphthyl carbene complexes with alkynes. Even in those cases where linear annulation has been tried to be forced by appropriate *ortho*-substitution, angular annulation is strongly preferred [74] and only a single exemption has been reported [75].

It has been established only recently that linear benzannulation may be equally feasible. The benzannulation of a dibenzofuran carbene complex **46** afforded two regioisomeric benzonaphthofuran benzannulation products among which the linear benzonaphthofuran **48** was slightly preferred over the angular isomer **47** (Scheme 18) [76]. An extension of this study to the homologous dibenzothienylcarbene chromium complex resulted in a reverse product distribution [77].

Whereas benzannulation of the naphthalene skeleton can – in principle – generate either phenanthrene or anthracene derivatives which considerably differ in their relative energies (by ca. 25 kJ mol^{-1} for the parent compounds) in favor of the angular product, the energy gap between angular and linear annulation products derived from dibenzofuran and -thiophene must be distinctly smaller since both isomers of benzo[*b*]naphtho[*d*]furan and -thiophenes are obtained. Arguments to rationalize the preference of angular

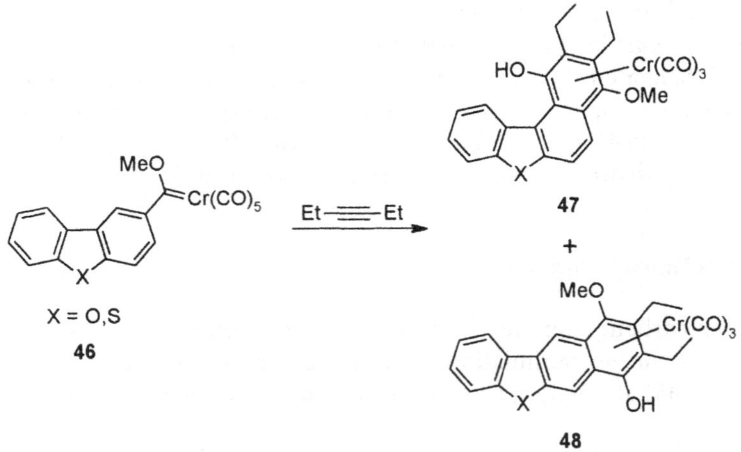

Scheme 18 Competition between linear and angular benzannulation

annulation in the thiophene series compared to the furan case may be based on the relative electron densities in 1- vs 3-position of the starting carbene complexes as well as in the ketene intermediates. Calculations [78] showed an increasing preference towards the angular annulation product in the sequence naphthalene≫thiophene>furane, according to the electrostatic charges in the ketene intermediates. This is confirmed by the experimental results, which gave no evidence for a linear annulation product in case of naphthalene- or naphthalene-homologue structures. A second argumentation may be based on the interring angle θ (Scheme 19) which decreases upon replacement of oxygen for sulfur and favors a conformation putting the sterically less demanding linear ketene moiety in close proximity to C-1 above or below the arene plane (**49**) instead of the more bulky methoxy substituent in **50**. The difference in atomic

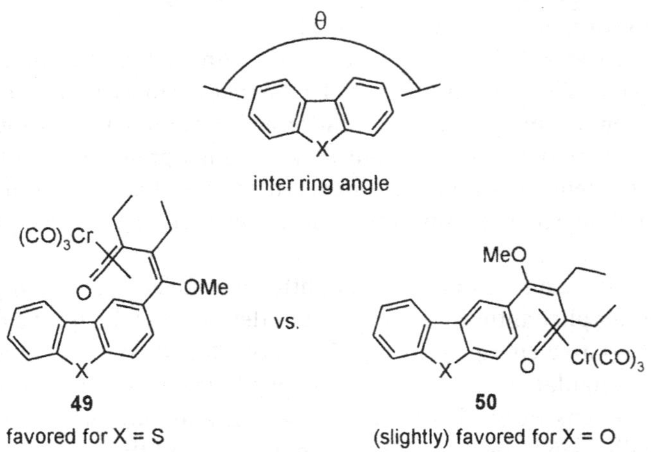

Scheme 19 Possible influence of the inter-ring angle on product development

size of sulfur vs oxygen is reflected in the helical distortion of both angular annulation products which is twice as much in the benzo[*b*]naphtho[1,2-*d*]-thiophene tricarbonylchromium complex [79].

Linear benzannulation is not restricted to dibenzofuran and -thiophene systems. It has been recently observed as well with mono- and bidirectional benzannulation reactions of chromium carbene modified [4]- and [6]-helicenes containing pyrrol rings [80].

2.3 Mechanistic Aspects

Benzannulation is the synthetically most useful but not the only option for chromium carbenes to react with alkynes. Besides hydroxyarenes, cyclobutenones, indenes and furans have also been observed as major or even predominating by-products as well, reflecting – in part – the stereochemistry and the tendency of CO-labilization in the intermediate formed upon alkyne insertion (Scheme 20). The chemoselectivity is mainly governed by the nature of the

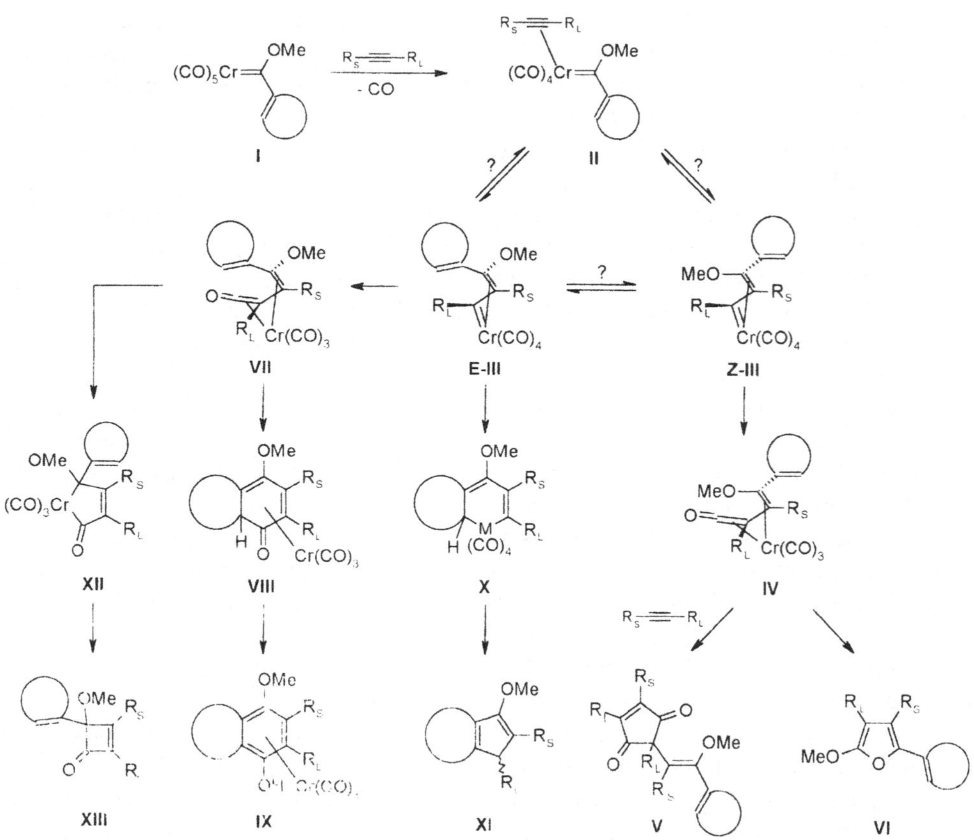

Scheme 20 Mechanism and product development of the [3+2+1]-benzannulation

metal and the donor ability of the carbene substituents; moreover, the solvent, concentration and temperature may influence the product distribution to a minor extent [47, 48]. Potent heteroatom donor stabilization of the carbene such as in aminocarbene complexes favors the formation of five-membered rings; an incorporation of CO is hampered in this case due to the increased back-donation from the metal to the carbonyl ligands. As expected, *N*-acceptor substitution weakens the metal-CO bonds and benzannulation may be the predominating pathway again [45]; steric bulk in the amino substitution pattern was found to play a role as well [81]. A generally accepted mechanistic scheme for chromium carbene-alkyne reactions, supported by experimental and theoretical studies [59, 82], is depicted in Scheme 20.

A primary decarbonylation from the pentacarbonyl chromium carbene complex has been recognized as the rate-determining step of the benzannulation reaction [83, 84]. Subsequent coordination of the alkyne followed by insertion into the chromium-carbene bond generates a η^1,η^3-vinylcarbene intermediate. It may be formed either as E-η^1,η^3-vinylcarbene complex (**E-III**) or its *Z*-isomer (**Z-III**) and is considered as a common intermediate for the variety of products observed. Cyclopentenedione (**V**) and furan (**VI**) side products may be traced back to **Z-III** while the hydroxyarenes and cyclobutenones are formed via vinylketene **VII** originating from **E-III**. A recent study points to an isomerization of vinylcarbene intermediates **E-III** and **Z-III** occurring in an alkyne de- and insertion equilibrium [48].

CO-insertion into the E-η^1,η^3-vinylcarbene produces η^4-vinylketene intermediate **VII** which undergoes subsequent ring closure and tautomerization to result in phenol formation (**IX**). The cyclobutenone side product observed in rare cases even may arise from intermediate **VII** by forming a metallacyclopentenone **XII** that, with subsequent extrusion of chromium, leads to **XIII**.

If the CO-insertion step in **E-III** is hampered as a result of increased stability of the Cr-CO bond, the formation of a chromacyclohexadiene intermediate **X** may give rise to indene formation (**XI**).

Over the years all relevant types of key intermediates have been isolated and structurally characterized as model compounds. Alkyne-alkoxycarbene tetracarbonyl complexes revealed a rather labile alkyne to chromium bond [85]. Tetracarbonyl chromium (amino)vinylcarbenes **52** generated upon photodecarbonylation from **51** and stabilized as aminocarbene complexes underwent insertion of electron-poor alkynes to give 1-chroma-1,3,5-hexatrienes **53** stabilized by coordination of the terminal C=C bond (Scheme 21) [43]. Even though coordination of the α,β-double bond has been shown to be generally favored over coordination of the γ,δ-double-bond [86], the uncoordinated vinylcarbene moiety reflects the electronic situation which requires a compensation of an electron-deficient carbene substituent on one hand by an enamine donor group on the other hand. Cyclization occurs upon warming to afford either a benzene (**54**) or a cyclopentadiene ring (**55**) depending on the substitution of the alkyne used.

Vinylketenes as the final acyclic intermediates have been first isolated from reactions of chromium carbenes and silylalkynes **56** (Scheme 22). A vicinal

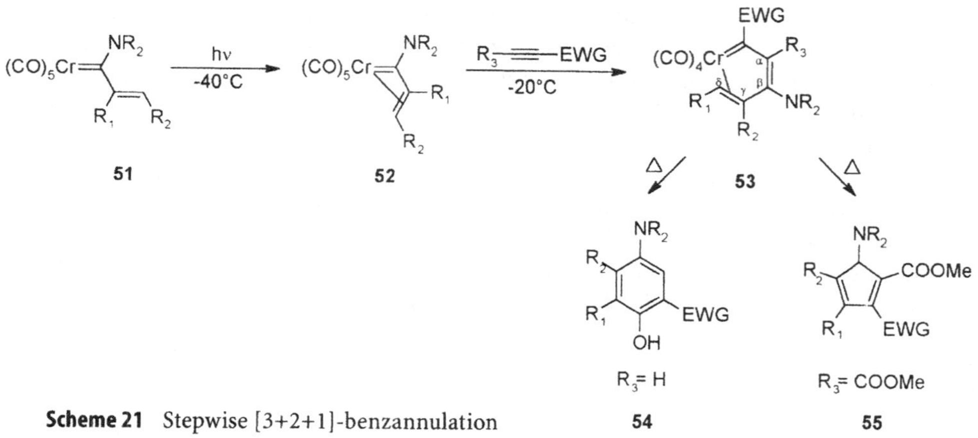

Scheme 21 Stepwise [3+2+1]-benzannulation

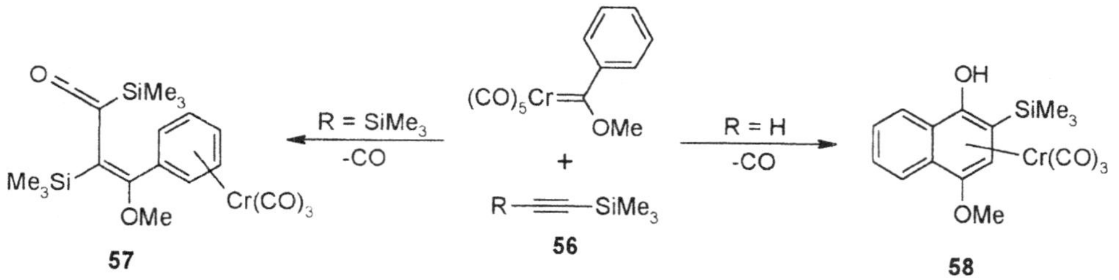

Scheme 22 Suppressed electrocyclization in case of a vicinal disilyl substituted vinylketene

disilyl substitution in **57** disfavors a vinylketene conformation required for electrocyclization; in contrast, trimethylsilylethyne allows for a vinylketene intermediate bearing the phenyl ring in close proximity to the ketene group which results in clean benzannulation to **58** [87].

The 6π-electrocyclization may be influenced even by a more remote substitution pattern as demonstrated for vinylketene complex **59** (Scheme 23). Ring closure has been observed only after photochemical or oxidative removal of the $Cr(CO)_3$-fragment leading to the uncoordinated vinylketene **60** that underwent ring closure with concomitant silyl migration under more rigorous thermal conditions to give **61** [88].

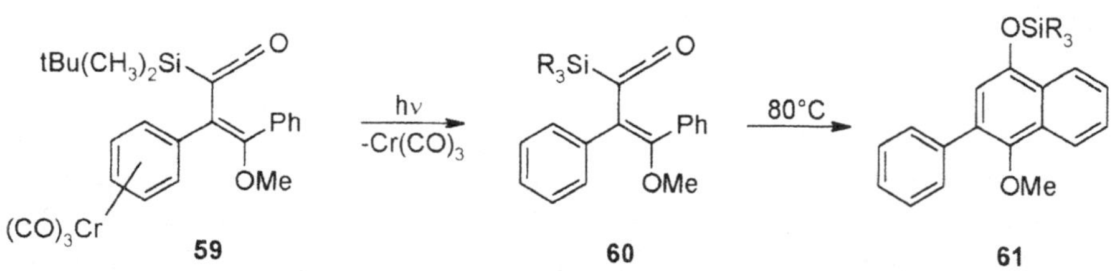

Scheme 23 Suppressed electrocyclization by a remote substitution pattern

Recently, the range of products accessible by chromium-templated benzannulation has been extended by electrocyclization of 1-chroma-1,3,5-hexatrienes available, e.g., from insertion of dienynes into the chromium-carbene bond of saturated carbene complexes [89, 90].

Since heteroalkynes undergo benzannulation only in rare examples [55], heteroatom-containing chromahexatrienes **63** were applied to the formation of heterocyclic systems (Scheme 24). Pyrone formation from 2-alkynyl-benzaldehyde **62** failed; the CO-insertion step was suppressed leading to the formation of isobenzofuran derivative **64** [91]. Exclusive pyrone formation to **65** was observed if 3-alkynyl-2-heteroaromatic carboxaldehyde was used [92].

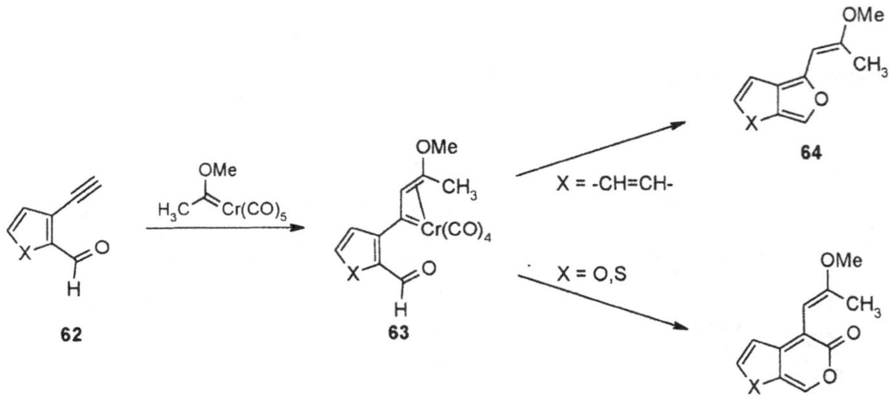

Scheme 24 Heteroanalogous benzannulation

3 Haptotropic Migration of $Cr(CO)_3$-Fragments in Fused Arenes

Half-sandwich $Cr(CO)_3$ complexes of fused arenes are suitable compounds for haptotropic rearrangements along extended π-ligand systems and have been studied in this context by several groups. In general, the haptotropic metal shifts are induced upon warming the arene complex to temperatures ranging from 40 °C to 100 °C; this protocol allows one to transform the thermodynamically less stable to the more stable regioisomer. If the coordination of one aromatic ring is strongly favored over the other, the haptotropic isomerization is virtually irreversible; in cases in which both isomers possess comparable thermodynamic stabilities, a dynamic equilibrium is observed. Four different mechanisms may account for the rearrangement as illustrated for a model complex in Scheme 25.

A dissociative mechanism involves the release of the metal fragment from the π-ligand followed by recomplexation (a). This path will be assisted by solvents or additives featuring donor abilities which allow a stabilization of the coordinatively unsaturated metal fragment intermediate. Bimolecular processes could be based either on the mutual exchange of the $Cr(CO)_3$ fragments between

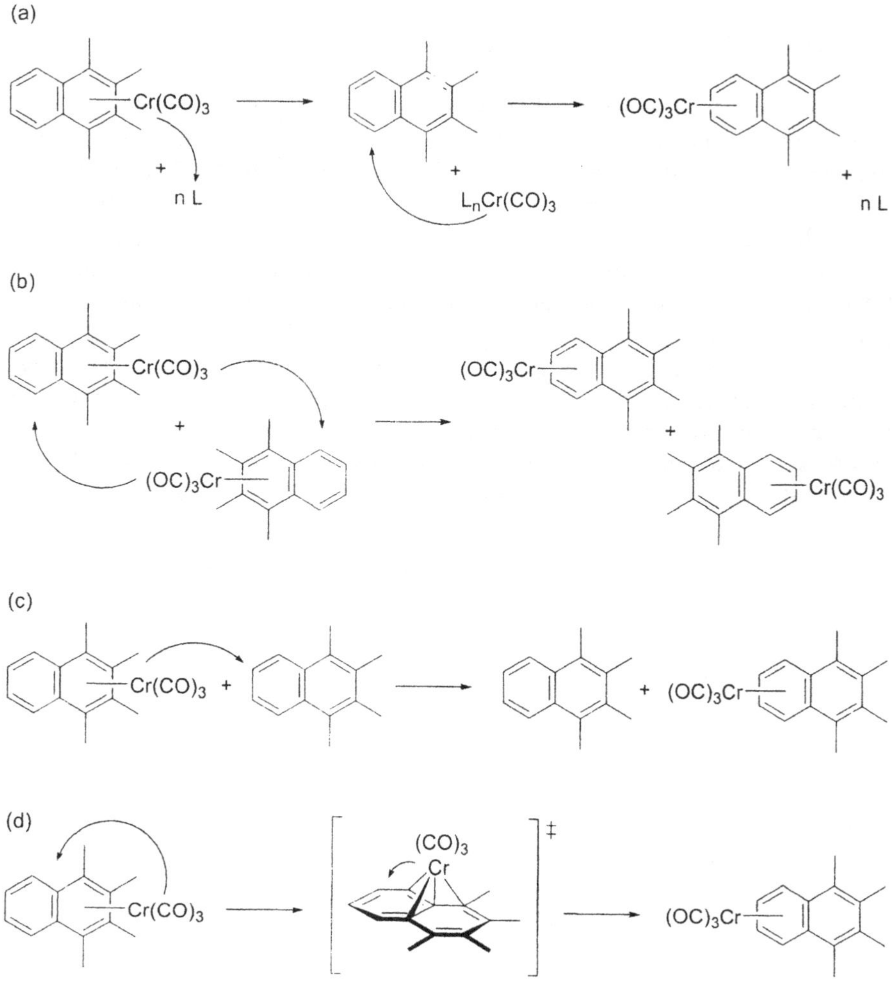

Scheme 25 Mechanisms of haptotropic rearrangements

two complex molecules (b) or on the transfer of the organometallic moiety to a non-coordinated π-ligand (c). Finally, the isomerization may result from an intramolecular metal shift during which the metal fragment remains coordinated to the π-electron-system (d). This implies that the metal moiety remains attached to the same face of the arene ligand throughout the whole process, and, thus, this mechanism fulfills the requirements of a haptotropic metal migration which has to proceed strictly *intra*molecularly. The extent to which *inter*molecular mechanisms contribute to the isomerization may increase with increasing polarity and donor properties of the solvent and with increasing temperature. Theoretical studies suggest that the metal migration proceeds close to the periphery of the aromatic ligand rather than by the direct way across the central C-C-bond [3, 98].

3.1 Complexes of Naphthalenes and Other Bicyclic Systems

A majority of studies deals with haptotropic rearrangements involving π-ligands derived from naphthalene which represents the most simple fused aromatic system available. The first example was reported by Deubzer and Öfele for $Cr(CO)_3$ complexes of methyl-substituted naphthalenes [6, 93]. Subsequent research by Treichel and Kündig provided rate constants and thermodynamic activation parameters for π-complexes in which the degeneration of the haptotropic rearrangement was removed by selective mono- or polydeuteration [94, 95]. The metal migration was found to be a first order process using perdeuterated benzene or cyclohexane as solvents demonstrating the intramolecular character of the metal shift. Further support for a true haptotropic rearrangement was provided for diastereomerically pure methylacenaphthene complexes **66–69** which underwent metal migration free of racemization in decane at 130 °C (Scheme 26) [96]. In specific cases the haptotropic metal

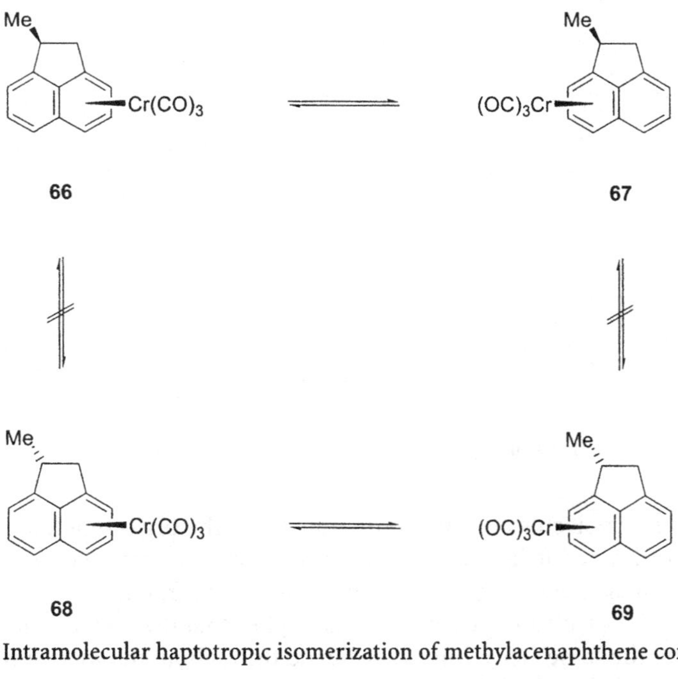

Scheme 26 Intramolecular haptotropic isomerization of methylacenaphthene complexes

migration may have to compete with other rearrangement reactions; for instance, the thermal isomerization of phenalene complexes **70–73** involves both haptotropic metal migration and sigmatropic hydrogen shift reactions. As determined in D_8-toluene, the metal migration is faster than the [1,5]-H shift of the *endo*-hydrogen atom; decomplexation proceeds by two orders of magnitude slower than the metal shift (Scheme 27) [97].

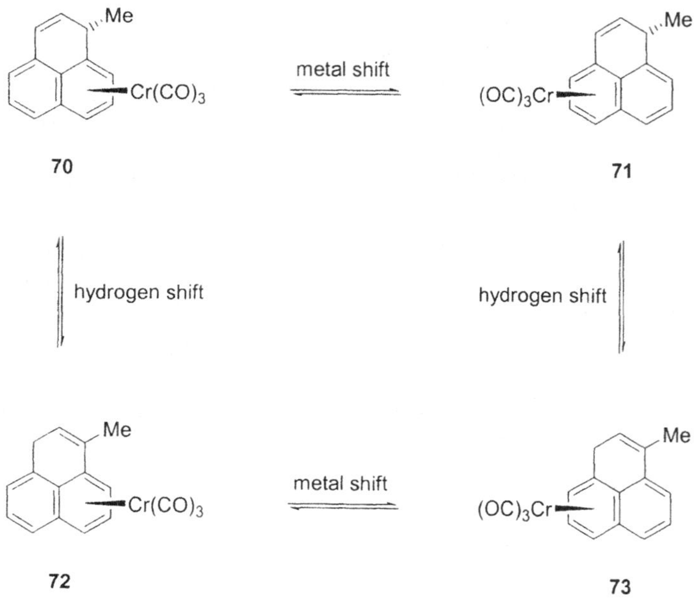

Scheme 27 Haptotropic metal shift versus sigmatropic hydrogen shift in phenalene complexes

The rate of the metal shift depends on the substitution pattern of the arene and on the solvent used. A systematic kinetic study of monosubstituted naphthalene $Cr(CO)_3$ complexes revealed that the metal shift is slightly slower in 1-substituted complexes compared with their 2-substituted analogues; the metal shift is accelerated if perdeuterocyclohexane is replaced as a solvent by hexafluorobenzene (for details above) [98].

Besides a single report using gas-chromatographic techniques, which also supports an intramolecular metal shift [99], the most convincing evidence for the mechanism of the complex isomerization originates from stereoselective transformations of enantiopure metal complexes. In this respect, the haptotropic metal shift in the (–)-menthyloxy-substituted naphthalene complex **74** – synthesized by diastereoselective chromium-templated [3+2+1]-benzannulation – at 90 °C in di-*n*-butyl ether is indicative for an intramolecular metal migration, regardless of the elevated temperatures applied and the coordinative properties of the etheral solvent used (Scheme 28) [68, 100].

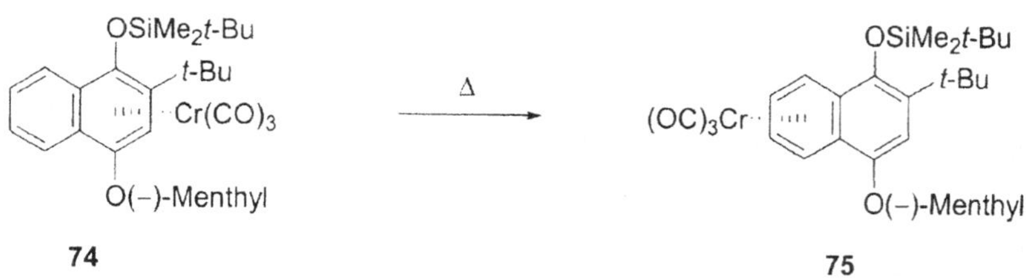

Scheme 28 Intramolecular haptotropic rearrangement of an enantiomerically pure complex

In contrast, nearly complete racemization (>90%) was observed when monosubstituted naphthalene complexes were heated in a 1:1 mixture of hexafluorobenzene and di-*n*-butyl ether [101]. The investigations have been carried out with enantiopure complexes obtained by HPLC on a chiral stationary phase (Daicel, Chiralcel OD). The prolonged reaction time of 48 h seems to be responsible for the considerable degree of racemization observed for the rearrangement products. When the same reaction was performed in toluene, the loss in optical purity decreased to ca. 20%; in non-coordinating solvents such as hexafluorobenzene or decane, however, no racemization of the sample at all could be detected after heating for 48 h.

Haptotropic shifts of $Cr(CO)_3$ fragments in bicyclic ligands other than naphthalenes have been studied for eight-membered rings [102, 103] as well as in heptalene complexes **76–79** where the inter-ring migration of the organometallic fragment has to compete with a 1,2-intra-ring shift (Scheme 29) [27, 104].

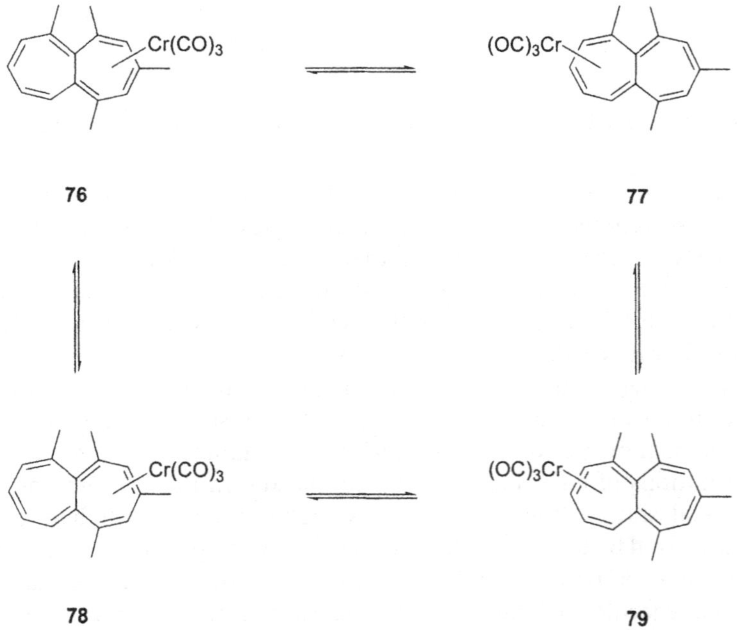

Scheme 29 Inter-ring haptotropic migration versus intra-ring 1,2-shift in heptalene complexes

3.2
Complexes of Oligocyclic Fused Arenes

Haptotropic metal migrations are not restricted to bicyclic π-systems; they may also proceed along more extended π-ligands. In the biphenylene system the aromatic rings suited for coordination of the $Cr(CO)_3$ moiety are formally separated by two σ-bonds. In order to avoid the inherent degeneracy of a metal shift in this system, the biphenylene $Cr(CO)_3$ complex was deuterated or methylated via metalation of the π-ligand with *n*-butyllithium followed by deuteration (**80**) or methylation [105]. The free activation enthalpy ($\Delta G^{\ddagger}$= 134.8 kJ mol^{-1}, T=403 K) for the metal migration is higher than that found along naphthalene ligands, but considerably lower as observed for biphenyl complexes which is due to geometrical arguments (Scheme 30). Evidence for haptotropic metal migrations across single σ-bonds was established for phenyl-substituted anthracenes [106]. The authors concluded that the process is based on an intermolecular reaction, but later an intramolecular mechanism of this shift has been discussed as well [107].

80 **81**

Scheme 30 Dynamic equilibrium between isomeric chromium biphenylene complexes

The haptotropic rearrangement reported for the starphenylene complex **82** is also based on the shift of the chromium fragment across a cyclobutanoid spacer; despite a bulky silyl substitution it occurs more facile to the terminal arene ring than the metal shift in the biphenylene complex ($\Delta G^{\ddagger}$=114.1 kJ mol^{-1}, T=360 K) (Scheme 31) [108].

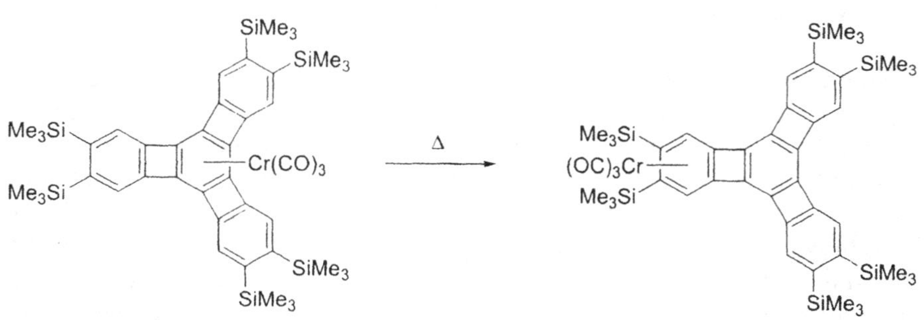

82 **83**

Scheme 31 Haptotropic isomerization of a starphenylene complex

The fluoranthene ligand features a benzoid and a naphthoid unit connected by two σ-bonds. Its $Cr(CO)_3$ complex **84** undergoes a degenerate metal shift within the naphthoid part of the molecule as well as a non-degenerate migration along the σ-bonds; the kinetics of the latter process have been determined in hexafluorobenzene solution and reveal a free enthalpy of activation in between those observed for the biphenylene and starphenylene complexes ($\Delta G^{\ddagger}$=125.4 kJ mol^{-1}, T=363 K) (Scheme 32) [109].

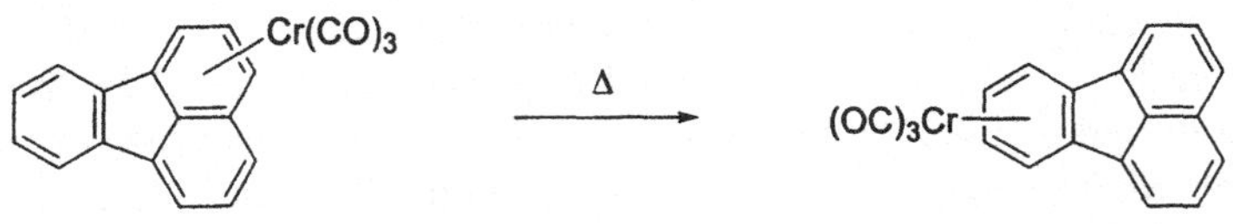

84 **85**

Scheme 32 Shift of a $Cr(CO)_3$ moiety across the fluoranthene ligand

Whereas thermodynamic conditions strictly prefer complexation of the terminal and more aromatic ring in fused arenes, a single exception has been reported for labeling an internal benzene ring with a $Cr(CO)_3$ fragment [110]. The reaction of 9,10-dimethylanthracene **86** with chromium hexacarbonyl under reflux in a solvent mixture of di-*n*-butyl ether and tetrahydrofuran afforded a 37% yield of the (η^6-4a,9,9a,8a,10,10a)-9,10-dimethylanthracene complex **87** surprisingly stable towards metal migration under these conditions (Scheme 33).

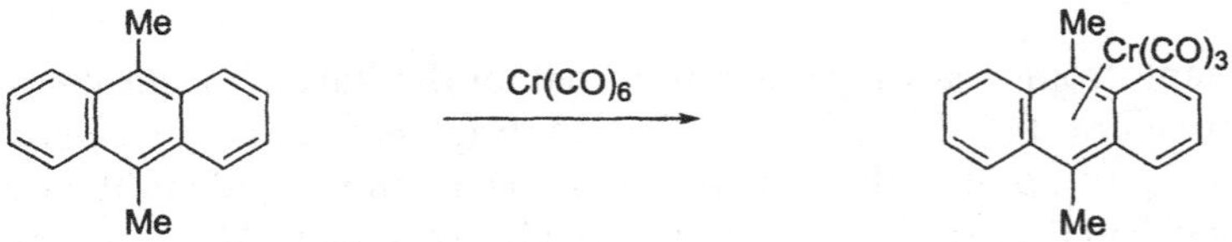

86 **87**

Scheme 33 Coordination of an internal hexacyclic ring of an anthracene ligand

This coordination behavior is quite unique and so far limited to the linear fused tricyclic arene skeleton. Attempts to label the central benzene ring by a $Cr(CO)_3$ fragment in angular tri- or tetracyclic systems failed. In a straightforward and highly flexible approach the benzannulation/haptotropic migration sequence was applied to phenanthrene complexes. The reaction of 1- and 2-naphthylcarbene complexes with alkynes generates chromium-coordinated phenanthrenes bearing a tunable arene substitution pattern under mild kinetic conditions. This approach aims at an increased versatility for the metal shift process which is intended to overcome the restriction of the haptotropic metal migration to a mono-directional one-step process – even in cases where no dynamic equilibrium between the two regioisomers is involved.

First results have been obtained by a set of phenanthrene complexes bearing different substitution patterns of the two primary "targets" of a haptotropic migration, the internal and the non-coordinated terminal ring [111, 112]. While the modification of the terminal ring allowed to tune the kinetics of the haptotropic metal shift, a variation of the substitution pattern of the internal ring led to complexes substantially differing in their capability of undergoing thermally induced haptotropic metal shifts. Upon warming to 90 °C the phenanthrene complexes **88–93** rearrange to their regioisomers **96–101** in which the $Cr(CO)_3$ fragment is bound to the non-hydroquinoid terminal ring [111] (Scheme 34, Table 1). In contrast, no evidence for a haptotropic isomerization could be detected neither for the Boc-aminosubstituted derivative **95** nor for the deprotected species **94** [112] which may reflect either an enhanced activation barrier for the metal shift or a reversal of the thermodynamic stability of the complex regioisomers.

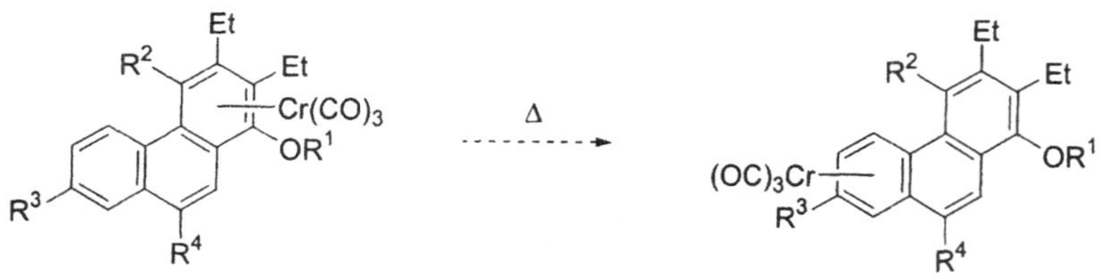

88 – 95 **96 – 103**

Scheme 34 Haptotropic rearrangements of phenanthrene complexes

Table 1 Substitution pattern of the phenanthrene chromium complexes **88–103**

	R^1	R^2	R^3	R^4
88, 96	Me	$SiMe_2{}^tBu$	H	H
89, 97	Me	$SiMe_2{}^tBu$	OMe	H
90, 98	Me	$SiMe_2{}^tBu$	Br	H
91, 99	$SiMe_2{}^tBu$	Me	H	H
92, 100	$SiMe_2{}^tBu$	Me	H	Me
93, 101	$SiMe_2{}^tBu$	Me	H	Br
94, 102	Et	Me	H	NHMe
95, 103	Et	Me	H	NMeBoc

The results obtained with the phenanthrene complexes prompted us to extend further the aromatic system as a π-pathway for the metal migration. A major interest focused on whether heteroarenes are compatible with the haptotropic migration or rather block the metal shift [76, 113]. The tetracyclic benzonaphthofuran and benzonaphthothiophene complexes **104** and **105** have been prepared by benzannulation of carbene complex precursors (Scheme 18). The haptotropic isomerization of the oxygen-based heterocycle proceeds under elevated temperatures of 90 °C and provides the complex isomer in which the chromium coordinates the non-hydroquinoid terminal ring (Scheme 35) [76]. This result indicates that the metal migration is not limited to benzene rings and even tolerates heteroatoms in five-membered ring systems. The molecular structures of both regioisomers **104** and **106** have been established by

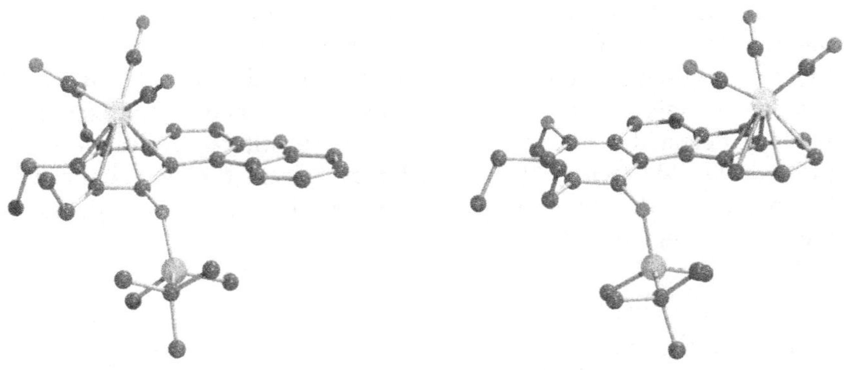

Fig. 1 Molecular structures of regioisomeric benzonaphthofuran complexes **104** and **106**

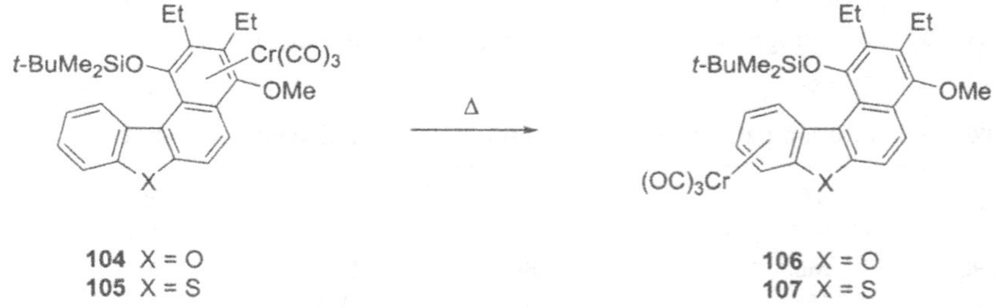

Scheme 35 Extended aromatic π-systems for hapotropic metal shifts

NMR spectroscopy and single crystal X-ray analysis (Fig. 1). In the homologous thiacyclic system **105** the metal shift is slowed down and only small amounts of rearrangement product are isolated [113]. An explanation why the metal shift proceeds more readily in the oxacyclic system has to take into account the extent of the helical twist present in the tetracyclic skeleton. The torsion angles observed in the thiophene derivative **105** amount to twice as much as those found for the fused furan analogue **104** (Fig. 2). Recently, a shift of a $Cr(CO)_3$ moiety across a tricyclic methylsubstituted dibenzothiophene ligand has been reported [114].

The angular-linear annulation pattern present in the pentacyclic dibenzo-[*b*,*g*]carbazole complex allows one to test long-range haptotropic metal shifts and to study a 1,5-interring migration across a central carbazole (Scheme 36) [115]. After warming complex **108** to 90 °C in di-*n*-butyl ether small amounts of rearrangement product **109** are isolated; the structures of both regioisomeric complexes were elucidated from their NMR spectra.

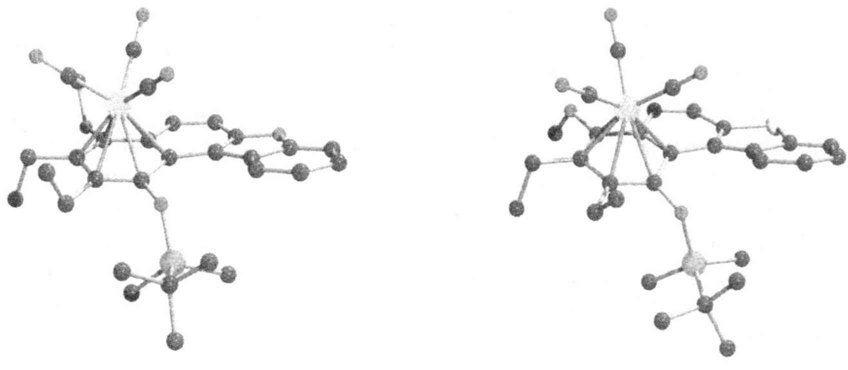

Fig. 2 Molecular structures of **104** and **105**, highlighting the helical twist of the tetracyclic arene ligand

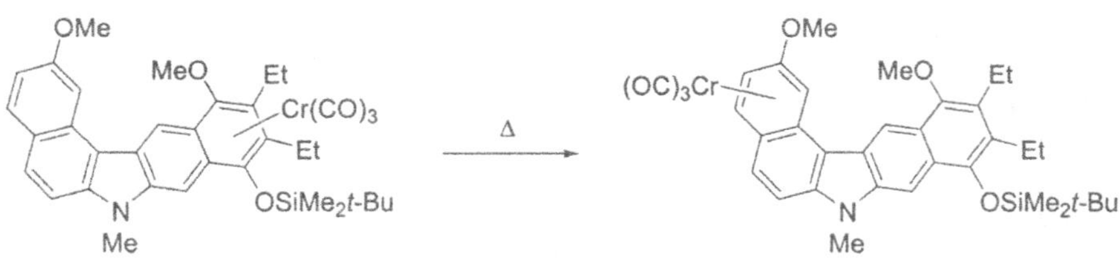

108 **109**

Scheme 36 Long-range metal migration along a pentacyclic arene ligand

4 Controllable Haptotropic Rearrangement

4.1 Tuning by the Arene Substitution Pattern

A systematic study addressing the impact of substituent influences on haptotropic metal shifts in arene chromium complexes can be based on a ligand tuning of this reaction. In this respect, four different substituents were incorporated in the naphthalene ligand in either the 1- or 2-position and the reactivity towards haptotropic rearrangement was examined (Scheme 37, Tables 2 and 3) [98]. While the metal shift of the 2-trimethylsilyl substituted naphthalene complex **113** was accelerated compared to the 1-substituted analogue **112**, for the methyl (**110** and **111**), trimethylstannyl (**114** and **115**) and the chloro (**116** and **117**) substituted complexes the haptotropic rearrangement was slowed down for the 1-substituted derivatives.

Studies of haptotropic metal shifts in polyfunctionalized naphthalene complexes are limited. The [3+2+1]-benzannulation (see above) is a versatile approach to such complexes, featuring the additional benefit of selectively providing the kinetic complex product suitable for subsequent thermal haptotropic isomerization. The substitution pattern of both naphthalene rings can be modified by prefunctionalization of the aromatic carbene substituent and the choice of the alkyne and the protective group for the phenolic benzannulation product (Table 2). The substituent effects have been investigated in terms of the relative thermodynamic stability of the two coordinative isomers and – whenever possible – the kinetics of the thermally induced haptotropic metal shift (Scheme 37, Table 3) [116].

110 – 130 131 – 151

Scheme 37 Haptotropic metal shifts in a series of naphthalene complexes

Table 2 Substitution pattern of the naphthalene chromium complexes 110–151

	R^1	R^2	R^3	R^4	R^5	R^6	R^7	R^8
110, 131	Me	H	H	H	H	H	H	H
111, 132	H	Me	H	H	H	H	H	H
112, 133	$Si(Me)_3$	H	H	H	H	H	H	H
113, 134	H	$Si(Me)_3$	H	H	H	H	H	H
114, 135	$Sn(Bu)_3$	H	H	H	H	H	H	H
115, 136	H	$Sn(Bu)_3$	H	H	H	H	H	H
116, 137	Cl	H	H	H	H	H	H	H
117, 138	H	Cl	H	H	H	H	H	H
118, 139	$SiMe_2{}^tBu$	Et	Et	Me	H	H	H	H
119, 140	$SiMe_2{}^tBu$	Et	Et	Me	OMe	H	H	H
120, 141	$SiMe_2{}^tBu$	Et	Et	Me	OMe	OMe	H	H
121, 142	$SiMe_2{}^tBu$	Et	Et	Me	OMe	H	H	OMe
122, 143	$SiMe_2{}^tBu$	Et	Et	Me	H	OMe	OMe	H
123, 144	$SiMe_2{}^tBu$	Et	Et	Me	Me	H	H	Me
124, 145	$SiMe_2{}^tBu$	Et	Et	(–)-Menthyl	H	H	H	H
125, 146	$SiMe_2{}^tBu$	^tBu	H	(–)-Menthyl	H	H	H	H
126, 147	$SiMe_2{}^tBu$	^tBu	H	(–)-Menthyl	OMe	H	H	H
127, 148	Me	Et	Et	Me	H	H	H	H
128, 149	Ac	Et	Et	Me	H	H	H	H
129, 150	$SiMe_2{}^tBu$	$SiMe_3$	H	Me	H	H	H	H
130, 151	$SiMe_2{}^tBu$	H	$SnBu_3$	Me	H	H	H	H

Table 3 Kinetic parameters and equilibrium constants of haptotropic rearrangements of the chromium naphthalene complexes **110–130** in hexafluorobenzene solution

	T [K]	k_1 [s^{-1}]	k_2 [s^{-1}]	K	$\Delta G^{\ddagger}$ [kJ mol^{-1}]
110	358	9.0×10^{-6}	9.0×10^{-6}	1.77	132.3
111	358	3.02×10^{-5}	1.45×10^{-5}	2.09	128.4
112	358	2.40×10^{-5}	5.0×10^{-6}	4.80	129.2
113	358	1.93×10^{-5}	1.1×10^{-6}	17.26	129.9
114	358	7.0×10^{-6}	1.4×10^{-6}	5.00	133.1
115	358	2.24×10^{-5}	1.7×10^{-6}	13.26	129.4
116	358	1.5×10^{-6}	4.36×10^{-5}	0.034	138.0
117	358	4.0×10^{-6}	3.46×10^{-5}	0.12	134.9
118	343	5.31×10^{-4}	–	–	105.9
119	343	7.01×10^{-4}	2.09×10^{-4}	3.35	105.1
120	343	1.26×10^{-3}	5.41×10^{-4}	2.33	103.5
121	343	–	–	0.02	–
122	343	–	–	0.08	–
123	343	6.31×10^{-4}	–	–	105.4
124a	333	8.98×10^{-4}	–	–	101.3
124b	333	1.55×10^{-3}	–	–	99.8
125	333	8.46×10^{-4}	–	–	101.5
126	333	3.04×10^{-3}	1.60×10^{-4}	19.0	97.9
127	348	1.38×10^{-4}	–	–	111.4
128	333	1.18×10^{-3}	–	–	100.5
129	343	1.59×10^{-4}	–	–	109.4
130	343	7.35×10^{-4}	–	–	105.0

The reaction of carbene complexes with internal aliphatic alkynes typically produces naphthohydroquinone complexes bearing two alkyl groups. If the non-hydroquinoid ring of the naphthalene remains unsubstituted, the complex isomers possess significantly different thermodynamic stability, and the metal shift proceeds quantitatively to the non-functionalized ring generating the thermodynamically more stable regioisomer. The complexes **118–123** display an identical substitution pattern of the hydroquinoid ring, originating from benzannulation of arylmethoxy carbene complexes with 3-hexyne followed by silylation with *tert*-butyldimethylsilylchloride, but they differ in the functionalization of the adjacent ring. The effect of methoxy substituents has been investigated in detail, revealing that both the number and the position of the methoxy groups is essential to the state of equilibrium. The difference in thermodynamic stability is reduced if a methoxy group is introduced in the 5-position (**119**) and, in consequence, a dynamic equilibrium (K=3.35) between the two regioisomers is observed. If a second methoxy substituent is added in the 6-position (**120**), the preference of the organometallic fragment for the non-hydroquinoid ring is further reduced, shifting the equilibrium to K=2.33. The two constitutional isomers of **120** (**121** and **122**) are much less susceptible to a haptotropic metal shift, the complex regioisomers prepared by benzannulation are the thermodynamically favored complexes in these cases. As observed for the parent complex **118** of the series, the haptotropic isomerization is quantitative for the 5,8-dimethylated ligand (**123**). This indicates that the metal migration is in principle compatible with 1,4,5,8-tetrasubstituted naphthalene ligands. In conclusion, the reason why complex **121** is not capable of a thermal haptotropic metal shift is related to electronic properties of the substituents rather than steric interactions, and has to be discussed in terms of the energy of the transition state, influenced by the substitution pattern at the arene carbon atoms next to the ring junction.

In the second part of this study, the emphasis was laid on the diversification of the hydroquinoid half of the naphthalene, while leaving the adjacent ring non-functionalized (except for **126**). For this purpose different carbene oxygen substituents, various alkynes and a set of protecting groups were employed, providing complexes **124–130**. The haptotropic rearrangements of these complexes proceed irreversibly due to the large difference in the complex isomers' thermodynamic stabilities. The kinetic parameters for the metal migration were correlated to the steric and electronic properties of the substituents present in the arene ligands. If the steric demand of the carbene oxygen substituent is increased by exchanging the methyl group for a (–)-menthyloxy substituent, the haptotropic isomerization is accelerated. As a result of the inherent planar chirality of the arene complexes synthesized via chromium-templated benzannulation, the incorporation of a the chiral terpenoid alcohol generates two diastereomeric complexes **124a** and **124b** with selectivities up to 80% d.e. [68]. In one of the stereoisomers the metal shift is twice as fast to the unsubstituted ring as in the other one. A *tert*-butyldimethylsilyl substituent is a popular protection group for the phenolic function directly resulting from CO-incorporation during the benzannulation, since it very effectively prevents oxidative decomposition of the complexes. If it is exchanged for a smaller methyl substituent, the rate constant of the metal shift is considerably decreased. On the contrary, the haptotropic isomerization is strongly accelerated when the electron-withdrawing acetyl protecting group is employed. It should be noted that the densely functionalized naphthohydroquinone complexes obtained by benzannulation display a remarkably low free activation enthalpy which amounts to about 20 kJ mol^{-1} less than that observed with monosubstituted naphthalene chromium complexes. This may be rationalized by additive steric effects of the persubstituted hydroquinoid ring.

As illustrated in the previous discussion a variety of the arene ligand substituents can be utilized for a reactivity tuning of haptotropic metal shifts. However, for each isomerization experiment a properly functionalized starting material has to be prepared. A possibly more elegant approach makes use of functional group interconversions to alter the reactivity of the arene complex. This strategy has been successfully applied to an aminosubstituted naphthalene complex in which the electron donating properties of the ligand may be varied by the nitrogen substituents: The methylaminonaphthalene complex **153** provides a more electron-rich arene ligand than present in its Boc-protected analogue **152** (Scheme 38) [112]. A completely different behavior of the complexes is observed upon heating: While any evidence for a haptotropic rearrangement is absent in the case of the Boc-protected aminonaphthalene **152** up to temperatures of 100 °C, cleavage of the Boc group enhances the electron donating properties of the amino substituent and facilitates the metal shift which now occurs readily at temperatures as low as 40 °C and results in nearly complete isomerization to complex **155**.

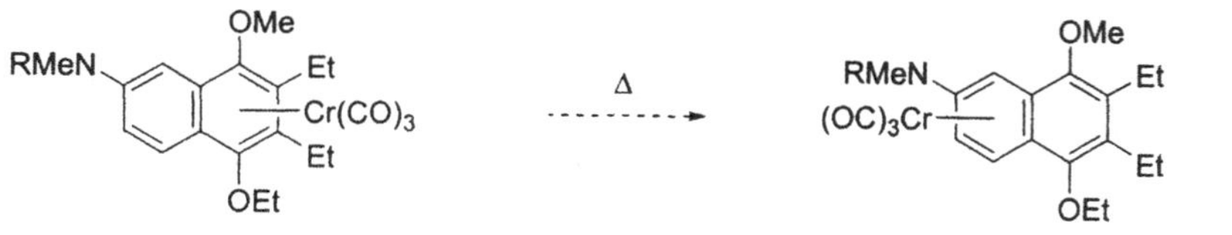

152 R = Boc
153 R = H

154 R = Boc
155 R = H

Scheme 38 Electronic substituent tuning based on a protection group variation

Analogously, the electron donating properties of the amino group can be influenced by protonation [52]. By treating the haptotropic rearrangement product **156** with tetrafluoroboric acid the precipitation of an orange colored salt indicated quaternization of the nitrogen by protonation. Subsequent heating to 70 °C monitored by IR spectroscopy suggested a retro migration of the chromium fragment to the hydroquinoid ring (Scheme 39). Unfortunately, the lability and the weak solubility of the reaction product hampered its isolation and complete characterization.

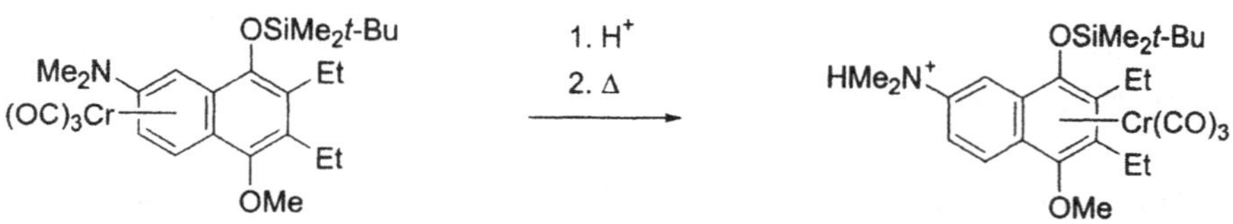

156 **157**

Scheme 39 Protonation of an aminonaphthalene complex as an electronic substituent tuning

4.2
Tuning by the Metal Coligand Sphere

4.2.1
Kinetic Tuning by Phosphorus Coligands

A modification of the arene ligand substitution pattern allows one to tune the kinetic and thermodynamic properties of haptotropic metal shifts. A different approach complementing these studies is based on a tailored diversification of the coligand sphere. Phosphorus coligands seemed to be adequate for this purpose due to the variety of phosphines and phosphites available for a template tuning. We aimed at a study which allows to evaluate the steric and electronic influences of the phosphorus coligands in a quantitative manner by investigating the haptotropic metal migration of a set of $Cr(CO)_2(PR_3)$ moieties along a naphthalene ligand.

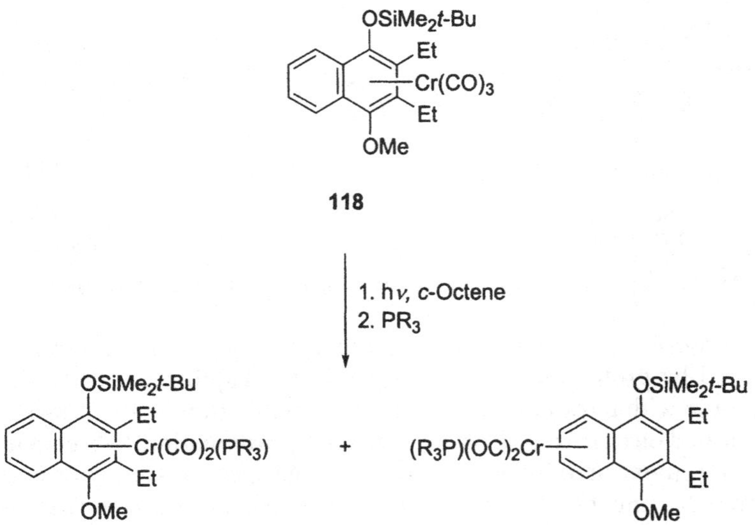

158 R = Ph
159 R = OPh
160 R = Me
161 R = OMe

162 R = Ph
163 R = OPh
164 R = Me
165 R = OMe

Scheme 40 Synthesis of $Cr(CO)_2(PR_3)$ complexes by photo-induced ligand exchange

The $Cr(CO)_2(PR_3)$ complexes **158–165** were synthesized by low-temperature photosubstitution of a carbonyl ligand from the benzannulation product **118** in the presence of cyclooctene (Scheme 40) [117]. The η^2-alkene chromium complex was subsequently reacted with the respective phosphorus ligand in the dark at room temperature to afford the regioisomers **158–161** as the major products; the complexes **162–165** already resulting from haptotropic rearrangement were isolated in minor amounts. Their formation involves a metal shift accompanying the cyclooctene/PR_3 exchange which is performed at conditions remarkably mild for a haptotropic isomerization. The mixture of regioisomers was separated by preparative HPLC for each pair of complexes. Both pairs of regioisomers bearing the triphenylphosphine (**158** and **162**) or the triphenylphosphite ligand (**159** and **163**) were characterized by single crystal X-ray analyses (Fig. 3). The thermodynamically less favored isomers **158–161** were studied in haptotropic rearrangement experiments which were performed in hexafluorobenzene solution at 60 °C (Scheme 41, Table 4).

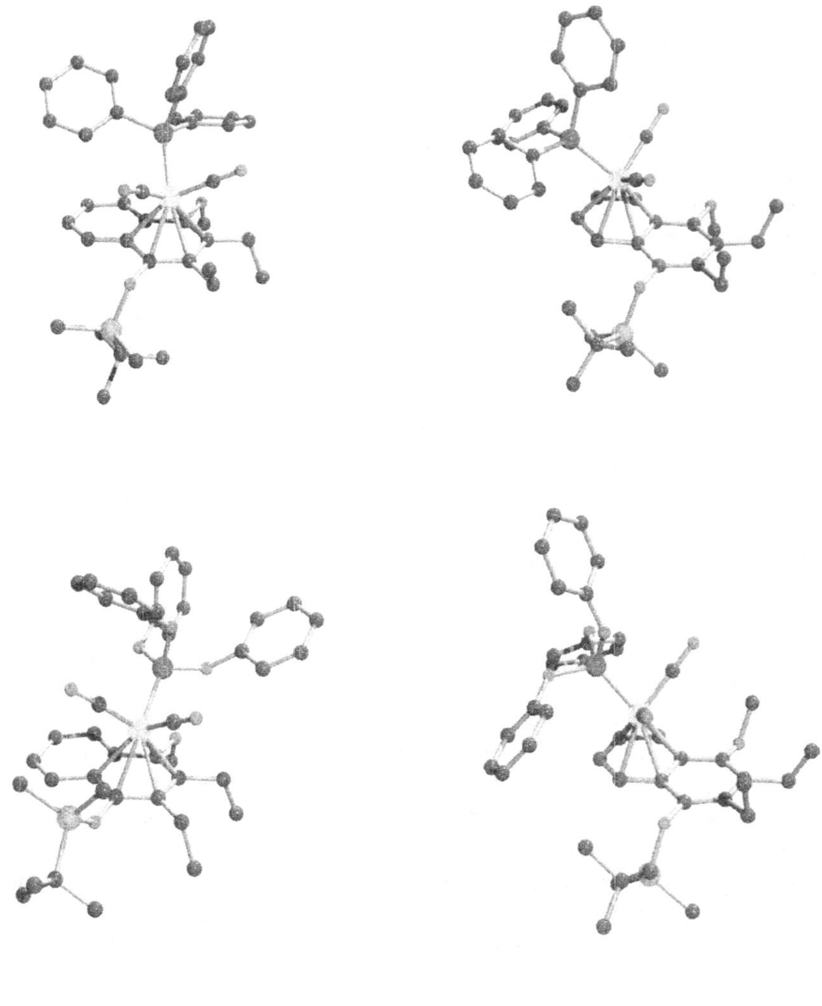

Fig. 3 Molecular structures of regioisomeric PPh_3 (**158**, **162**) and $P(OPh)_3$ (**159**, **163**) complexes

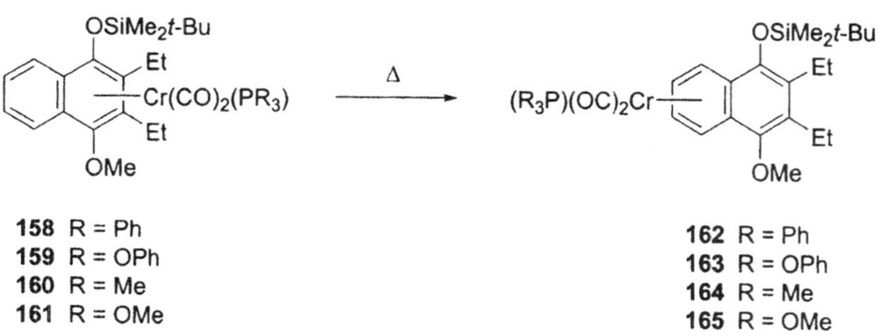

Scheme 41 Haptotropic shifts of $Cr(CO)_2(PR_3)$ moieties

Table 4 Kinetic parameters for the isomerization of $Cr(CO)_2(PR_3)$ complexes **158–161**, cone angels and values for the electronic parameter χ

	L	T [K]	k_1 [s^{-1}]	$\Delta G^{\ddagger}$ [kJ mol^{-1}]	θ [°]	χ [cm^{-1}]
118	CO	333	$(6.5\pm0.7)\times10^{-5}$	108.5±0.3	–	–
118	CO	348	$(4.3\pm0.4)\times10^{-4}$	108.1±0.3	–	–
158	PPh_3	333	$(1.6\pm0.2)\times10^{-4}$	106.0±0.3	145	13.25
159	$P(OPh)_3$	333	$(1.9\pm0.2)\times10^{-4}$	105.6±0.3	128	30.20
160	PMe_3	348	$(7.2\pm0.7)\times10^{-6}$	119.9±0.3	118	8.55
161	$P(OMe)_3$	333	$(6.1\pm0.6)\times10^{-6}$	115.1±0.3	107	24.10
161	$P(OMe)_3$	348	$(3.4\pm0.3)\times10^{-5}$	115.5±0.3	107	24.10

The haptotropic rearrangement of the triphenylphosphine (**158**) and the triphenylphosphite complex **159** displayed a reactivity two to three times enhanced compared to that observed for the parent $Cr(CO)_3$ complex **118**. In contrast, a trimethylphosphite coligand (**161**) reduces the rate constant of the metal shift by an order of a magnitude. The trimethylphosphine ligand further increases the thermostability towards haptotropic isomerization; for this reason, the rearrangement of complex **160** was examined at elevated temperature; at 75 °C the reactivity of the trimethylphosphine complex **160** is about 60 times lower than that of the parent tricarbonyl complex **118**. Figure 4 illustrates the reactivity tuning by logarithmic concentration plots.

We discussed the reactivity tuning by the phosphorus coligands in terms of their steric and electronic properties. By utilizing a standard empirical approach we performed a quantitative analysis of ligand effects (QALE) [118, 119]. For the fit of the experimentally obtained $\Delta G^{\ddagger}$ values we employed Tolman's cone angles θ as a measure of the ligands' steric demand [120] and χ – which is the hypsochromic shift of the IR ν_{CO} band of $Ni(CO)_3(PR_3)$ compared to that of the P^tBu_3 complex – as the electronic parameter [121]. A multiple regression according to Eq. (1) revealed that both the steric and the electronic properties

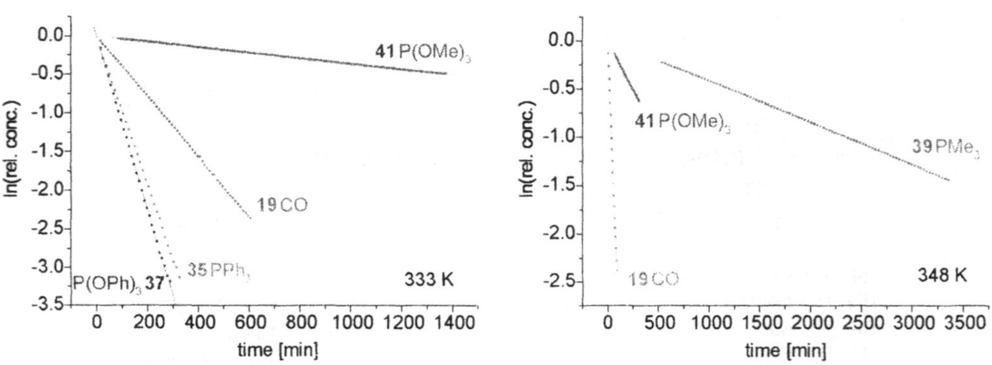

Fig. 4 Kinetic tuning of haptotropic rearrangements by phosphorus coligands

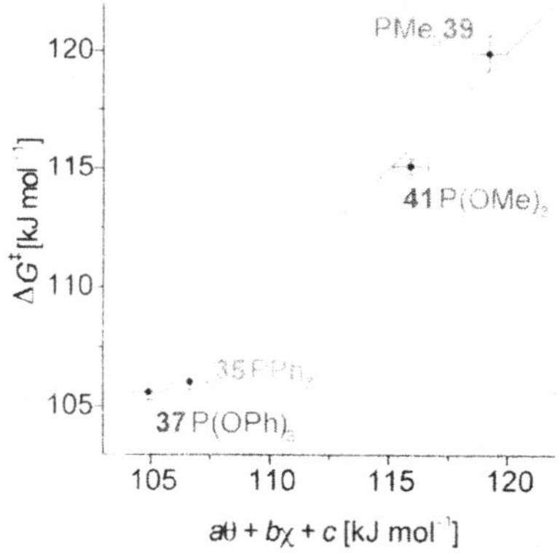

Fig. 5 Plot of the experimentally determined $\Delta G^{\ddagger}$ values against $a\theta+b\chi+c$

of the phosphorus ligands display a significant impact onto the kinetics of the metal shift. Bulky and electron-deficient ligands enhance the reactivity towards isomerization while the metal shift is slowed down by small and electron-rich phosphines. These results as well as the goodness of fit are illustrated by Fig. 5; the experimentally measured $\Delta G^{\ddagger}$ values are plotted against the data obtained by the fit according to Eq. (1):

$$\Delta G^{\ddagger} = a\theta + b\chi + c \qquad (1)$$

In order to avoid the necessity of using HPLC techniques to obtain pure coordinative isomers, the photolytic ligand exchange procedure was applied to the thermodynamically favored tricarbonyl complex **139**. Upon treatment of the intermediate cyclooctene complex with triphenylphosphine, complex isomer **162** was expected as the only product. However, both regioisomeric complexes **158** and **162** were isolated after chromatographic work-up, paralleling the results received from the analogous reaction of the thermodynamically less stable $Cr(CO)_3$ complex **118**. This surprising outcome has to be rationalized as a metal shift against the direction favored by thermodynamics and seemed to be a promising starting point for gaining further control concerning haptotropic complex rearrangements.

4.2.2 Thermo-Optical Switchable Devices

By observing a reverse metal shift accompanying the formation of the triphenylphosphine complex **158** we felt encouraged to test whether a photochemical procedure is suitable for reconverting the product of the thermally induced haptotropic rearrangement to the regioisomer provided by benzannulation [122]. In this context we first studied the UV irradiation of complex **139** under CO atmosphere in order to stabilize potential dicarbonyl intermediates. However, decomposition was observed under these conditions with no evidence for a haptotropic metal shift indicating that a metal migration at the stage of a coordinatively unsaturated dicarbonyl species seems to be unlikely.

Mimicking the synthesis of the triphenylphosphine complexes **158** and **162** we then added cyclooctene before irradiating the sample in order to trap the dicarbonyl intermediate as a η^2-alkene complex (Scheme 42). Interestingly, we observed an identical set of A_1 and the B_1 ν_{CO} stretching frequencies, irrespective whether complex **118** or its regioisomer **139** was subjected to photolysis suggesting the formation of the same reaction intermediate in both cases. The irradiation was stopped upon complete removal of one equivalent of carbon monoxide, and the solution was subsequently flushed with a slight stream of CO to refurnish a tricarbonyl chromium complex. IR and NMR analyses revealed that this protocol exclusively generates the thermodynamically less favored complex isomer **118** along with a small amount of demetalated hydroquinone as a by-product. A crystallographic study of the tricarbonyl complex independently established the formation of complex **118** by a haptotropic rearrangement against the direction favored by thermodynamics.

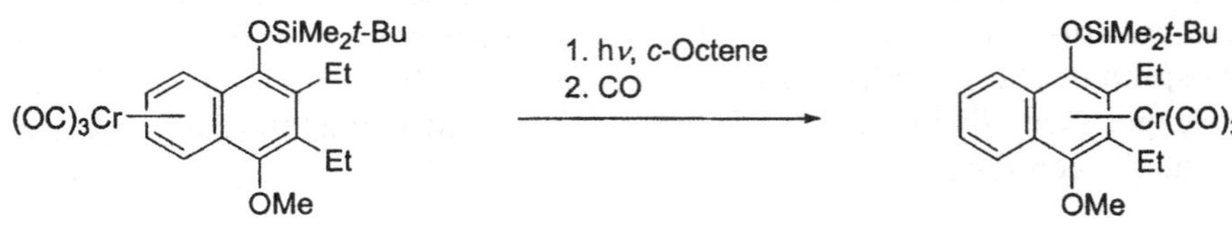

139 **118**

Scheme 42 Photo-induced reverse metal shift to the thermodynamically less favored isomer

The planar chirality of $Cr(CO)_3$ complexes **118** and **139** arising from the unsymmetrical substitution pattern in the hydroquinoid ring generally allows the construction of an organometallic molecular switch transferring chiral information. This aim requires that the metal shift occurs under retention of configuration in both directions. The thermally initiated haptotropic metal migration is known to proceed via an intramolecular process [68, 94–100], which is supported by the kinetic data collected for the isomerization of complex **118** to **139** ($\Delta G^{\ddagger}$=108.5 kJ mol^{-1}, T=333 K) (Fig. 6). In order to investigate the mechanism of the photo-induced metal shift the enantiomers of complex **139** have been separated using HPLC on a chiral stationary phase (Daicel, Chiralcel OD). The absolute configuration of the complex enantiomers could be established by X-ray analysis, assigning the (R)-configuration to the (–)-enantiomer of **139**. The (+)-enantiomer (S)-**139** was subjected to the photochemical haptotropic isomerization, and the rearrangement product was analyzed by chiral HPLC and compared to the racemic mixture of complex **118**. The formation of a single enantiomer indicated the intramolecularity of the photo-induced haptotropic rearrangement; a stepwise release of the organometallic moiety followed by recomplexation by the arene ligand could be ruled out on the basis of this experiment.

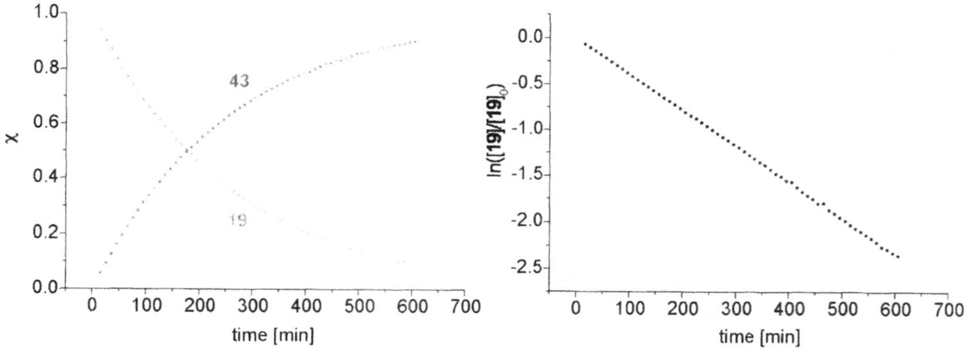

Fig. 6 Kinetic analysis of the rearrangement of complex **118** to **139**

Organometallic molecular switches are rare. Some reports refer to dinuclear cyclopentadienyl complexes of ruthenium, molybdenum and tungsten [14, 123, 124]; a single study deals with a switchable mononuclear molybdenum complex [125]. None of them has been tested for the stereospecificity of the switching cycle, which is partly due to the presence of achiral species, which would erase any stereoinformation incorporated. In consequence, the pair of enantiopure complexes (*R*)-**118** and (*S*)-**139** complemented by two stereospecific interconverting procedures represents the first example of a racemization-free organometallic switch (Scheme 43, Fig. 7).

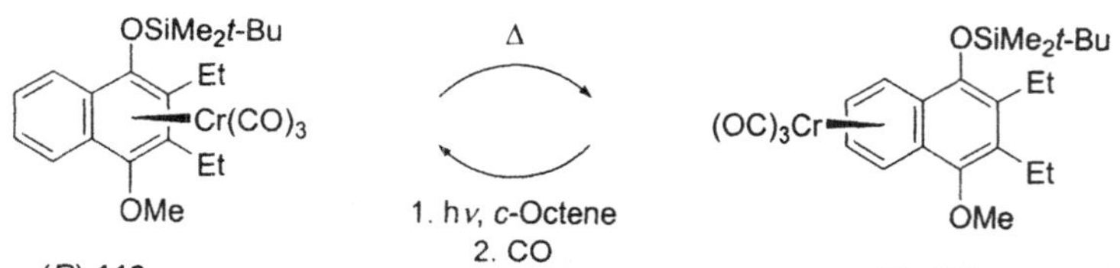

(*R*)-**118** (*S*)-**139**

Scheme 43 Stereospecific organometallic switch based on a reversible metal migration

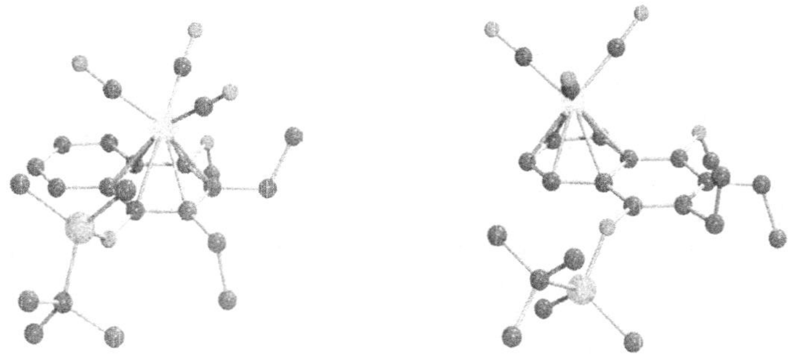

Fig. 7 Both positions of the molecular switch as molecular structures (**118**, **139**)

5
Conclusion

The chromium-templated [3+2+1]-benzannulation of Fischer carbene complexes allows for the regio- and stereoselective one-pot synthesis of densely functionalized hydroquinones obtained as (arene)$Cr(CO)_3$ complexes. The range of functional groups tolerated in the alkyne and in the carbene complex moiety along with the chemoselectivity arising from a different substitution pattern allows to tune the reaction towards a huge number of possible products, making the reaction an important tool in chemical synthesis in the last decades. Additionally, the mild reaction conditions led to numerous applications in bioorganic chemistry or in the synthesis of natural products.

Intensive research concentrated on the haptotropic rearrangement reactions with the focus set on the controllability of this kind of dynamic process, making it suitable for applications in molecular machines. The metal migration may be rendered switchable between two specifiable isomers by external influence, e.g. heating, immision of light or pH-change. The haptotropic rearrangement in $Cr(CO)_3$ complexed naphthalenes was tuned by variation of the substitution pattern of the organic ligand and of the coligand sphere of the chromium template. The migration is governed by both the steric and the electronic properties of the ligands, as demonstrated by substitution of carbon monoxide for phosphines or phosphites. In addition, research on enantiopure tricarbonyl chromium complexes proved the intramolecular reaction pathway culminating in the development of a racemization-free organometallic switch.

References

1. Anh NT, Elian M, Hoffmann R (1978) J Am Chem Soc 100:110
2. Cotton FA (1968) J Am Chem Soc 90:6230
3. Albright TA, Hofmann P, Hoffmann R, Lillya CP, Dobosh PA (1983) J Am Chem Soc 105:3396
4. Matsubara K, Oda T, Nagashima H (2001) Organometallics 20:881
5. Nesmeyanov AN, Kravtsov DN (1960) Dokl Akad Nauk SSSR 135:331
6. Deubzer B (1966) PhD Thesis, Technische Hochschule München
7. Waldbach TA, van Rooyen PH, Lotz S (1993) Angew Chem 105:795; Angew Chem Int Ed Engl 32:710
8. Nordlander JE, Roberts JD (1959) J Am Chem Soc 81:1769
9. Mikhailov BM, Bubnov YN (1977) Organoboron compounds in organic synthesis. Nauka, Moscow
10. Kwart H, Slutsky J (1972) J Am Chem Soc 94:2515
11. Lutsenko IF, Baukov YI, Dudykina OV, Kramarova EN (1968) J Organomet Chem 11:35
12. Lutsenko IF, Baukov YI, Belavin I, Tvorogov AN (1968) J Organomet Chem 14:229
13. Bennet MA, Bown M, Hockless DCR, McGrady JE, Schranz HW, Stranger R, Willis AC (1998) Organometallics 17:3784
14. Matsubara K, Mima S, Oda T, Nagashima H (2002) J Organomet Chem 650:96
15. Foxman B, Marten D, Rosan A, Raghu S, Rosenblum M (1977) J Am Chem Soc 99:2160

16. Amendola V, Fabrizzi L, Mangano C, Miller H, Pallavicini P, Perotti A, Taglietti A (2002) Angew Chem 114:14; Angew Chem Int Ed Engl 41:2553
17. Stanger A, Weismann H (1996) J Organomet Chem 515:183
18. White C, Thompson J, Maitlis PM (1977) J Chem Soc Dalton Trans 1654
19. Ustynyuk YA, Trifonova OI, Oprunenko YF, Mstislavsky VI, Gloriozov IP (1990) Organometallics 9:1707
20. Nakasuji K, Yamaguschi M, Murata I (1986) J Am Chem Soc 108:325
21. Keasey A, Bailey PM, Maitlis PM (1978) J Chem Soc Chem Commun 142
22. For a recent review see: Dötz KH, Jahr HC (2004) Chem Rec 4:61
23. Razuvaev GA, Artemov AN, Aladjin AA, Sirotkin NI (1976) J Organomet Chem 111:131
24. Dötz KH, Tomuschat P (1999) Chem Soc Rev 28:187
25. Whitlock HW Jr, Reich C, Woessner WD (1971) J Am Chem Soc 93:2483
26. Hails MJ, Mann BE, Spencer CM (1985) J Chem Soc Dalton Trans 693
27. Uebelhart P, Linden A, Hansen H-J, Ustynyuk YA, Trifonova OA, Akhmedov NG, Mstislavsky VI (1999) Helv Chim Acta 82:1930
28. Stanger A, Vollhardt KPC (1992) Organometallics 11:317
29. Oprunenko Y (2000) Russ Chem Rev 69:683 (and references therein)
30. Nagashima H, Fukahori T, Itoh K (1991) J Chem Soc Chem Commun 786
31. Amendola V, Fabrizzi L, Mangano C, Miller H, Pallavicini P, Perotti A, Taglietti A (2000) J Chem Soc Dalton Trans 185
32. Belle C, Pierre J-L, Saint-Aman E (1998) New J Chem 22:1399
33. Balzani V, Credi A, Raymo FM, Stoddart JF (2000) Angew Chem 112:3484; Angew Chem Int Ed 39:3348
34. Sauvage J-P (1998) Acc Chem Res 31:611
35. Dötz KH (1975) Angew Chem 87:672; Angew Chem Int Ed Engl 14:644
36. Dötz KH (1984) Angew Chem 96:573; Angew Chem Int Ed Engl 23:587
37. Vorogushin AV, Wulff WD, Hansen HJ (2001) Org Lett 3:2641
38. For a further example see: Dötz KH, Popall M (1987) Angew Chem 99:1220; Angew Chem Int Ed Engl 26:1158
39. Harrity JPA, Kerr WJ (1993) Tetrahedron Lett 34:2995
40. Choi YH, Rhee KS, Kim KS, Shin GC, Shin SC (1995) Tetrahedron Lett 36:1871
41. Harrity JPA, Kerr WJ, Middlemiss D (1993) Tetrahedron 49:5565
42. Hutchinson EJ, Kerr WJ, Magennis EJ (2002) Chem Commun 2262
43. Barluenga J, Aznar F, Gutiérrez I, Martín A, García-Granda S, Llorca-Baragaño MA (2000) J Am Chem Soc 122:1314
44. Anderson JC, Cran JW, King NP (2002) Tetrahedron Lett 43:3849
45. Barluenga J, López LA, Martínez S, Tomás M (2000) Tetrahedron 56:4967
46. Liptak VP, Wulff WD (2000) Tetrahedron 56:10229
47. Bos ME, Wulff WD, Miller RA, Chamberlin S, Brandvold TA (1991) J Am Chem Soc 113:9293
48. Waters ML, Bos ME, Wulff WD (1999) J Am Chem Soc 121:6403
49. Wulff WD, Gilbert AM, Hsung RP, Rahm A (1995) J Org Chem 60:4566
50. Barluenga J, López LA, Martínez S, Tomás M (1998) J Org Chem 63:7588
51. Yamashita A (1986) Tetrahedron Lett 27:5915
52. Wenzel B (2001) Diploma thesis, Universität Bonn
53. Dötz KH, Grotjahn D, Harms K (1989) Angew Chem 101:1425; Angew Chem Int Ed Engl 28:1384
54. Wenzel B, Dötz KH (publication in preparation)
55. Dötz KH, Tiriliomis A, Harms K (1993) Tetrahedron 46:5577
56. Barluenga J, Aznar F, Palomero MA (2002) Chem Eur J 8:4149

57. Barluenga J, Aznar F, Palomero MA (2000) Angew Chem 112:4514; Angew Chem Int Ed Engl 39:4346
58. Dötz KH, Mühlemeier J, Schubert U, Orama O (1983) J Organomet Chem 247:187
59. Hofmann P, Hämmerle M, Unfried G (1991) New J Chem 15:769
60. Chamberlin S, Waters ML, Wulff WD (1994) J Am Chem Soc 116:3113
61. Davies MW, Johnson CN, Harrity JPA (2001) J Org Chem 66:3525
62. Woodgate PD, Sutherland HS, Rickard CEF (2001) J Organomet Chem 627:206
63. Wang H, Huang J, Wulff WD, Rheingold AL (2003) J Am Chem Soc 125:8980
64. Bitterwolf TE, Hubler TL, Todime R (1990) J Macromol Sci Chem A 27:1437
65. Hsung RP, Wulff WD, Rheingold AL (1994) J Am Chem Soc 116:6449
66. Hsung RP, Wulff WD, Challener CA (1996) Synthesis 773
67. Dötz KH, Otto F, Nieger M (2001) J Organomet Chem 621:77
68. Stinner C, Dötz KH (1997) Tetrahedron Asymmetry 8:1751
69. Hsung RP, Wulff WD, Chamberlin S, Liu Y, Liu RY, Wang H, Quinn JF, Wang SLB, Rheingold AL (2001) Synthesis 2:200
70. Janes C, Nieger M, Saarentko P, Dötz KH (2003) Eur J Org Chem 2276
71. Fogel L, Hsung RP, Wulff WD (2001) J Am Chem Soc 123:5580
72. Dötz KH, Mittenzwey S (2002) Eur J Org Chem 39
73. Dötz KH, Gerhardt A (1999) J Organomet Chem 578:223
74. Semmelhack MF, Ho S, Cohen D, Steigerwald M, Lee MC, Lee G, Gilbert AM, Wulff WD, Ball RG (1994) J Am Chem Soc 116:7108
75. Dötz KH, Popall M (1985) J Organomet Chem 295:C1
76. Jahr HC, Nieger M, Dötz KH (2002) J Organomet Chem 641:185
77. Jahr HC, Dötz KH (publication in preparation)
78. Calculations were performed on a semiempirical stage with a PM3(tm) basis set. DFT calculations are still in progress; Wenzel B, Schneider J, Dötz KH (publication in progress)
79. Jahr HC (2003) PhD Thesis, Universität Bonn
80. Schneider J, Dötz KH (publication in preparation)
81. Flynn BL, Schirmer H, Duetsch M, de Meijere A (2001) J Org Chem 66:1747
82. Gleichmann MM, Dötz KH, Hess BA (1996) J Am Chem Soc 118:10551
83. Dötz KH, Fischer H, Mühlemeier J, Märkl R (1982) Chem Ber 115:1355
84. Fischer H, Hofmann P (1999) Organometallics 18:2590
85. Dötz KH, Schäfer T, Kroll F, Harms K (1992) Angew Chem 104:1257; Angew Chem Int Ed Engl 31:1236
86. Jackson TJ, Herndon JW (2001) Tetrahedron 57:3859
87. Dötz KH, Fügen-Köster B (1980) Chem Ber 113:1449
88. Moser WH, Sun L, Huffman JC (2001) Org Lett 3:3389 (and references therein)
89. Herndon JW, Zhang Y, Wang H, Wang K (2000) Tetrahedron Lett 41:8687 (and references therein)
90. Barluenga J, Aznar F, Palomero MA (2003) J Org Chem 68:537
91. Jiang D, Herndon JW (2000) Org Lett 2:1267
92. Zhang Y, Herndon JW (2002) J Org Chem 67:4177
93. Deubzer B, Fritz HP, Kreiter CG, Öfele K (1967) J Organomet Chem 7:289
94. Kirss RU, Treichel PM Jr (1986) J Am Chem Soc 108:853
95. Kündig EP, Desobry V, Grivet C, Rudolph B, Spichiger S (1987) Organometallics 6:1173
96. Oprunenko YF, Malyugina SG, Ustynyuk YA, Ustynyuk NA, Kravtsov DN (1988) J Organomet Chem 338:357
97. Akhmedov NG, Malyugina SG, Mstislavsky VI, Oprunenko YF, Roznyatovsky VA, Ustynyuk YA, Batsanov AS, Ustynyuk NA (1998) Organometallics 17:4607

98. Oprunenko YF, Akhmedov NG, Laikov DN, Malyugina SG, Mstislavsky VI, Roznyatovsky VA, Ustynyuk YA, Ustynyuk NA (1999) J Organomet Chem 583:136
99. Marriott PJ, Lai Y-H (1986) Inorg Chem 25:3680
100. Dötz KH, Stinner C, Nieger M (1995) J Chem Soc Chem Commun 2535
101. Oprunenko Y, Malyugina S, Nesterenko P, Mityuk D, Malyshev O (2000) J Organomet Chem 597:42
102. Müller J, Göser P, Elian M (1969) Angew Chem 81:331; Angew Chem Int Ed Engl 8:374
103. Berno P, Ceccon A, Gambaro A, Venzo A, Ganis P, Valle G (1987) J Chem Soc Perkin Trans 2:935
104. Ochertyanova EA, Hansen H-J, Ustynyuk YA (2002) Helv Chim Acta 85:1166
105. Oprunenko Y, Gloriozov I, Lyssenko K, Malyugina S, Mityuk D, Mstislavsky V, Günther H, von Firks G, Ebener M (2002) J. Organomet Chem 656:27
106. Cunningham SD, Öfele K, Willeford BR (1983) J Am Chem Soc 105:3724
107. Traylor TG, Stewart KJ (1986) J Am Chem Soc 108:6977
108. Nambu M, Mohler DL, Hardcastle K, Baldridge KK, Siegel JS (1993) J Am Chem Soc 115:6138
109. Oprunenko Y, Malyugina S, Vasil'ko A, Lyssenko K, Elschenbroich C, Harms K (2002) J Organomet Chem 641:208
110. Own ZY, Wang SM, Chung JF, Miller DW, Fu PP (1993) Inorg Chem 32:152
111. Stendel J Jr, Dötz KH (publication in preparation)
112. Wenzel B, Dötz KH (publication in preparation)
113. Jahr HC, Nieger M, Dötz KH (publication in preparation)
114. Zabalov MV, Gloriozov IP, Oprunenko YF, Lemenovskii DA (2003) Russ Chem Bull Int Ed 52:1567
115. Schneider J, Dötz KH (publication in preparation)
116. Dötz KH, Szesni N, Nieger M, Nättinen K (2003) J Organomet Chem 671:58
117. Jahr HC, Nieger M, Dötz KH (publication in preparation)
118. Golovin MN, Rahman MM, Belmonte JE, Giering WP (1985) Organometallics 4:1981
119. Rahman MM, Liu HY, Prock A, Giering WP (1987) Organometallics 6:650
120. Tolman CA (1977) Chem Rev 77:313
121. Bartik T, Himmler T, Schulte H-G, Seevogel K (1984) J Organomet Chem 272:29
122. Jahr HC, Nieger M, Dötz KH (2003) Chem Commun 2866
123. Boese R, Cammack JK, Matzger AJ, Pflug K, Tolman WB, Vollhardt KPC, Weidman TW (1997) J Am Chem Soc 119:6757
124. Burger P (2001) Angew Chem 113:1971; Angew Chem Int Ed Engl 40:1917
125. Zhu G, Tanski JM, Churchill DG, Janak KE, Parkin G (2002) J Am Chem Soc 124:13658

Topics in Current Chemistry (2004) 248: 105–139
DOI 10.1007/978-3-540-44505-0

Supramolecular Templating in the Formation of Helicates

Markus Albrecht (✉)

Institut für Organische Chemie, RWTH-Aachen, Professor-Pirlet-Strasse 1, 52074 Aachen, Germany
markus.albrecht@oc.rwth-aachen.de

Abstract This chapter discusses the influence of templates on the self-assembly and on the structure of helicates. Hereby, templating in the chemistry of linear double- and triple-stranded helicates as well as of circular helicates will be described.

Keywords Self-assembly · Helicate · Host-guest-chemistry · Template · Coordination Compound

1 Introduction

The manufacturing or grafting of macroscopic structures very often proceeds by use of appropriate supports or of stencils. Those help to fix independently preformed parts of the structure in a relative orientation to each other and by finally connecting them provide the basis to obtain a desired shape and function. Chemists are manufacturers on a microscopic level, transforming and building up molecules. Hereby, molecular stencils, so-called templates, can be used to arrange reactants, direct bond forming processes, and to obtain a desired molecular structure. Templating not only can be used to specifically prepare covalently linked molecules but the formation of big, non-covalently connected supramolecular aggregates can also be influenced [1]. This chapter gives some insight into helicate chemistry and shows that in many cases the formation of helicates and related compounds is controlled by the action of highly specific templates [2, 3].

1.1 Templating in Supramolecular Chemistry and the Concept of "Dynamic Combinatorial Chemistry"

Templating very often is found not only in chemical but also in biological processes. The formation of the tobacco mosaic virus (TMV) is a prominent example where template-directed self-assembly occurs in nature [4]. The TMV is composed of one molecule of RNA containing 6400 bases and of 2130 protein subunits. The TMV spontaneously forms from its components if RNA is added to the protein. Hereby the protein units assemble around the RNA strand, which acts as a template and controls the overall shape of the resulting virus. Molecular recognition and non-covalent interactions between the protein units result in a stable rod like molecule with the RNA in the interior and the protein subunits forming a spindle around it [4].

In the early days of supramolecular chemistry templating already entered the game. In the 1960s Pedersen observed that the ring closure of the open chain molecule **1** to form 18-crown-6 **2** is favoured if potassium cations are present as templates. As depicted in Scheme 1, the precursor wraps around the templating potassium cation and brings the reactive termini in close spatial

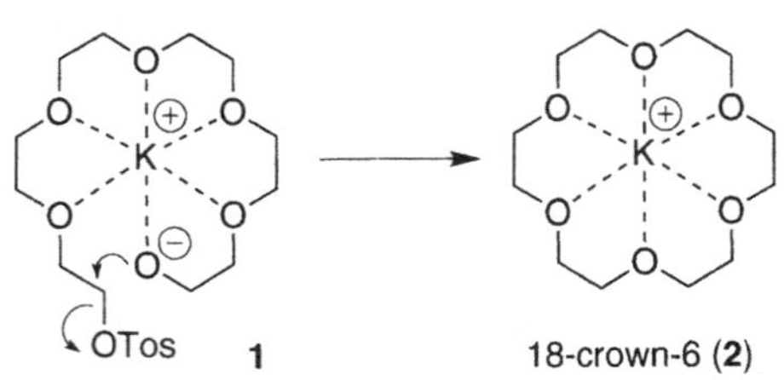

Scheme 1 Formation of 18-crown-6 **2** templated by a potassium cation

contact. Thus an intramolecular Williamson-type ether formation is favoured over an intermolecular oligomerization process [5].

Non-covalent interactions between complementary molecular building blocks can also be directed by a template so that a specific supramolecular structure is obtained either by covalent or non-covalent bond forming processes [6]. As an example, the formation of catenanes proceeds by slipping of a linear molecule through a macrocycle and the linear unit then has to be cyclized. The crucial step during this process is the slipping of the linear moiety through the ring which has to be induced by some "supramolecular" interactions. Different approaches have been used for the formation of catenanes [7]; e.g. charge transfer interactions between aromatic moieties were used in Stoddarts approach [8], while Vögtle and Leigh control the slipping by hydrogen bonding interactions [9].

In Scheme 2 the approach by Sauvage is shown. Metal coordination fixes the linear molecule inside of the ring and enables the formation of the "catenate" by macrocyclization. The metal acts as a template which fixes the two molecular components in a special spatial arrangement which is ideally predisposed for the final ring closure. In a last step, the templating metal ion can be removed to obtain the "free" catenane [10].

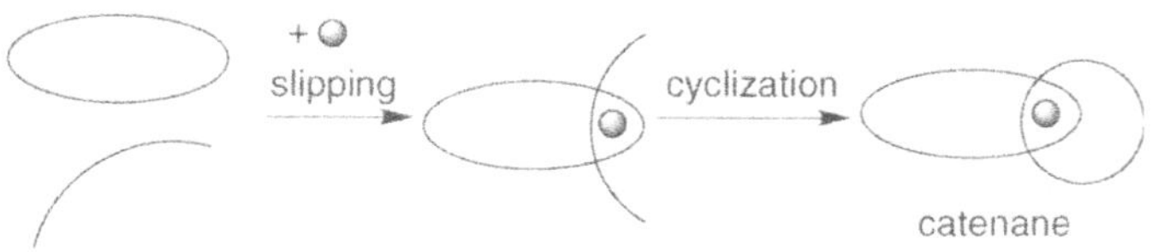

Scheme 2 Metal template-directed formation of catenanes as described by Sauvage et al. [10]

In contrast to the templating in covalently bonded systems, non-covalently linked supramolecular aggregates show a high degree of dynamic behaviour (which is important for proof reading and error correction during the self-assembly process) [1, 11]. Mixtures of different supermolecules which are formed from molecular building blocks can show fast interconversion and high fluxionality. Addition of an appropriate template might interfere with the equilibrium mixture and one of the components then might be favoured.

As one of many, an example from Fujitas group will be discussed here. Mixing of the tridentate ligand **3** with (en)$Pd(NO_3)_2$ (en=diaminoethane) in water leads to the formation of oligomeric coordination compounds $\{[(\mathbf{3})_2Pd_3]^{6+}\}_n$ (n=1,2,...) [12].

The same reaction is performed in the presence of 4-methoxyphenylacetate as a template, and the trinuclear complex $[(\mathbf{3})_2Pd_3]^{6+}$ is obtained in high yield (Scheme 3). It is even possible to transform slowly a preformed mixture of oligomers $\{[(\mathbf{3})_2Pd_3]^{6+}\}_n$ into the desired trinuclear Pd-complex by addition of the template [12]. A two-step reaction takes place, which follows the concept of "Dynamic Combinatorial Chemistry". In contrast to "traditional" combinatorial chemistry, each single step of the reaction sequence has to be reversible. This initially generates molecular diversity by formation of a mixture of oligomers ("dynamic combinatorial library"). Addition of a template leads in a subsequent step to a selection process which due to the reversibility of the processes accumulates the thermodynamically favoured supramolecular species.

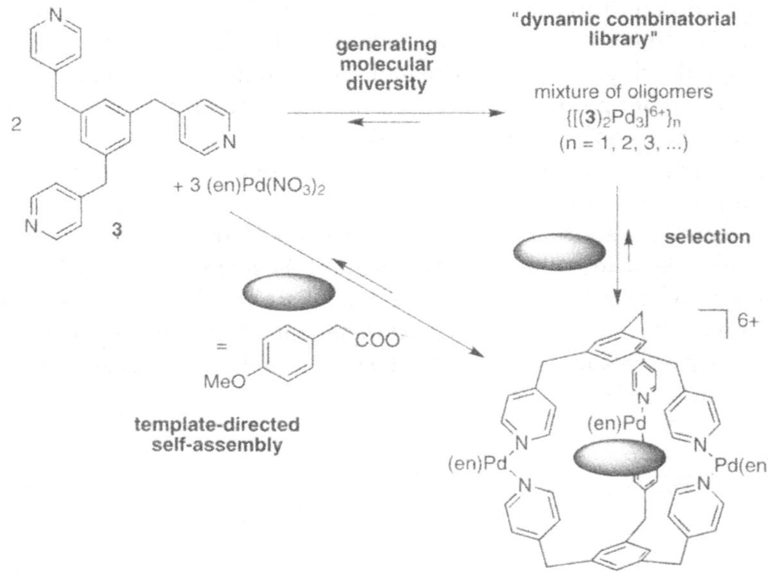

Scheme 3 Formation of $[(3)_2Pd_3]^{6+}$ in a template-directed self-assembly process or in two steps following the principle of "Dynamic Combinatorial Chemistry"

The principle of "Dynamic Combinatorial Chemistry" is used in covalent synthesis as well. Here receptors with labile covalent bonds (imines, ester, disulfides) have to be present [13, 14].

As will be discussed in here, templating and dynamic combinatorial processes are tightly connected and play an important role in the self-assembly of some helicate-type supramolecular coordination compounds. However, first we have to answer the question as to what "helicates" are.

1.2 Helicates: Definitions and Historical Background

In 1987 Lehn introduced the term "double-stranded helicate" for coordination compounds, in which two linear oligo-donor ligands wrap around two or more metal centres to form a double stranded helix. The formation of the helicates (e.g. 5) proceeds by self-assembly from the organic ligands in the presence of appropriate metal centres. Ether-linked bis- and tris(bipyridine) ligands **4** were used in combination with copper(I) or silver(I) ions (Fig. 1) [15].

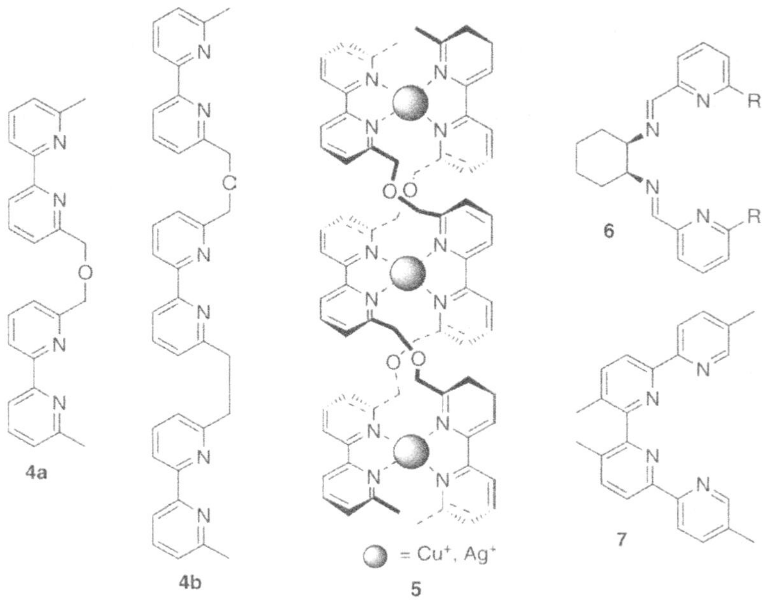

Fig. 1 Early ligands (**4**,**6**,**7**), which form double-stranded helicates (like **5**)

However, related helical structures were already observed in the 1970s with formylbilliverdine (**12**, see below) [16] or decamethylbiladiene as ligands in combination with zinc(II) ions [17]. In addition, the bis(iminopyridine) [18] and bis(bipyridine) [19] ligands **6** and **7** also form dinuclear double-stranded metal complexes $[(\mathbf{6})_2Ag_2]^{2+}$ and $[(\mathbf{7})_2Cu_2]^{2+}$.

A triple-stranded helicate was first introduced by Williams and Piguet [20]. In the dinuclear cobalt(II) complex **8**, three ligands **9** with two bidentate ligand moieties wrap around two metals which possess an octahedral coordination geometry (Fig. 2). Shortly after Lehn described a trinuclear triple-stranded helicate with tris(bipyridine) ligands (**58**, see below) [21].

Several years earlier, a triple-stranded dinuclear coordination compound had already been described by Raymond for the iron(III) complex $[(\mathbf{10})_3Fe_2]$ of Rhodoturulic acid **10**-H_2 [22] and the X-ray structure of the iron(III) complex $[(\mathbf{11})_3Fe_2]$ of **11**-H_2 revealed a similar helical motif [23].

In 1995 we described the first triple-stranded dinuclear meso-helicate [24] (compounds of this type were also termed "side-by-side complex" or "mesocate") [25]. The ligands are not wrapping around the metal centres, but an achiral complex is formed in which both metal complex units possess opposite configuration. The relative configuration of the metal complex units (helicate vs meso-helicate) can be controlled by introduction of appropriate spacers in the ligand. Due to the preferred zigzag conformation of alkyl chains, a spacer with an odd number of methylene units will lead to the meso-helicate while a spacer with an even number of methylene units prefers the formation of a

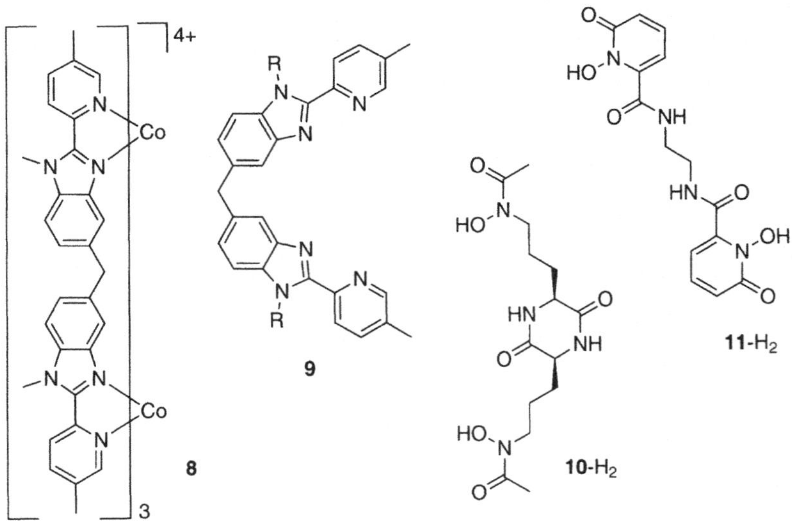

Fig. 2 Early ligands (**9**,**10**,**11**) for triple-stranded helicates (like **8**)

helicate [26]. This still works if the spacer is not only composed of alkyl chains but of a combination of rigid moieties and short "stereo-controlling" alkyl units [27].

Besides the linear helicates, a series of circular helicates were described, in which metal centres are arranged in a circular fashion with the ligands wrapping around them [28].

In many cases, guest species are bound in the cavities which are formed in the interior of linear or circular helicates; e.g., in the solid state, the iron complex $[(\mathbf{11})_3Fe_2]$ bears a water molecule bound in its interior [23]. The guest species are often responsible for the specific formation of the helical complexes and are sometimes necessary templating units. In cases where no direct templating takes place, other "secondary" stabilizing effects are often present (e.g., rigidity or predisposition of ligands). In fact, in the 1987 paper on helicates by Lehn, he already stated that the helicate formation is supported by favourable π-π interactions between the bipyridine units of ligands **4** [15].

2 Templating in the Formation of Linear Helicates

2.1 Double-Stranded Helicates

Examples for templating in the formation of double-stranded helicates are rare. However, the formylbiliverdine **12**, which was described in the 1970s, forms a mononuclear zinc(II) complex $[H_2O(\mathbf{12})Zn]^{2+}$, which under acidic conditions

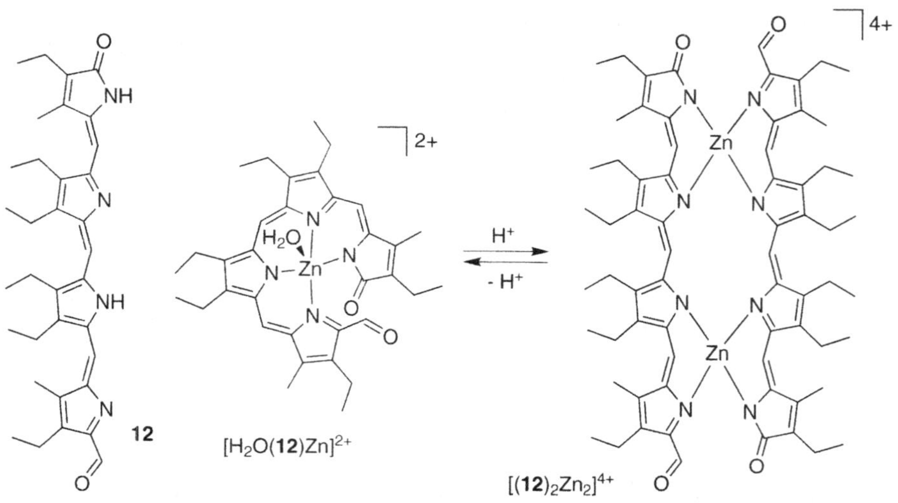

Scheme 4 Transformation of a water-templated mononuclear zinc(II) complex into a double-stranded helicate by acidification

is transformed into a double-stranded helicate $[(\mathbf{12})_2Zn_2]^{4+}$ (Scheme 4). In this example water acts as template, which stabilizes the mononuclear complex rather than the helicate. Labilization of the water as ligand results in a structural rearrangement and in the formation of the double-stranded dinuclear helicate [16].

In the dinuclear double-stranded europium(III) complex $[(H_2O)_4(\mathbf{13})_2Eu_2][F_3CSO_3]_6$ (Fig. 3) four water molecules are bound in the interior of a helicate-type coordination compound, filling up the cavity and probably stabilizing the complex [29].

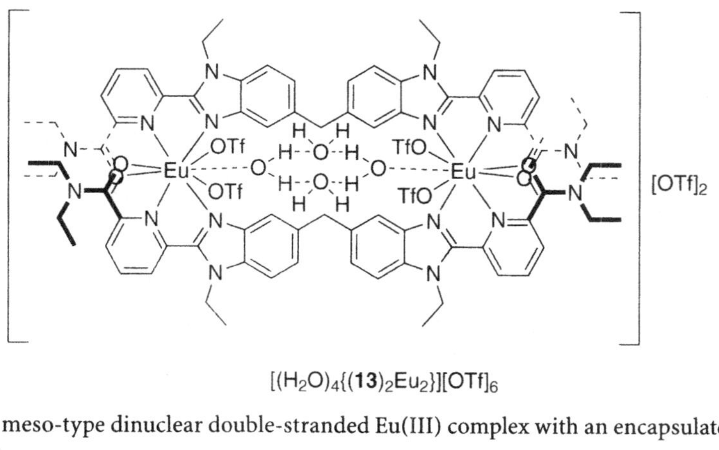

Fig. 3 A meso-type dinuclear double-stranded Eu(III) complex with an encapsulated water tetramer

The principle of "Dynamic Combinatorial Chemistry" could also be applied to obtain double-stranded helicates by templating (Scheme 5). It was even possible to obtain a catenane from a dynamic combinatorial library which incorporates a double-stranded helicate as a cyclic component.

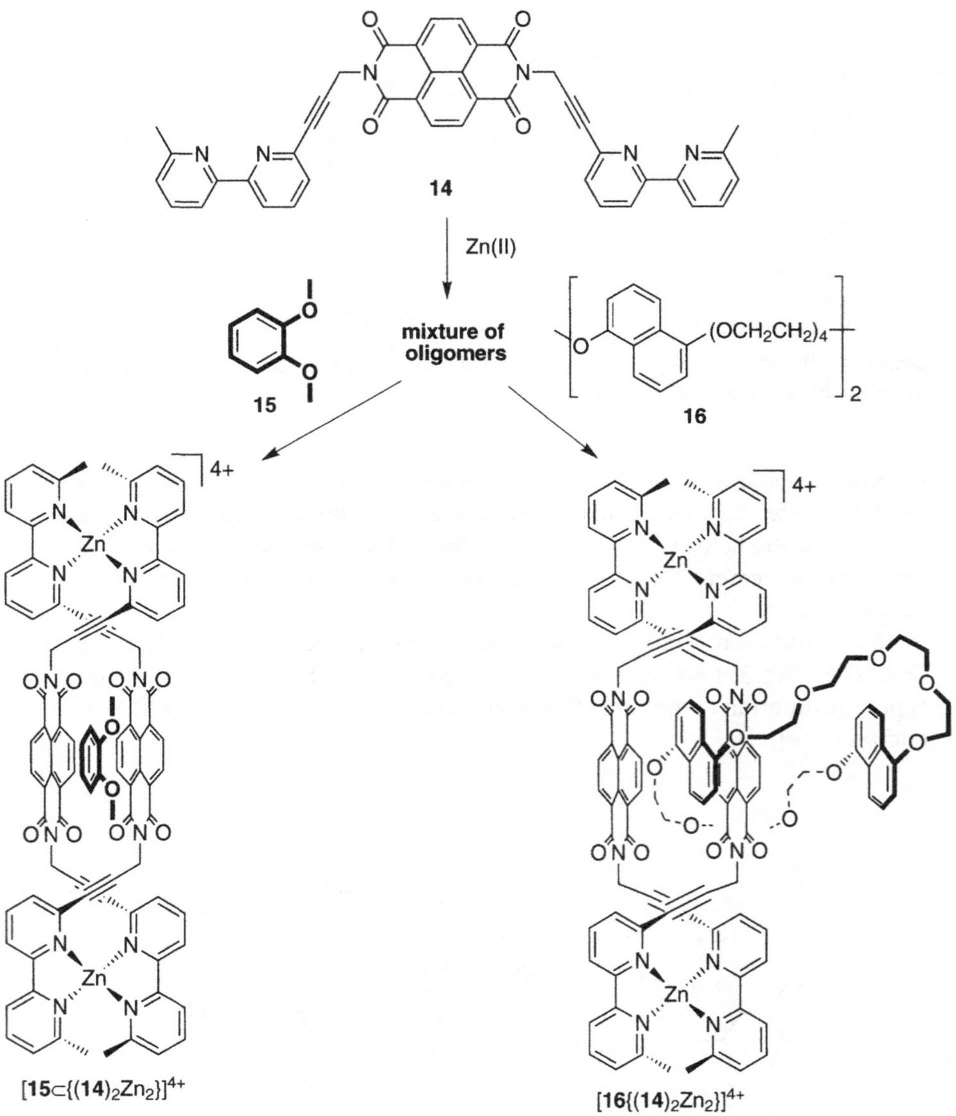

Scheme 5 Template directed formation of a double-stranded helicate following the principle of "Dynamic Combinatorial Chemistry"

The bis(bipyridine) ligand **14** reacts in the presence of zinc(II) ions to form bis(bipyridine) zinc complexes. However, no specific complex but a mixture of oligomers (dynamic combinatorial library) is obtained if no template is present. Addition of electron rich aromatic systems to the mixture leads to a selection and the dinuclear complex $[(\mathbf{14})_2Zn_2]^{4+}$ is formed. If veratrole **15** is used as the template, the formation of the dinuclear zinc(II) complex $[\mathbf{15}\subset\{(\mathbf{14})_2Zn_2\}]^{4+}$ is supported by attractive π-π-interactions between the electron rich **15** and the electron poor aromatic units of the spacer of ligand **14** [30, 31]. Addition of the macrocycle **16** by templating leads to the catenane $[\mathbf{16}\{(\mathbf{14})_2Zn_2\}]^{4+}$. In a dynamic process, the dinuclear helicate is formed and cleaved and upon addition of **16**, slipping of the ligand **14** through the ring **16** has to occur. The thermodynamically most stable complex $[\mathbf{16}\{(\mathbf{14})_2Zn_2\}]^{4+}$, the catenane, is obtained; again due to favourable aromatic π-π interactions [32].

Mixing of the aryl-bridged bisbenzimidazole ligands **17** and **18** with silver(I) triflate in a 1:1 ratio results in the formation of double-stranded complexes $[(F_3CSO_3)_2\{(\mathbf{17})_2Ag_2\}]$ and $[\mu\text{-}(F_3CSO_3)\subset\{(\mathbf{18})_2Ag_2\}]^+$ (Fig. 4). The ligands are orientated in a linear fashion, so that a helicate-type complex is formed which, however, does not show helical twisting [33].

Each of the silver(I) centres of $[(F_3CSO_3)_2\{(\mathbf{17})_2Ag_2\}]$ bridges between two benzimidazole units and additionally binds a triflate anion. In contrast to this the Ag-Ag distance in $[\mu\text{-}(F_3CSO_3)\subset\{(\mathbf{18})_2Ag_2\}]^+$ is larger and there is enough space for only one triflate anion to fit into the cavity and to bridge the two metals. A second triflate acts as a non-bonding anionic counter cation. In both compounds the binding of the anions to the metals seems to be essential for a stabilization of the dinuclear complexes $[(F_3CSO_3)_2\{(\mathbf{17})_2Ag_2\}]$ and $[\mu\text{-}(F_3CSO_3)\subset\{(\mathbf{18})_2Ag_2\}]^+$ [33].

A related situation, as observed for $[\mu\text{-}(F_3CSO_3)\subset\{(\mathbf{18})_2Ag_2\}]^+$, can be found for the biscatechol ligands **19**-H_4 [34] and **20**-H_4 [35, 36] and the catechol/benzene-1,2-thiol ligand **21**-H_4 [37]. Upon complexation of **19** with iron(III) ions

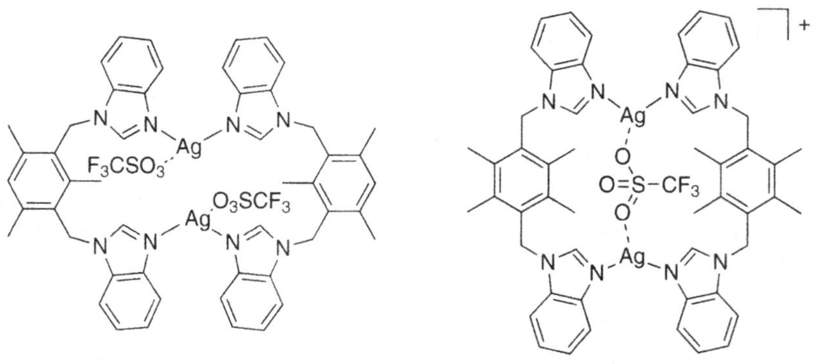

$[(F_3CSO_3)_2\{(\mathbf{17})_2Ag_2\}]$ $[\mu\text{-}(F_3CSO_3)\{(\mathbf{18})_2Ag_2\}]$

Fig. 4 The double-stranded complexes $[(F_3CSO_3)_2\{(\mathbf{17})_2Ag_2\}]$ and $[\mu\text{-}(F_3CSO_3)\subset\{(\mathbf{18})_2Ag_2\}]^+$

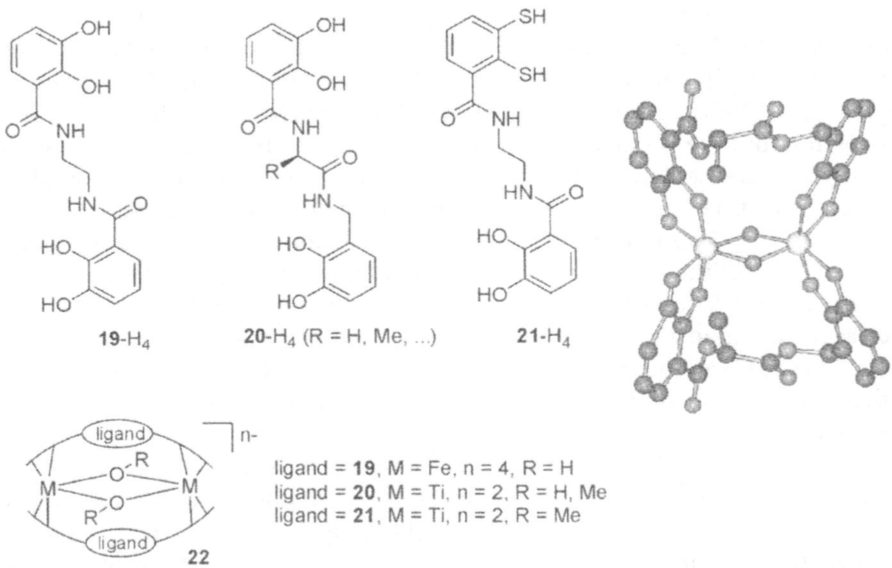

Fig. 5 Alkoxide/hydroxide-bridged complexes **22** from dicatechol-type ligands **19–21**-H_4 and the X-ray structure of the alanine-bridged complex $[(\mathbf{20})_2(OH)_2Ti_2]^{2-}$

or **20/21** with titanium(IV), double-stranded complexes **22** are formed (Fig. 5), which in the thermodynamically favoured isomer possess a meso-relation between the complex units. Additional bridging coligands (hydroxide, methoxide) act as template and stabilize the double-stranded structure **22**.

With the amino acid-derived ligands **20**-H_4 it was tested how tolerant the system is towards alkoxide coligands with different sterical demands. In the presence of ethanol or allyl alcohol the formation of compounds **22** still proceeds, while isopropanol or *tert*-butanol destabilize the double-stranded complex **22** and lead to mixtures of complexes which contain triple-stranded dinuclear coordination compounds as one component. Similar results are obtained if the coordination studies of ligands **20** are performed in non-protic solvents like acetonitrile or dmf [38].

In the formation of **22** the alcoholates act as essential templates. In the presence of an appropriate coligand, the compound **22** is always obtained independent of the ligand to metal ratio [36].

For ligand **19**-H_4 and iron(III) ions the OR^- groups have to be present to stabilize a compound like **22**. However, complexation with a 3:2 ligand to metal stoichiometry leads to the exclusive formation of the triple-stranded helicate [34].

The ligand **23**-H_2 with two 1,3-dicarbonyl binding sites leads with copper(II) acetate in the presence of a large excess of alkali metal acetate ($M{=}K^+, Rb^+, Cs^+$) to the double-stranded helicate-type complex with the templating cation bound internally $[M{\subset}\{(\mathbf{23})_2Cu_2\}]^+$ [39] (Scheme 6).

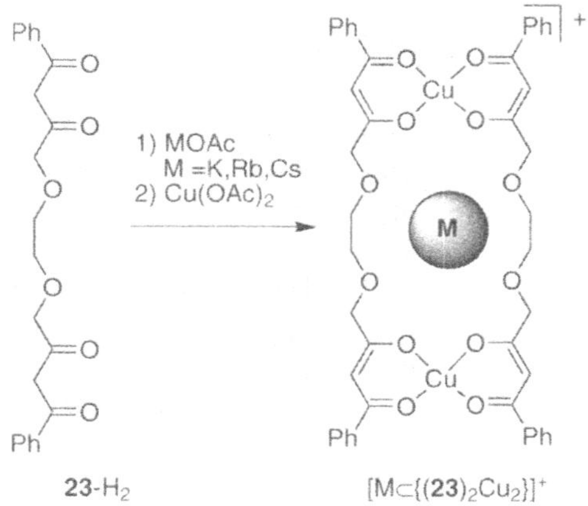

Scheme 6 Template-directed formation of $[M{\subset}\{(23)_2Cu_2\}]^+$

2.2 Triple-Stranded Helicates

Triple-stranded helicates are formed from three linear ligand strands and two or more metal ions. To enable an effective templating, appropriate binding sites for the template have to be present inside of the helicate.

However, interaction of templates with the helicate also can occur from the "outside".

As an example, the transfer of chiral information from stereochemically well defined chiral counterions (Fig. 6) onto the bipyridine-based complex **24** or the catecholamide-based helicate **25** have to be mentioned. The chiral counterions interact with the helicate from the "outside" and induce one preferred twist of the helix [40–42].

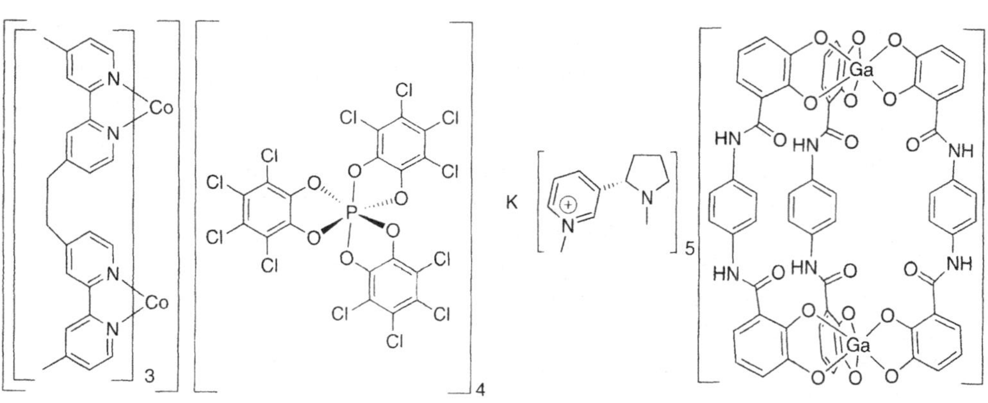

Fig. 6 Chiral induction with chiral counterions as templates

2.2.1
Triple-Stranded Helicates as Host-Species

Guests can be bound in the interior of triple-stranded dinuclear helicates. Thereby, binding of small templates in the interior of triple-stranded helicates can control the relative configuration at the two metal centres, leading either to a helicate or a meso-helicate. It was already described, that ligand **11**-H_2 forms a triple-stranded helicate [(**11**)$_3$$Fe_2$] which encapsulates one water molecule in the solid state. The spacer of the related ligand **26**-H_2 is somewhat longer than the one of **11**-H_2 and in its "relaxed" form is predisposed for the formation of a meso-helicate. Indeed triple-stranded meso-helicates [(**26**)$_3$$Al_2$] and [(**26**)$_3$$Ga_2$] are formed if no water is present [43] (Scheme 7).

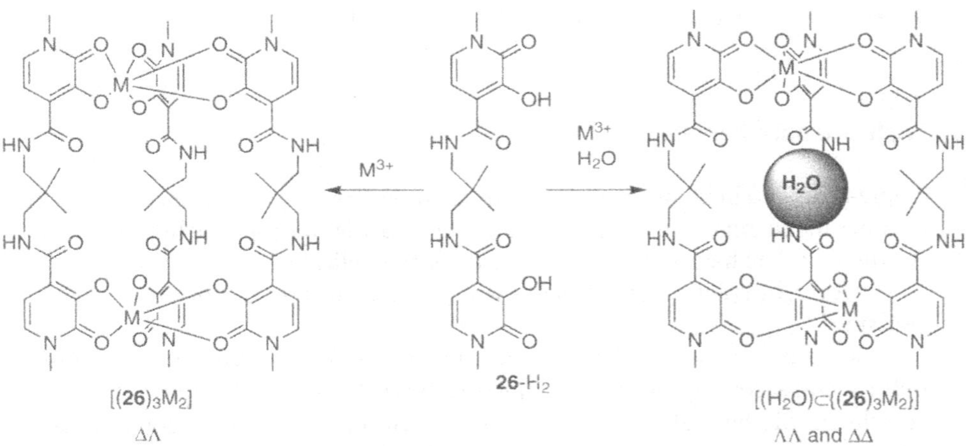

Scheme 7 Formation of the meso-helicate [(**26**)$_3$$M_2$] vs the template-controlled formation of the helicate [(H_2O)⊂{(**26**)$_3$$M_2$}] (M=Al, Ga)

If the coordination study with ligand **26**-H_2 is performed in the presence of water as template, the more condensed helicate [(H_2O)⊂{(**26**)$_3$$M_2$}] is obtained. By twisting of one complex unit, the relatively big cavity adjusts its size to the small template water [43].

A very similar template-control of the relative stereochemistry in dinuclear helicates is observed in the case of the ligands **27**-H_2 and **28**-H_2 (Fig. 7). The rigid linear ligand **27**-H_2 with a 1,3-phenylene spacer forms the helicate [(**27**)$_3$$Fe_2$]. The related ligand **28**-H_2 with a pyridine spacer, on the other hand, is able to encapsulate a potassium cation. This potassium stabilizes the meso-helicate structure for [K⊂{(**28**)$_3$$Fe_2$}]$^+$ due to a better size complementarity between host and template ("lock-and-key principle"). If the complex is formed in the absence of potassium cations, the triply N-protonated species [(**28**-H)$_3$$Fe_2$]$^{3+}$ is obtained [44]. (A derivative related to **28** which bears thiophene instead of the *tert*-butyl groups leads to similar results in coordination studies) [45].

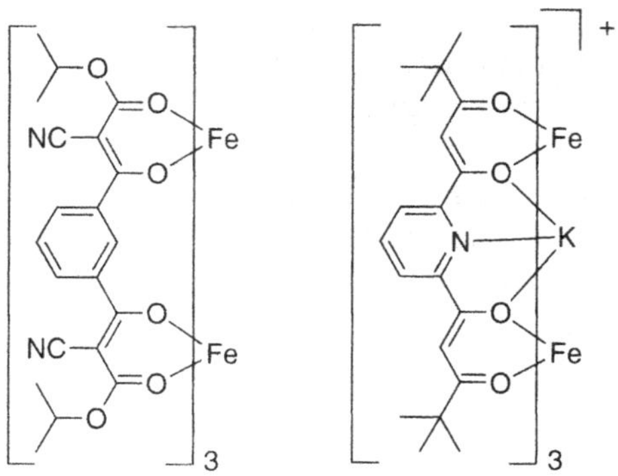

helicate
[$(\mathbf{27})_3Fe_2$]

meso-helicate
[$K\subset\{(\mathbf{28})_3Fe_2\}]^+$

Fig. 7 Formation of a helicate [$(27)_3Fe_2$] and template-controlled formation of the meso-helicate [$K\subset\{(\mathbf{28})_3Fe_2\}]^+$

Many triple-stranded helicates provide internal binding sites for guest species. However, the guests are not always essential for the assembly of the coordination compound, but can be introduced after complex formation. [$K\subset\{(\mathbf{28})_3Fe_2\}]^+$ represents an example for such a system [44]. Others were described for the related β-dicarbonyl derived ligands **29**-H_2, **30**-H_2 and **31**-H_2 [39, 46, 47] (Fig. 8).

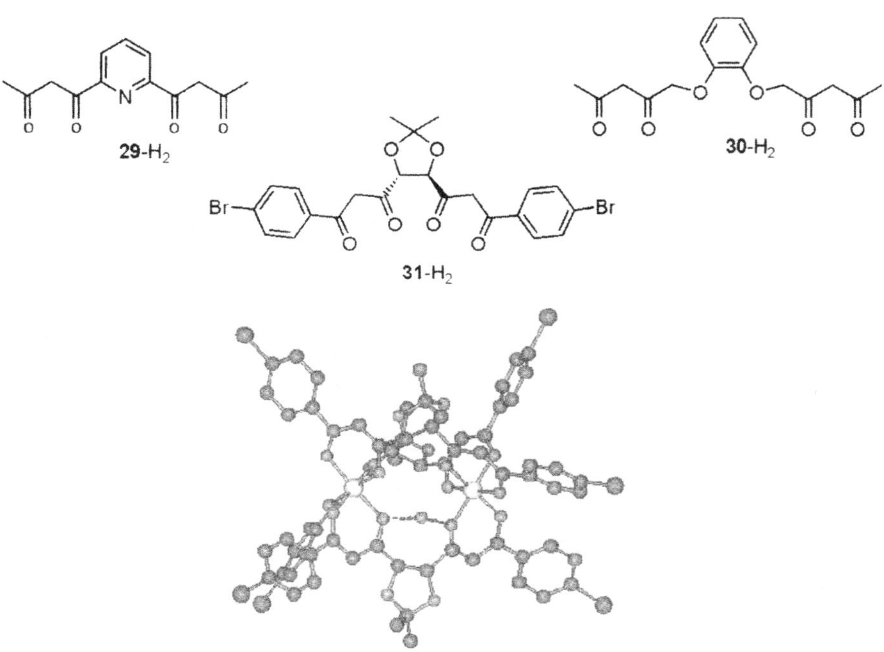

Fig. 8 Ligands **29–31**-H_2 and the X-ray structure of [$(H_2O)\subset\{(\mathbf{31})_3Ga_2\}$]

Ligand **29**-H_2 forms a triple-stranded dinuclear complex [$(\mathbf{29})_3Fe_2$] which is able to encapsulate lanthanum(III) ions in its interior to obtain [$\{(THF)(H_2O)La\}\subset\{(\mathbf{29})_3Fe_2\}]^{3+}$. The related complex [$\{(H_2O)Sr\}\subset\{(\mathbf{28})_3Fe_2\}]^{2+}$ could also be characterized [46].

Coordination studies of ligand **30**-H_2 with nickel(II) ions leads in the presence of cesium acetate to the triple-stranded helicate with the templating cation bound in the interior of the self-assembled dinickel-cryptand [$Cs\subset\{(\mathbf{30})_3Ni_2\}]^-$ [39].

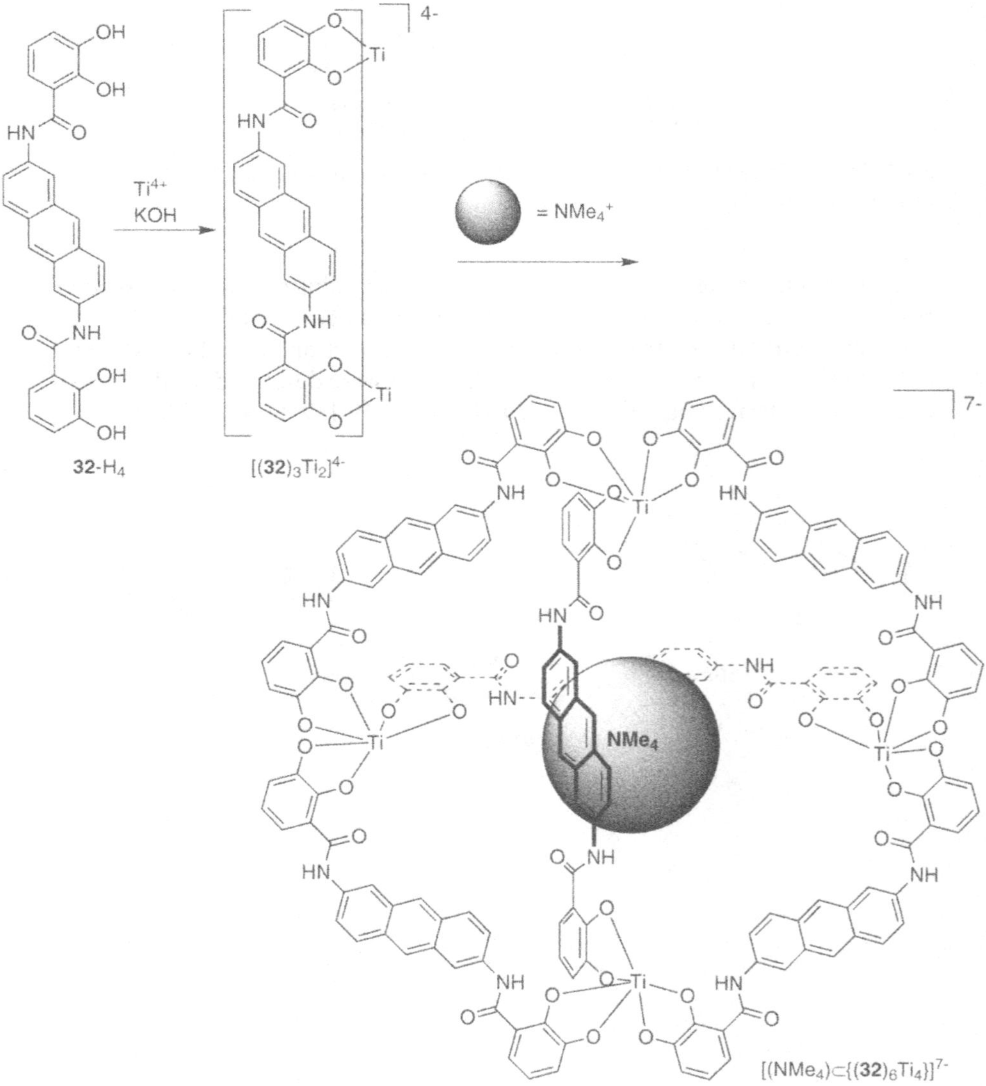

Scheme 8 Formation of the helicate [$(\mathbf{32})_3Ti_2]^{4-}$ and its tetramethylammonium-templated destabilization to form a molecular tetrahedron [$(NMe_4)\subset\{(\mathbf{32})_6Ti_4\}]^{7-}$

The chiral tartaric acid-derived ligand **31**-H_2 forms enantiomerically pure helicates with gallium(III) and iron(III) ions. For the gallium complex [(H_2O) $\subset${($\mathbf{31}$)$_3Ga_2$}] it was found that one water is encapsulated in the internal cavity in the solid state. In solution this compound is able to act as a cryptand which shows a high selectivity for the binding of lithium cations [47].

An example, were addition of a guest species does not stabilize the triple-stranded helicate, but templates the formation of a molecular tetrahedron, is represented in Scheme 8. Coordination chemistry of ligand **32**-H_4 with titanium(IV) ions results in the formation of the triple-stranded helicate $[(\mathbf{32})_3Ti_2]^{4-}$, which upon addition of tetramethylammonium cations is destabilized and the supramolecular tetrahedron $[(NMe_4)\subset\{(\mathbf{32})_6Ti_4\}]^{7-}$ is formed. The ammonium cation acts as template and is encapsulated in the interior of the tetrahedron [48].

2.2.2 Helicates from Alkyl- and Imino-Bridged Dicatechol-Ligands

Alkyl- (**A**) and imino-bridged (**B**) triple-stranded helicates with dicatechol ligands provide an internal cavity to bind guest species. The imino groups possess the ability to favour additionally the binding of the template (**B2**) [49]. This is not the case, if catecholamides are used, because here the internal oxygen donor-sites are blocked by hydrogen bonding (**C**) [50] (Fig. 9).

Alkyl-bridged dicatechol ligands **33**-H_4 form either the triple-stranded helicate (even spacer length **33b, 33d, 31f**) or meso-helicate (odd spacer length **33a, 33c, 33e**) $M_4[(\mathbf{33})_3Ti_2]$ with titanium(IV) cations. This self-assembly process highly depends on the templating by the counter cations M which are present [26].

With the small ligand **33a**-H_4 the templating could be impressively demonstrated. If a coordination study of **33a**-H_4 (Scheme 9) with titanium(IV) ions is performed in the presence of potassium carbonate, a red solid with the formula "$K_4[(\mathbf{33a})_3Ti_2]$" is obtained. However, NMR spectroscopy as well as mass spec-

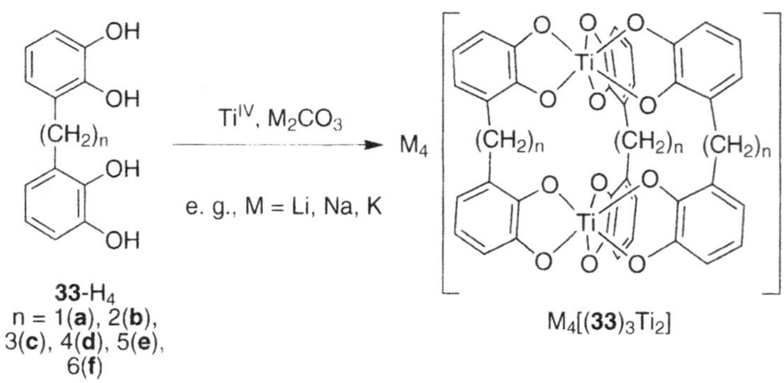

Scheme 9 Formation of dinuclear triple-stranded helicates from alkyl-bridged dicatechol ligands **33**-H_4

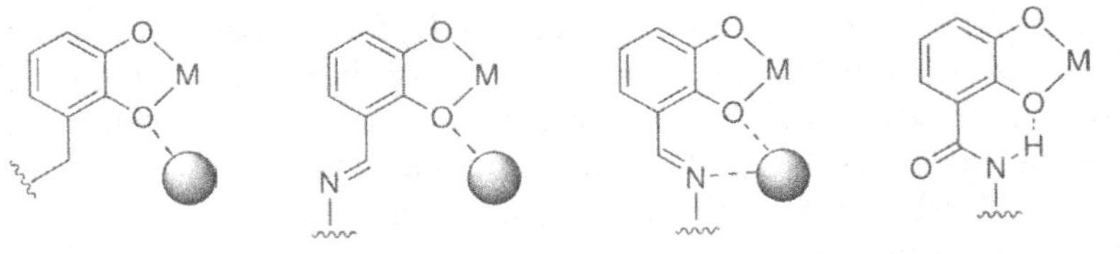

Fig. 9 Binding of acceptors in the interior of catecholate-derived helicates with alkyl- (**A**) or imine-spacers (**B**). In case of catecholamides (**C**) the internal binding-site is blocked

trometry show that here a mixture of oligomeric complexes $\{K_4[(\mathbf{33a})_3Ti_2]\}_n$ (n=1,2,...) is present [51].

If the same experiment is performed with lithium carbonate or sodium carbonate as base, only the dinuclear titanium meso-helicates $Li/Na_4[(\mathbf{33a})_3Ti_2]$ are observed. Strong templating by alkali metal cations takes place. Small lithium or sodium cations are able to stabilize the relatively small dinuclear titanium complex while the big potassium destabilizes this species. Consequently, if sodium or lithium perchlorate is added to a solution of the oligomer $\{K_4[(\mathbf{33a})_3Ti_2]\}_n$, the mixture is transformed into the dinuclear complex with the small templates bound to the self-assembled titanium complexes [52].

However, this templating by the small cations is not as simple as it on the first sight seems to be. The X-ray structure of $Li_4[(\mathbf{33a})_3Ti_2]$ (Fig. 10) reveals that no lithium cation is bound in the interior of the helicate. However, three lithium cations are binding from the "outside" and are coordinatively saturated by one dmf per cation [51].

In contrast to the results in the solid state, Li NMR spectroscopic investigations in solution reveal that only two lithium cations are bound to the meso-helicate unit [52].

In case of the corresponding larger helicate-type complexes $M_4[(\mathbf{33b{-}f})_3Ti_2]$, the binding of alkali metal cations could be shown in solution as well as in the solid state. Representative solid state structures are shown in Fig. 11.

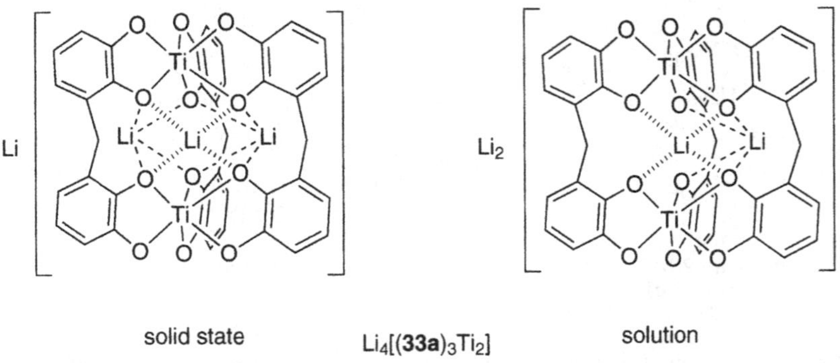

Fig. 10 Structure of $Li_4[(\mathbf{33a})_3Ti_2]$ as found in the solid state and in solution

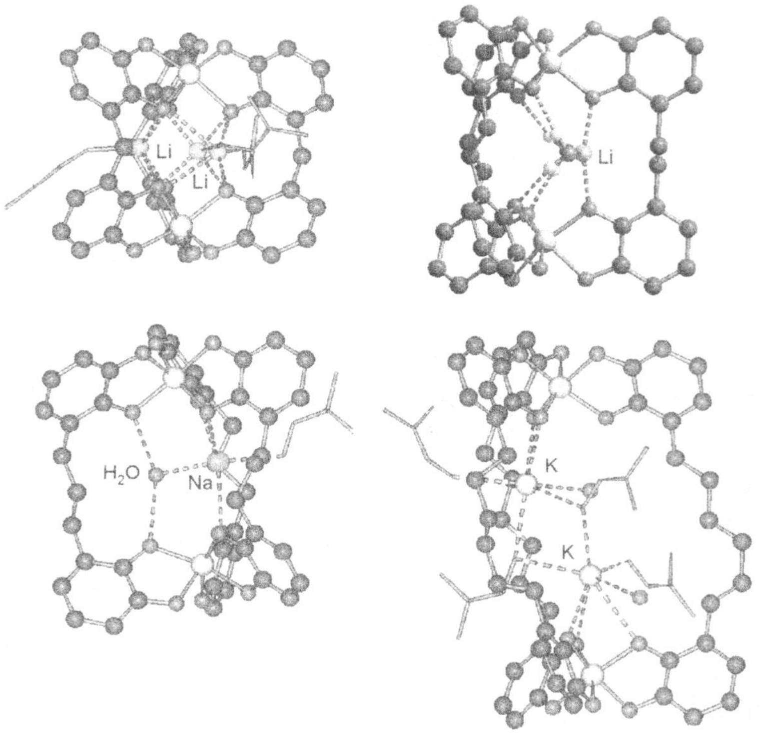

Fig. 11 Solid state structures of $[\{(dmf)Li\}_3\{(\mathbf{33a})_3Ti_2\}]^-$, $[\{(H_2O)_2Li\}\subset\{(\mathbf{33b})_3Ti_2\}]^{3-}$, $[\{(H_2O)(dmf)Na\}\subset\{(\mathbf{33c})_3Ti_2\}]^{3-}$, and $[\{(H_2O)(dmf)_2K\}_2\subset\{(\mathbf{33f})_3Ti_2\}]^{2-}$ (dmf is only indicated)

In case of $[(\mathbf{33b})_3Ti_2]^{4-}$ and $[(\mathbf{33c})_3Ti_2]^{4-}$ water molecules can be found in the crystal which are encapsulated in the dinuclear titanium complexes in addition to the alkali metal cations [24, 53]. The internal cavity of $[(\mathbf{33f})_3Ti_2]^{4-}$ is so big, that two potassium cations are bound in the interior [54].

The templating effect of the alkali metal cations can be used in ligand self-recognition during the self-assembly of the dinuclear complexes. The outcome of a coordination study of a 1:1 mixture of ligand **33a**-H_4 and **33b**-H_4 with titanium(IV) ions highly depends on the template which is present [55].

With lithium carbonate as base, as expected the two homoleptic, $[(\mathbf{33a})_3Ti_2]^{4-}$ and $[(\mathbf{33b})_3Ti_2]^{4-}$, but only one heteroleptic complexes $[(\mathbf{33a})_2(\mathbf{33b})Ti_2]^{4-}$ are obtained. The other possible heteroleptic complex $[(\mathbf{33a})(\mathbf{33b})_2Ti_2]^{4-}$ cannot be observed. With sodium carbonate, on the other hand, only homoleptic coordination compounds $[(\mathbf{33a})_3Ti_2]^{4-}$ and $[(\mathbf{33b})_3Ti_2]^{4-}$ are formed. In the presence of sodium a strong ligand self-recognition occurs [55].

As already stated, potassium cations are not able to stabilize the compound $[(\mathbf{33a})_3Ti_2]^{4-}$. Therefore, with K^+ the homoleptic dinuclear coordination compound $[(\mathbf{33b})_3Ti_2]^{4-}$ is obtained together with oligomeric titanium complexes of ligand **33a** [55] (Scheme 10).

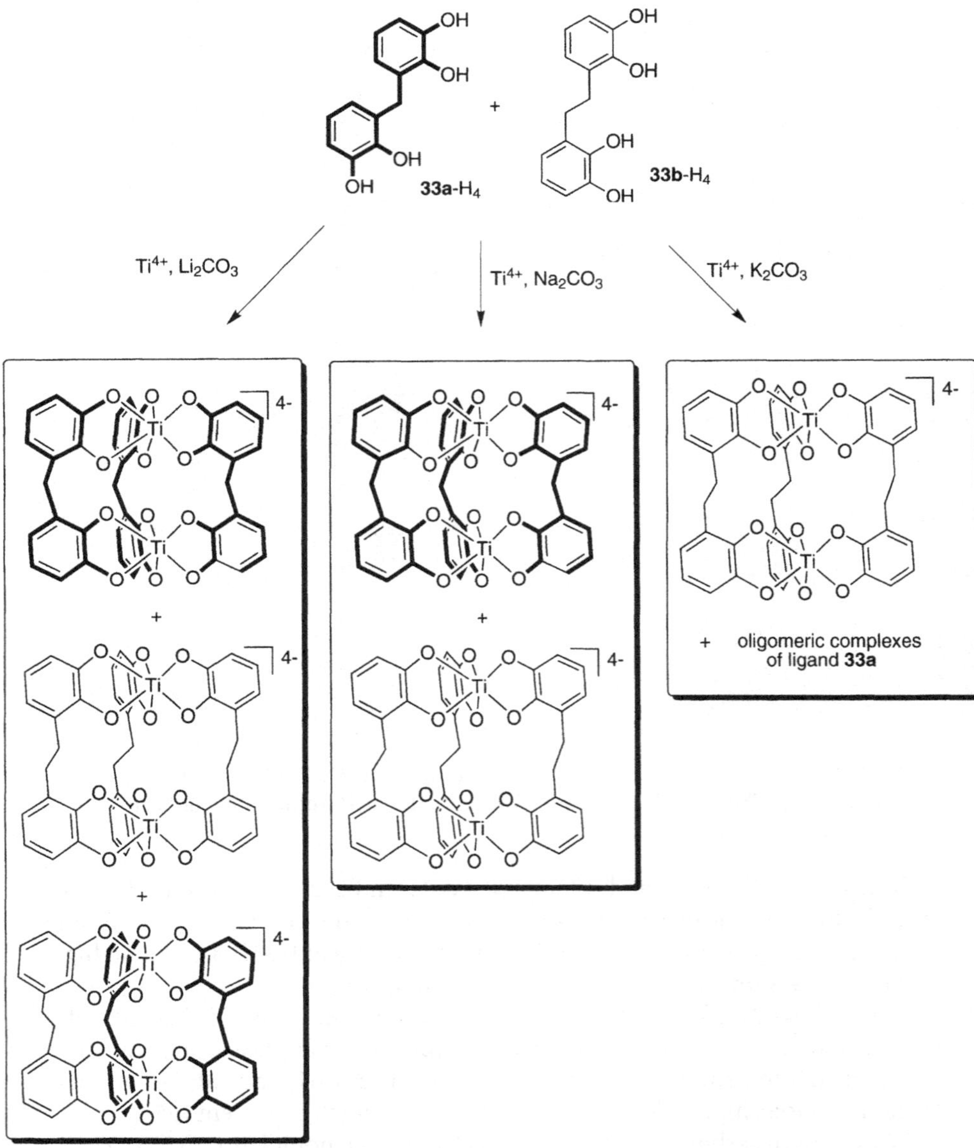

Scheme 10 Template-directed self-recognition experiments with a 1:1 mixture of **33a**-H_4 and **33b**-H_4

In contrast to **33a**-H_4, the rigid ligand **34**-H_4 possesses a direct bond between the catechol units and forms triple-stranded dinuclear titanium complexes $[(\mathbf{34})_3Ti_2]^{4-}$ even in the presence of the large potassium cations. Probably, due to the rigidity and the predisposition of the ligand **34**-H_4, it does not have the option to form oligomeric compounds. The dinuclear complexes are favoured without templating by the cations. However, it could be shown, that lithium, sodium as well as potassium coordinate to the dinuclear complexes from the "outside" [56].

The elongated rigid ligands **35a**-H_4 and **35b**-H_4 (Fig. 12) lead to dinuclear triple-stranded helicates $[(\mathbf{35a/b})_3Ti_2]^{4-}$ with titanium(IV) ions. In case of the smaller ligand **35a**-H_4 this works in the presence of lithium, sodium and potassium, while, surprisingly, the larger $[(\mathbf{35b})_3Ti_2]^{4-}$ is formed only in the presence of lithium or of sodium, but not of potassium cations [56].

Due to the size of the ligands **35a/b**-H_4, there is enough space in the interior of the helicates $[(\mathbf{35a/b})_3Ti_2]^{4-}$ for more than one counter cation. The solid state structure of $K_4[(\mathbf{35a})_3Ti_2]$ (Fig. 13) reveals, that two of the potassium cations are bound in the interior of $[(\mathbf{35a})_3Ti_2]^{4-}$. Dimerization with bridging solvent molecules is found in the crystal [57].

Formation of imines is a simple entry to obtain dicatechol ligands which possess different lengths and flexibilities at the spacers. Figure 14 shows a

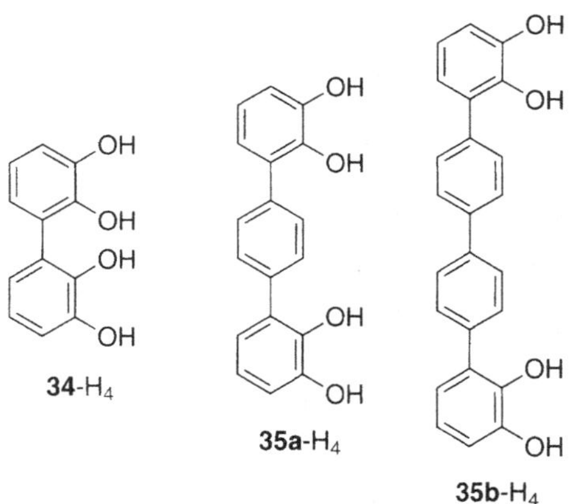

Fig. 12 The rigid ligands **34**-H_4 and **35a,b**-H_4

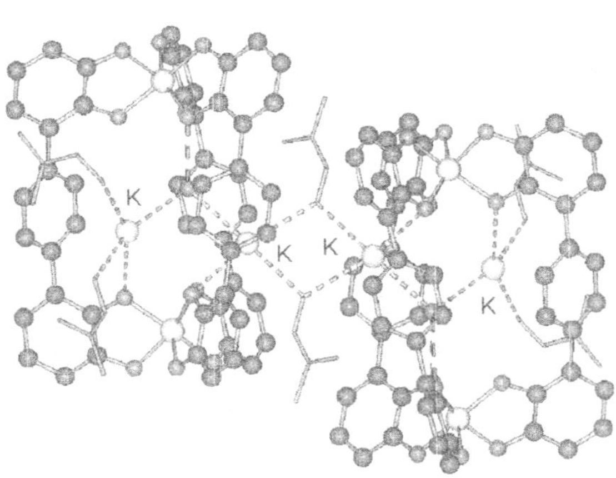

Fig. 13 The dimer $[\{(H_2O)(dmf)_3K_2\}\{(\mathbf{35a})_3Ti_2\}]_2^{4-}$ in the solid state

representative selection of ligands which were thus prepared. Hereby ligands **36–38**-H_4 represent rigid compounds with different length of the spacer. In **39**-H_4 and **40**-H_4 some flexibility is introduced by short alkyl units in addition to aromatic moieties. **41**-H_4 and **42**-H_4 are related ligands with heteroatoms introducing flexibility and in **43–45**-H_4 alkyl spacers are present, either butyl, *trans*-1,2- or *trans*-1,4-cyclohexyl units.

All of the ligands **36–45**-H_4 form triple-stranded helicate-type complexes with titanium(IV) ions [27, 49, 58] (Scheme 11). In all cases a binding of alkali metal cations is observed in the interior of the compounds and with the alkyl bridged ligands **43–45**-H_4 a templating effect by the cations is shown.

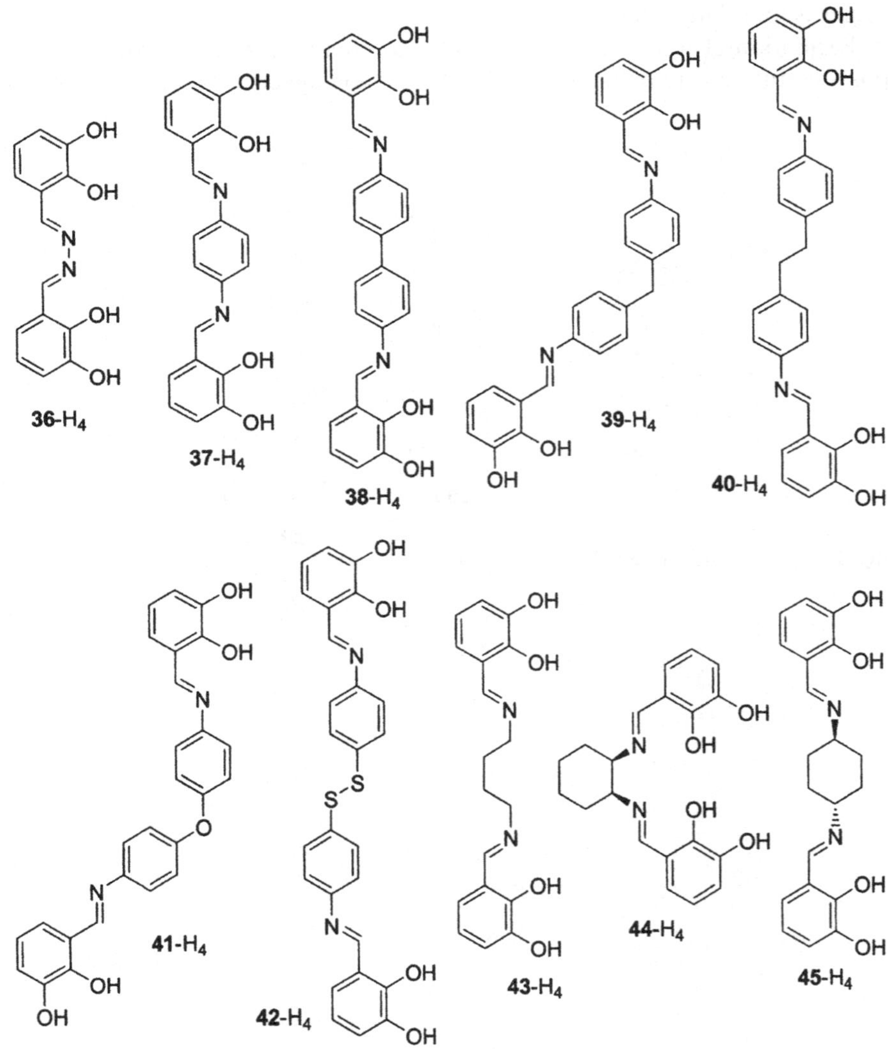

Fig. 14 Imino-bridged dicatechol ligands 36–45-H_4

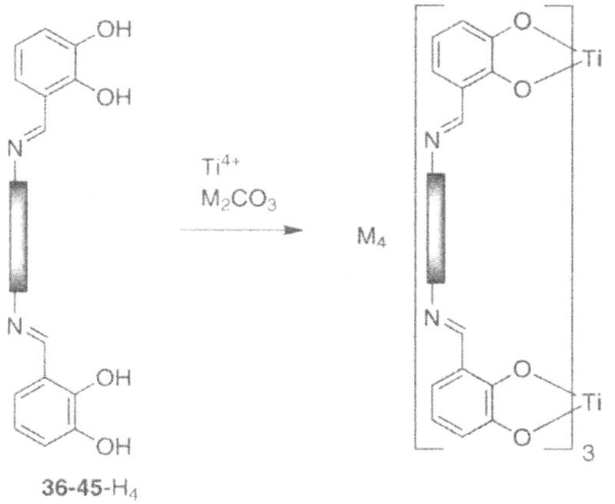

Scheme 11 Formation of dinuclear helicate-type complexes from imino-bridged dicatechol ligands **36–45**-H_4

Solid state structures of the titanium(IV) complexes of the rigid ligands **36–38**-H_4 and of a corresponding vanadium(IV) complex $[(\mathbf{36})_3V_2]^{4+}$ are shown in Fig. 15.

All the structural analyses show the presence of triple-stranded dinuclear helicates with encapsulated counter cations. In all four shown examples two cations are bound per cavity of a helicate. Only in the case of the short azine bridged ligand **36** do the imine nitrogen atoms partially bind to the cations [58]. Usually (e.g., $K_4[(\mathbf{37})_3Ti_2]$ or $Na_4[(\mathbf{38})_3Ti_2]$) the imines are orientated to the outside of the cavity and binding of the cation occurs as is indicated in Fig. 9 **B1** [27].

In the larger compounds $K_4[(\mathbf{37})_3Ti_2]$ or $Na_4[(\mathbf{38})_3Ti_2]$ as well as in Na_4-$[(\mathbf{39})_3Ti_2]$ or $Na_4[(\mathbf{40})_3Ti_2]$, one cation is bound to the three internal oxygen

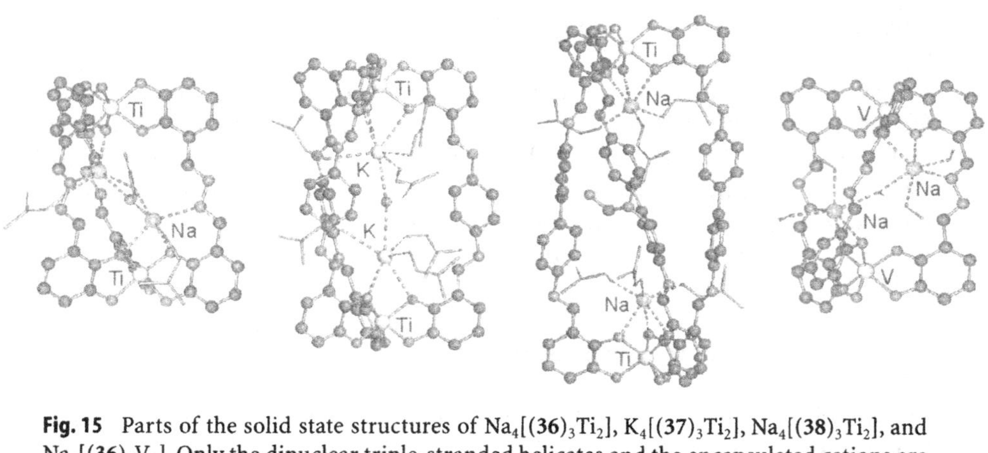

Fig. 15 Parts of the solid state structures of $Na_4[(\mathbf{36})_3Ti_2]$, $K_4[(\mathbf{37})_3Ti_2]$, $Na_4[(\mathbf{38})_3Ti_2]$, and $Na_4[(\mathbf{36})_3V_2]$. Only the dinuclear triple-stranded helicates and the encapsulated cations are shown. Solvent molecules are only indicated

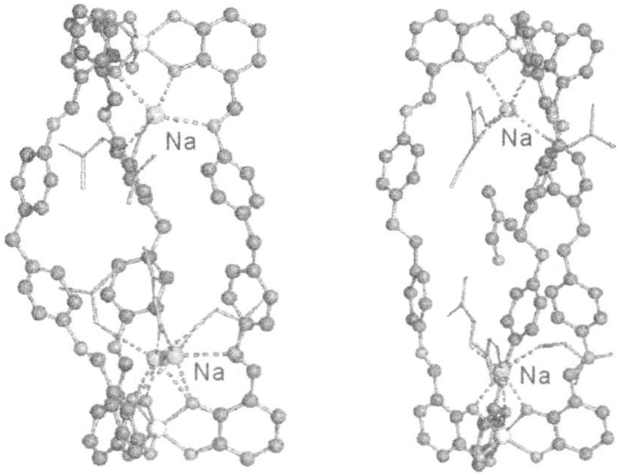

Fig. 16 Parts of the solid state structures of $Na_4[(\mathbf{39})_3Ti_2]$ and $Na_4[(\mathbf{40})_3Ti_2]$

atoms of the (triscatecholato)titanium complex units and is further coordinatively saturated by (at least) three solvent molecules. (In case of $Na_4[(\mathbf{39})_3Ti_2]$ a disordered sodium cation binds to one of the imine nitrogen atoms [27,49].)

$Na_4[(\mathbf{39})_3Ti_2]$ (Fig. 16) represents a meso-helicate. Here the relative stereochemistry is controlled by the small methylene unit, which is incorporated in the long and rigid spacer of ligand **39**.

In case of $Na_4[(\mathbf{40})_3Ti_2]$ a helicate is present with the ethylene groups acting as the stereo-controlling units. The relative configuration of the metal complex units is controlled over a distance of two nm with the control unit located about one nm apart from the stereocentres [27].

In case of the alkyl-bridged diimino-catechol ligands **41**–**45**-H_4, the importance of the counter cations as templates can be shown [49].

The described templating effect of alkali metal cations in the formation of dinuclear complexes can be used to obtain tetranuclear supramolecular tetrahedra from triscatechol ligands which possess an axis of C_3-symmetry [59].

The results discussed here show, how important templating can be in the formation of helicates. However, in the case of the catechol derivatives the templates are always counter cations which have to be present. Therefore it would be of interest to form neutral coordination compounds with the help of externally added cations (or neutral molecules) as templates. Unfortunately, with catechols the coordination compounds are always charged.

2.2.3 Helicates from 8-Hydroxyquinoline-Ligands

The 8-hydroxyquinoline ligand unit is geometrically very similar to catechol but forms neutral coordination compounds. Thus, templating by externally added species can be investigated.

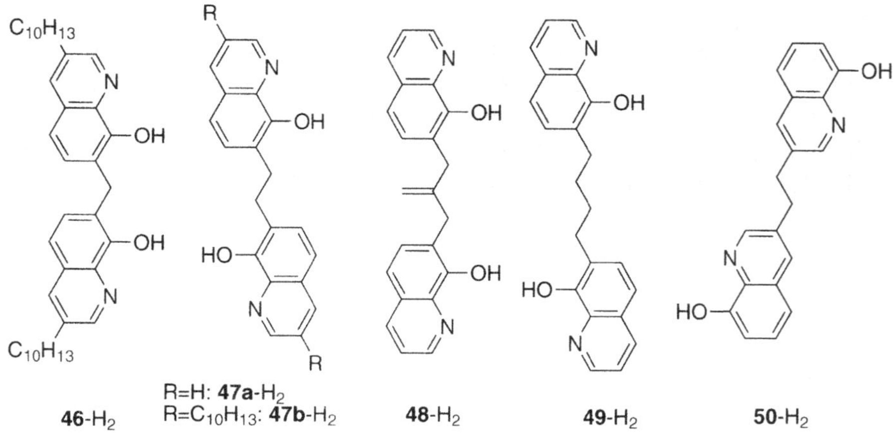

Fig. 17 Bis(8-hydroxyquinoline) ligands **46–50**-H_2

A series of di(8-hydroxyquinoline) ligands **46–50**-H_2 (Fig. 17) was prepared [60], which possess different spacer length and in case of **50**-H_2 different orientation of the ligand moiety. In **50**-H_2 the nitrogen and not the oxygen donors are located at the internal positions of the ligands. The oxygen atoms of **50**-H_2 cannot interfere with templating by small species like alkali metal cations [61].

Reaction of the ligands **46–50**-H_2 with M^{3+} cations ($M=Al^{3+}$, Ga^{3+},Fe^{3+}, Cr^{3+}) in the absence of appropriate templates produces coordination compounds of the composition "[(**46–50**)$_3$$M_2$]". This can be shown by elemental analysis. However, most of the compounds are insoluble and therefore cannot be investigated in solution. Due to the presence of long alkyl chains, "[(**47b**)$_3$$Ga_2$]" is soluble in, e.g., chloroform [62, 63]. NMR spectroscopy reveals that not a defined complex but a mixture of oligomers is formed. MALDI TOF MS or FT-ICR-MS show that, besides the dinuclear compound [(**47b**)$_3$$Ga_2$], oligomers thereof are present. A specific self-assembly process does not take place but unspecific mixtures of oligomers ("libraries") are formed. In analogy to the corresponding alkyl-bridged catecholate systems it should be possible to obtain defined helicate-type complexes by addition of appropriate templates [61, 62].

Ligand **50**-H_2 does not possess internal oxygen donor atoms and therefore no templating by cations is possible. Consequently, no defined dinuclear coordination compounds are obtained if alkali metal salts are added as template during the formation of gallium(III) complexes with **50**-H_2. Ligands **46–49**-H_2, on the other hand, possess internal donors and here templating is possible. However, this templating highly depends on the "size-selectivity" between the template and the cavity of the helicate ("lock-and-key principle") (Scheme 12). The short ligand **46**-H_2 only forms dinuclear complexes [M⊂{(**46**)$_3$$Ga_2$}]$^+$ if the small lithium or sodium cations are present as templates. With bigger cations, like potassium or ammonium, no defined gallium(III) complexes are obtained. The somewhat longer ligand **47**-H_2 forms dinuclear coordination compounds

[M⊂{$(\mathbf{47})_3Ga_2$}]$^+$ in the presence of sodium, potassium, rubidium or ammonium cations. Small lithium as well as the big cesium cannot act as templates in this case. Ligand **48**-H_2, with three carbon atoms in the spacer, forms dinuclear gallium(III) complexes [M⊂{$(\mathbf{48})_3Ga_2$}]$^+$ in the presence of potassium, rubidium and cesium, while sodium is already too small as a template. Formation of [M⊂{$(\mathbf{49})_3Ga_2$}]$^+$ with the longest alkyl spacer proceeds only in the presence of the large cesium template [64].

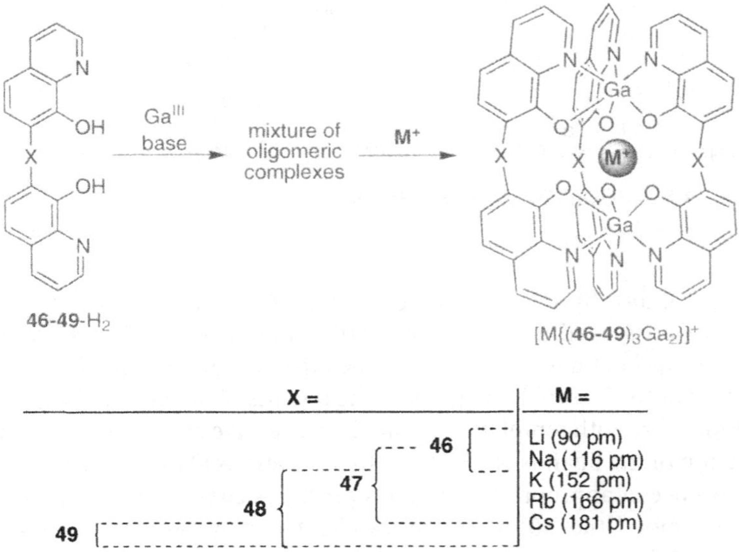

Scheme 12 "Size-selective" preparation of dinuclear complexes [M⊂{$(\mathbf{46–49})_3Ga_2$}]$^+$ following the principle of "Dynamic Combinatorial Chemistry"

There are two possible reaction paths for the specific formation of dinuclear [M⊂{$(\mathbf{46–49})_3Ga_2$}]$^+$. (1) Direct preparation is possible, if the templates are added during the complex formation. Here a template-directed self-assembly process occurs. (2) In a two step procedure, first the mixture of oligomers is formed which then by addition of an appropriate template in a dynamic process is transformed into the dinuclear complexes [M⊂{$(\mathbf{46–49})_3Ga_2$}]$^+$. Here we follow the principle of "Dynamic Combinatorial Chemistry" [64].

Not only the 8-hydroxyquinoline ligands **46–49**-H_2 can be used for the preparation of dinuclear helicate-type complex of gallium(III), aluminium(III), chromium(III), iron(III) as well as of zinc(II). In the case of zinc, charged complexes $M_2[(\mathbf{47})_3Zn_2]$ are formed which possess the templates as counter cations. The results of the coordination studies with aluminium(III), chromium(III) and iron(III) are similar to those, which were discussed for gallium(III) [64].

X-ray structures show that in the solid state the dinuclear helicate-type complexes bind templating cations in the internal cavity. Different coordination modes can be observed with different cations and different central metals (Ga, Al, Fe).

Figure 18 shows representative results of some structural analyses. In $[(dmf)_2K\subset\{(\mathbf{47a})_3Ga_2\}]^+$ a potassium cation is encapsulated in the interior of the dinuclear triple-stranded helicate and two dmf molecules are additionally binding from the outside to the templating cation. A very similar structure is observed for the corresponding $[(dmf)_2Na\subset\{(\mathbf{47a})_3Ga_2\}]^+$ [62]. Ligand **48**-H_2 shields the interior of the cavity by the vinyl groups of the spacer. Therefore, in $[Cs\subset(\mathbf{48})_3Ga_2\}]^+$ the templating cesium cation possesses an unusual low sixfold coordination. As shown with $[(NH_4)\subset\{(\mathbf{48})_3Al_2\}]^+$, the templating alkali metal cation can be substituted by ammonium. No coordinative bonds can be formed between the templating NH_4^+ and the meso-helicate $[(\mathbf{48})_3Al_2]$. The stabilizing interactions here are due to hydrogen bonding and electrostatic attraction. In iron complexes, as represented by $[ClK(H_2O)\subset\{(\mathbf{48})_3Fe_2\}]$, the skeleton of the dinuclear meso-helicate can be distorted and a water molecule is encapsulated in the cavity in addition to the potassium. Additionally chlorine coordinates

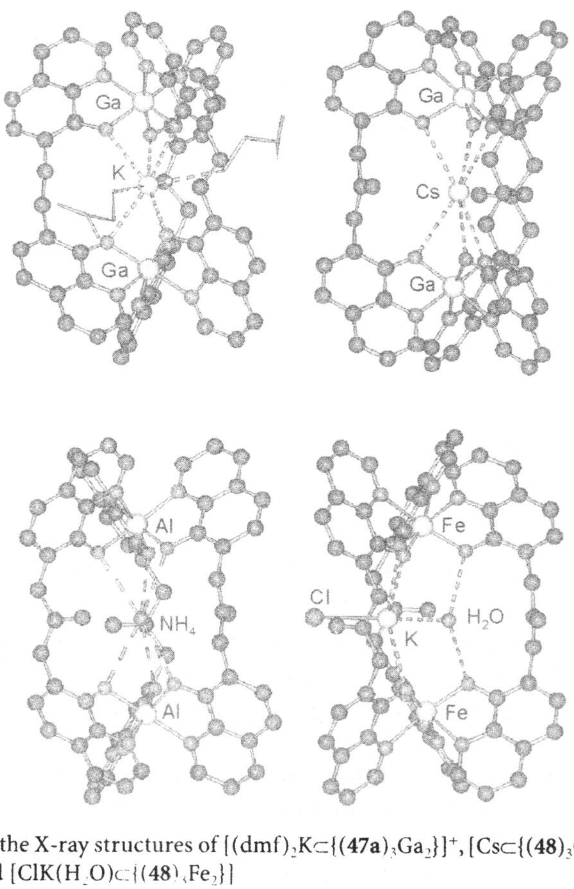

Fig. 18 Parts of the X-ray structures of $[(dmf)_2K\subset\{(\mathbf{47a})_3Ga_2\}]^+$, $[Cs\subset\{(\mathbf{48})_3Ga_2\}]^+$, $[(NH_4)\subset\{(\mathbf{48})_3Al_2\}]^+$, and $[ClK(H_2O)\subset\{(\mathbf{48})_3Fe_2\}]$

from the outside of the cavity to the internal potassium ion. A related structure is found for [ClRb(H_2O)⊂{(**48**)$_3$Fe$_2$}] [64].

As was discussed, the formation of dinuclear triple-stranded helicate-type coordination compounds from alkyl-bridged 8-hydroxyquinoline ligands **46–49**-H$_2$ depends heavily on appropriate templating. The templates are always encapsulated in the cryptand-type cavity of the complexes. Now the question remains, how strong this templating can be. Is it eventually possible, to remove the spacers of the ligands and still obtain helical complexes?

The simple 7-urea-8-hydroxyquinoline **50**-H (as well as related compounds bearing amide or urea moieties in the 7-position), forms a trinuclear zinc(II) complex [(**50**)$_6$Zn$_3$] (Scheme 13). Initially, tris(quinolinato) zinc complexes [(**50**)$_3$Zn]$^-$ are formed. In their facial structure two of them are able to bind to a third zinc atom by bridging coordination of the quinolinato oxygen atoms. Here the formation of a hexahelical [(**50**)$_6$Zn$_3$] is templated by a zinc(II) ion. This templating is supported by stabilizing hydrogen-bridges between ligands of the terminal complex units [(**50**)$_3$Zn]$^-$ [65, 66].

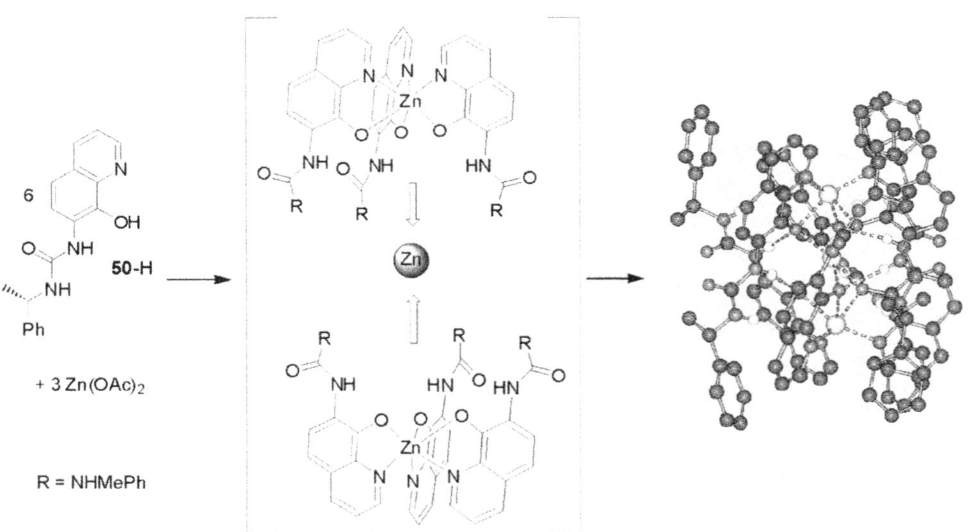

Scheme 13 Template directed formation of [(**50**)$_6$Zn$_3$]

The formation of a related trinuclear coordination compound **51** (Fig. 19) was also described. Nickel(II) acts as a template, which binds two octahedral ruthenium(II) complex units through N-oxide functions of the coordinating ligands [67].

The described results show, that a strong templating effect can lead to the assembly of helicate-type coordination compounds without a covalent connection between the complex moieties. Bidirectional coordinating abilities of ligands, together with electrostatic interactions and maybe hydrogen bonding,

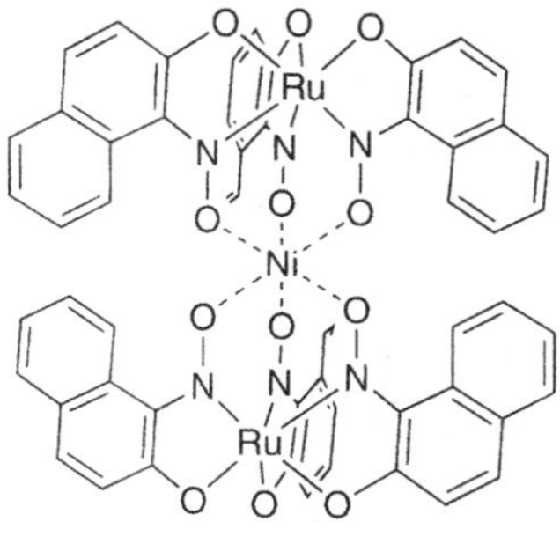

Fig. 19 Compound 51

51

can be sufficient for the formation of well defined oligonuclear coordination compounds.

3 Templating in the Formation of Circular Helicates

In circular helicates linear ligands (Fig. 20) are bound to a circular arrangement of metal ions and thus are forming a metallo-supramolecular ring or torus.

A trinuclear circular helicate $[(\mathbf{52})_3Ag_3]^{3+}$ is formed from ligand **52** with silver(I) cations, [68] while **53** forms a tetranuclear iron(II) circular helicate $[(\mathbf{53})_4Ag_4]^{8+}$ [69]. The chiral bipyridine derivative **54** in the presence of silver(I) ions leads to an enantiomerically pure hexanuclear circular helicate $[(\mathbf{54})_6Ag_6]^{6+}$.

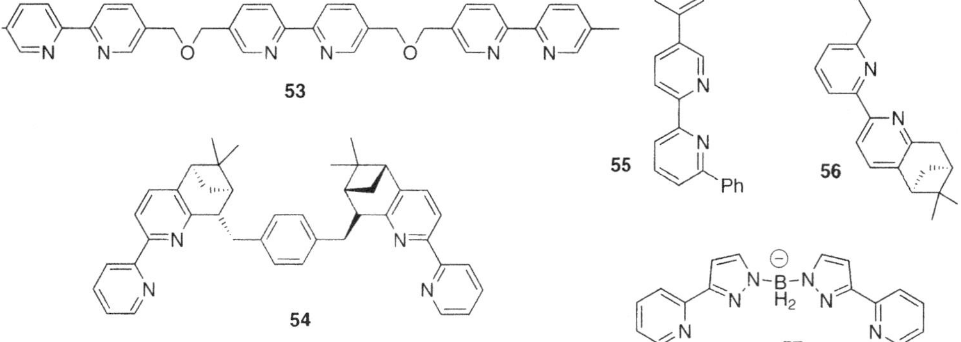

52 53 54 55 56 57

Fig. 20 Ligands for circular helicates

[70] Very often the self-assembly of circular helicates does not proceed smoothly, but mixtures of circular complexes with different ring size are formed. Thus, ligand **55** forms a mixture of copper(I) complexes $\{[(\mathbf{55})_2Cu_2]^{2+}\}_n$ (n=1,2,...) from which the dinuclear species $[(\mathbf{55})_2Cu_2]^{2+}$ can be isolated by crystallization [71]. Similarly, **56** forms mixtures of cyclic oligomers of copper(I) complexes. However, depending on the crystallization conditions, it is possible to obtain either $[(\mathbf{56})_2Cu_2]^{2+}$ or $[(\mathbf{56})_3Cu_3]^{3+}$ [72].

An octanuclear coordination compound $[(ClO_4)\subset\{(\mathbf{57})_8Co_8\}]^{7+}$, which is related to the circular helicates is formed from **57** by reaction with cobalt(II) perchlorate. In the solid state one of the perchlorate counterions is bound in the interior of the circular metal complex and probably is stabilizing the aggregate [73].

With the tris(bipyridine) ligand **58** it was shown that a templating influence of counterions on the formation of circular helicates can be important. Reaction of **58** with nickel(II) salts leads to the triple-stranded trinuclear helicate $[(\mathbf{58})_3Ni_3]^{6+}$ as the product [21] (Scheme 14).

Reaction of **58** with iron(II) salts reveals that the triple-stranded helicate $[(\mathbf{58})_3Fe_3]^{6+}$ is an initial product which is produced under kinetic control of the reaction. As thermodynamic product circular helicates are formed, depending on the anions which are present as templates [28, 69, 74].

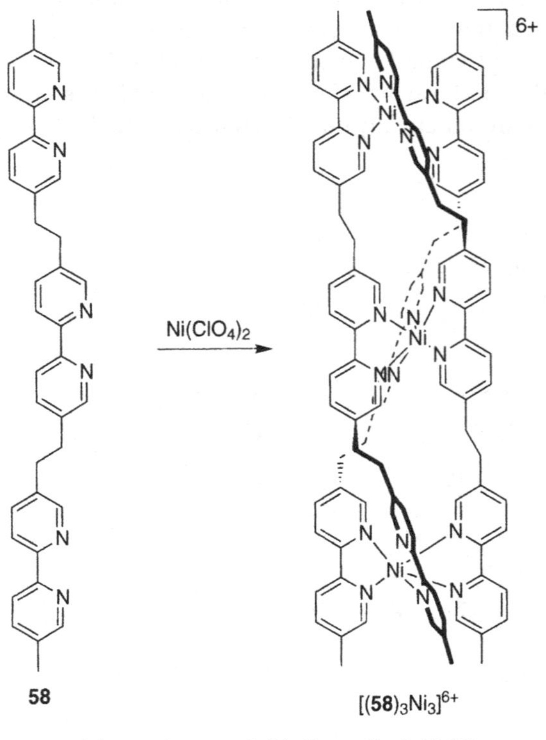

58 $[(\mathbf{58})_3Ni_3]^{6+}$

Scheme 14 Formation of the triple-stranded helicate $[(\mathbf{58})_3Ni_3]^{6+}$

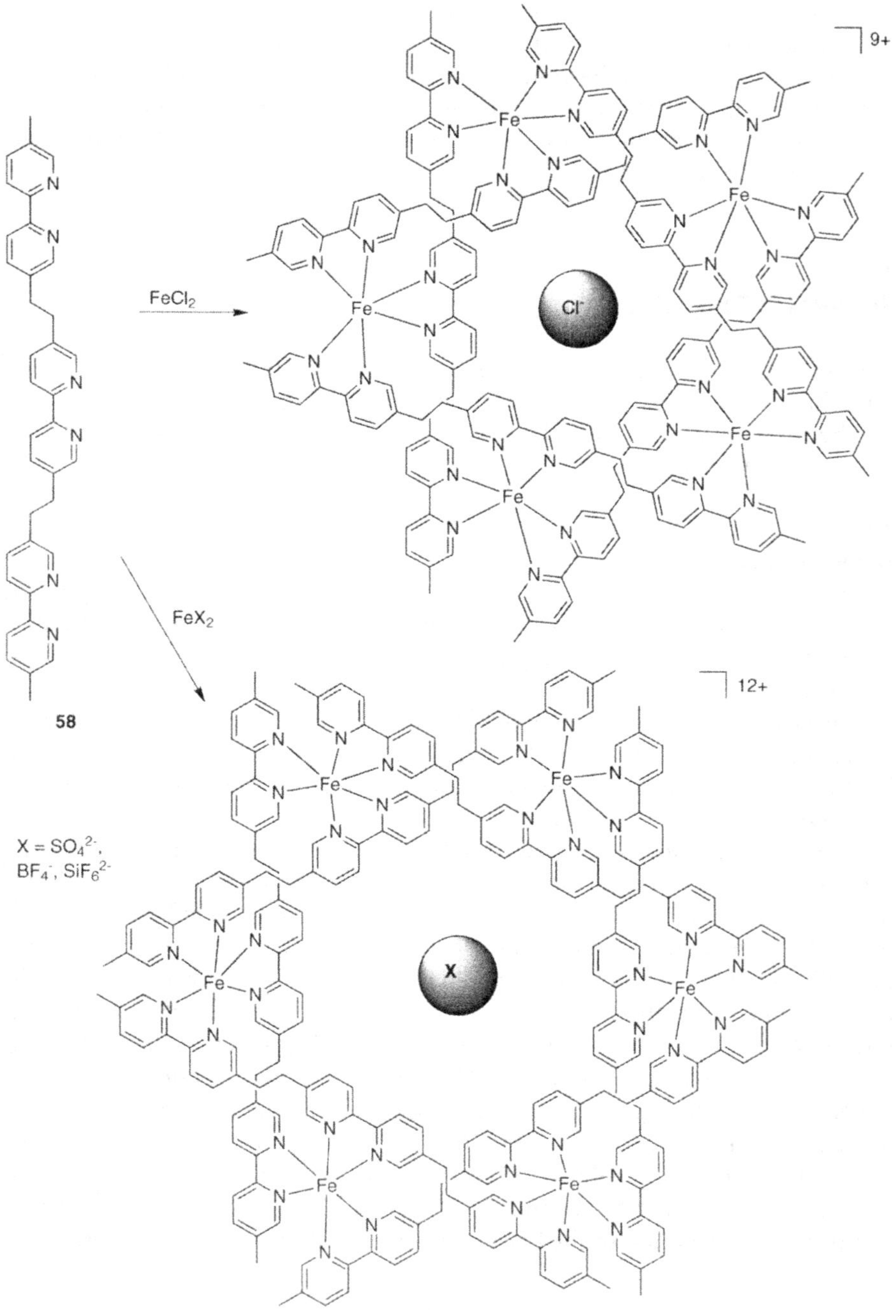

Scheme 15 Template-directed formation of pentanuclear vs hexanuclear circular helicates $[(\mathbf{58})_5Fe_5]^{10+}$ or $[(\mathbf{58})_6Fe_6]^{12+}$

In the presence of the relatively small chloride anion, the pentanuclear circular helicate $[Cl\subset\{(\mathbf{58})_5Fe_5\}]^{9+}$ is obtained. Medium sized anions like bromide support the formation of the pentanuclear as well as hexanuclear helicate, while big anions (sulfate, tetrafluoroborate, hexafluorosilicate) lead to the exclusive formation of the hexanuclear complex $[(\mathbf{58})_6Fe_6]^{12+}$ (Scheme 15). Templating allows the adjustment of the ring size of circular helicates [69].

4
Helicates as Templates

It is not only templating that plays an important role in the formation of helicates; helicates can also act as templates themselves. This will be demonstrated for the preparation of molecular knots via helicates as intermediates and of polymeric networks.

4.1
Molecular Knots

J.-P. Sauvage introduced helicates as templates for the formation of molecular trefoil knots, in which a macrocycle is wound through itself as it is schematically shown in Fig. 21.

The strategy is to form a double-stranded dinuclear helicate first, and then to crosswise connect the termini of the helicated ligands to obtain a molecular knot with two metals fixing its geometry. Removal of the metal ions yields the final trefoil knot-type molecule [75].

Several procedures were used for the final ring closure reactions, e.g., Williamson ether synthesis or Glaser-type coupling reaction. As one of the most successful approaches ring closure by olefin metathesis will be described here. The bisphenanthroline ligand **59** self-assembles in the presence of copper(I) ions to form the double-stranded dinuclear helicate $[(\mathbf{59})_2Cu_2]^{2+}$. Addition of an olefin-metathesis catalyst affords a coupling of the terminal olefinic units to obtain a *cis-trans* mixture of the corresponding macrocyclic olefin. The double bonds are reduced and the helicate **60** is obtained, in which the termini are bridged and the organic ligand possesses the structure of a trefoil knot. The metal ions can be removed to obtain the free knot **61** [76, 77].

In this example, the helicate $[(\mathbf{59})_2Cu_2]^{2+}$ acts as a template and orientates the reactive termini of the ligands **59** for a ring closure reaction. This finally results in the formation of the knotted organic molecule **61** (Scheme 16).

Fig. 21 A trefoil knot

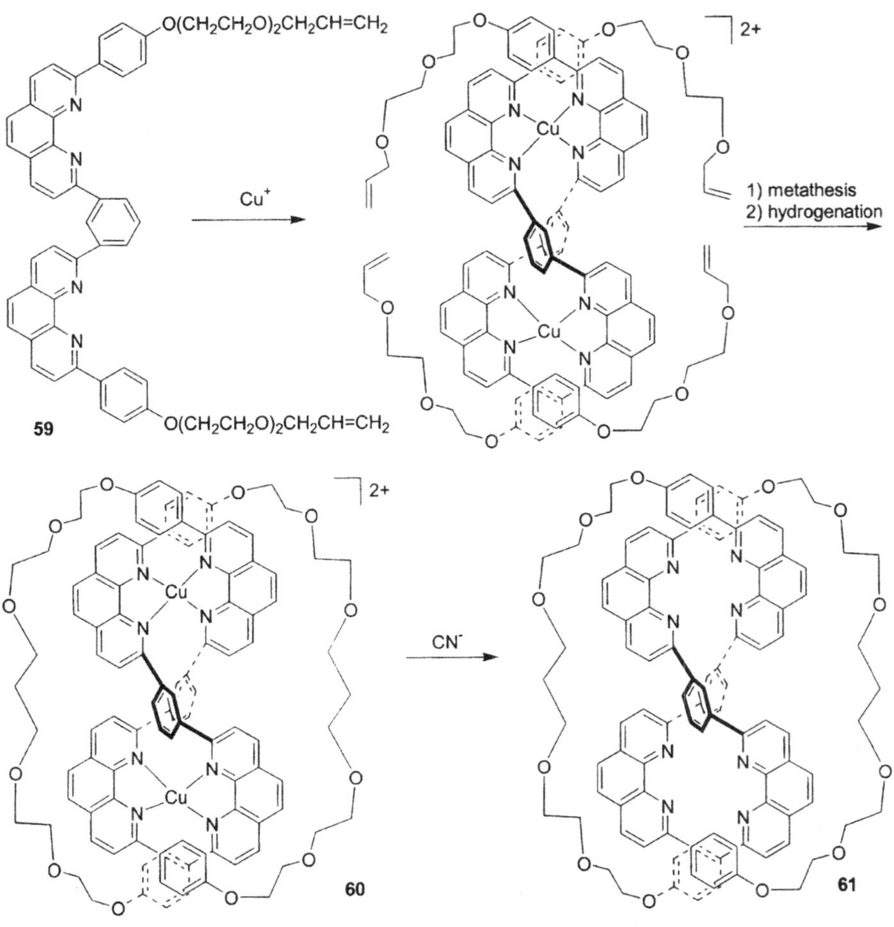

Scheme 16 Preparation of a molecular knot **61**

4.2 Helicates as Parts of Polymers

Due to their rigidity and their intrinsic chirality, helicates are interesting building blocks to be incorporated into polymers (organic as well as inorganic). The structure (and thus the properties) of a polymer will be influenced by the metallosupramolecular component [78].

In compound **62** two bipyridine units are part of a polymeric chain. Upon addition of copper(I) ions the bipyridines bind to the copper and double-stranded helicate-type domains are formed within the polymer [79].

Ligand **63** was used to form the triple-stranded helicate $[(\mathbf{63})_3Fe_2]^{4+}$, which then was copolymerized with methyl methacrylate, thus incorporating the triple-stranded helical coordination compound in the polymer network [80] (Fig. 22).

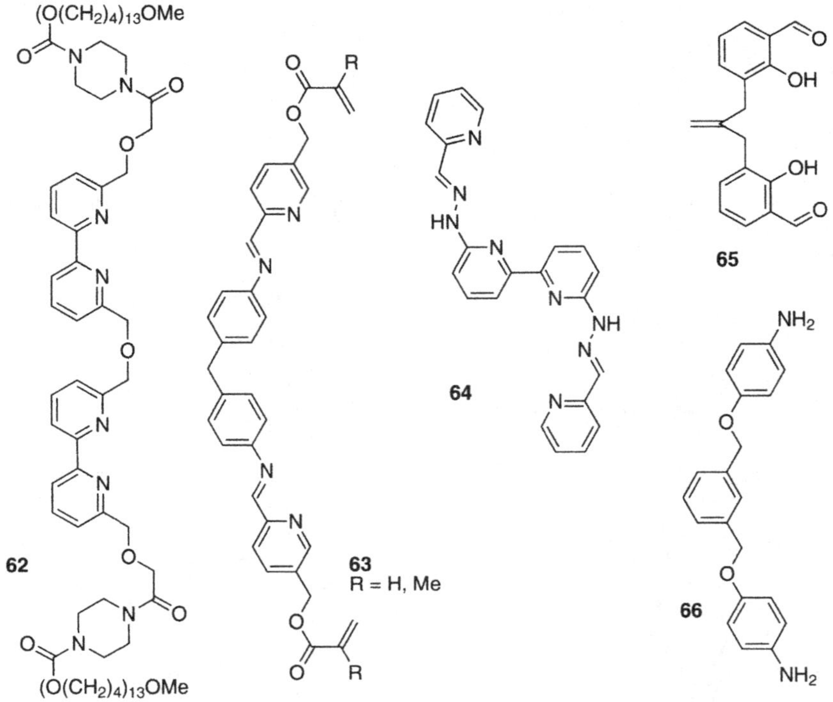

Fig. 22 Building blocks which were used for the incorporation of helicates into polymers

The formation of a cross linked inorganic double helical polyiodine is templated by the double-stranded dinuclear helicate $[(\mathbf{64})_2Zn_2]^{4+}$ [81].

Just recently the imine formation of dialdehyde **65** with the diamine **66** was performed in the presence of copper(II), nickel(II), zinc(II), or cobalt(II) salts. In this reaction a polymeric network which is made up from double-stranded helicates is formed. The polymer precipitates under formation of colloidal microspheres with a high degree of monodispersity. The size of the colloidal microspheres depends on the metal ion which is used in the reaction [82].

5 Conclusions

The self-assembly of supramolecular structures is a process which is not at all understood today and which only in some special examples can be predicted. Therefore, the understanding of fundamental processes which play a role in self-assembly is very important.

As shown in this chapter, templating can play an important role in the formation of such simple supramolecular coordination compounds as helicates. Due to their simplicity they can act as our "molecular drosophila" and we can use them to study basic principles of supramolecular chemistry. Important aspects are not only templating but also stereochemistry or regiochemistry. In the long term this will eventually help us to develop a rational approach for the build up of huge functional materials.

Acknowledgements Our part of this work was supported by the Deutsche Forschungsgemeinschaft. I am very grateful to all of my coworkers and collaborators who contributed to this work and who are mentioned in the references.

References

1. Lehn J-M (1995) Supramolecular chemistry. VCH, Weinheim
2. Piguet C, Bernardinelli G, Hopfgartner G (1997) Chem Rev 97:2005
3. Albrecht M (2001) Chem Rev 101:3457
4. Klug A (1983) Angew Chem Int Ed Engl 22:565
5. Pedersen CJ (1967) J Am Chem Soc 89:7017
6. Chapman RG, Sherman JC (1997) Tetrahedron 53:15911
7. Amabilino DB, Stoddart JF (1995) Chem Rev 95:2725
8. Nepogodiev SA, Stoddart FJ (1998) Chem Rev 98:1959
9. Vögtle F, Jäger R (1997) Angew Chem Int Ed Engl 36:931
10. Sauvage J-P (1990) Acc Chem Res 23:319
11. Lehn J-M (2000) Chem Eur J 6:2097
12. Hiraoka S, Fujita M (1999) J Am Chem Soc 121:10239
13. Huc I, Lehn JM (1997) Proc Natl Acad Sci USA 94:2106
14. Otto S, Furlan RLE, Sanders JKM (2002) Science 297:590
15. Lehn J-M, Rigault A, Siegel J, Harrowfield J, Chevrier B, Moras D (1987) Proc Natl Acad Sci USA 84:2565
16. Struckmeier G, Thewalt U, Fuhrhop J-H (1976) J Am Chem Soc 98:278
17. Sheldrick WS, Engel J (1980) J Chem Soc Chem Commun 5
18. Van Stein GC, Van der Poel H, Van Koten G, Spek AL, Duisenberg AJM, Pregosin PS (1980) J Chem Soc Chem Commun 1016
19. Lehn J-M, Sauvage J-P, Simon J, Ziessel R (1983) Nouv J Chim 7:413
20. Williams AF, Piguet C, Bernardinelli G (1991) Angew Chem Int Ed Engl 30:1490
21. Krämer R, Lehn J-M, DeCian A, Fischer J (1993) Angew Chem 105:764
22. Carrano CJ, Raymond KN (1978) J Am Chem Soc 100:5371
23. Scarrow RC, White DL, Raymond KN (1985) J Am Chem Soc 107:6540
24. Albrecht M, Kotila S (1995) Angew Chem Int Ed Engl 34:2134
25. Albrecht M (2000) Chem Eur J 6:3485
26. Albrecht M (1998) Chem Soc Rev 27:281
27. Albrecht M, Janser I, Houjou H, Fröhlich R (2004) Chem Eur J 10:2839
28. Hasenknopf B, Lehn J-M, Kneisel BO, Baum G, Fenske D (1996) Angew Chem Int Ed Engl 35:1208
29. Martin N, Bünzli J-CG, McKee V, Piguet C, Hopfgartner G (1998) Inorg Chem 37:577
30. Bilyk A, Harding MM (1995) J Chem Soc Chem Commun 1697
31. Houghton MA, Bilyk A, Harding MM, Turner P, Hambley TW (1997) J Chem Soc Dalton Trans 2725
32. Try AC, Harding MM, Hamilton DG, Sanders JKM (1998) Chem Commun 723
33. Su C-Y, Cai Y-P, Chen C-L, Smith MD, Kaim W, zur Loye H-C (2003) J Am Chem Soc 125:8595

34. Enemark EJ, Stack TDP (1996) Inorg Chem 35:2719
35. Albrecht M, Napp M, Schneider M, Weis P, Fröhlich R (2001) Chem Commun 409
36. Albrecht M, Napp M, Schneider M, Weis P, Fröhlich R (2001) Chem Eur J 7:3966
37. Hahn FE (personal communication)
38. Albrecht M, Nolting R (unpublished results)
39. Saalfrank RW, Maid H, Mooren N, Hampel F (2002) Angew Chem 114:323
40. Yeh RM, Ziegler M, Johnson DW, Terpin AJ, Raymond KN (2001) Inorg Chem 40:2216
41. Jodry JJ, Lacour J (2000) Chem Eur J 6:4297
42. Lacour J, Jodry JJ, Monchaud D (2001) Chem Commun 2302
43. Xu J, Parac TN, Raymond KN (1999) Angew Chem Int Ed Engl 38:2878
44. Saalfrank RW, Dresel A, Seitz V, Trummer S, Hampel F, Teichert M, Stalke D, Stadler C, Daub J, Schunemann V, Trautwein AX (1997) Chem Eur J 3:2058
45. Saalfrank RW, Seitz V, Heinemann FW, Göbel C, Herbst-Irmer R (2001) J Chem Soc Dalton Trans 599
46. Saalfrank RW, Seitz V, Caulder D, Raymond KN, Teichert M, Stalke D (1998) Eur J Inorg Chem 1313
47. Albrecht M, Schmid S, deGroot M, Weis P, Fröhlich R (2003) Chem Commun 2526
48. Scherer M, Caulder D, Johnson DW, Raymond KN (1999) Angew Chem Int Ed 38:1588
49. Albrecht M, Janser I, Kamptmann S, Weis P, Wibbeling B, Fröhlich R (2004) J Chem Soc Dalton Trans 37
50. Kersting B, Meyer M, Powers RE, Raymond KN (1996) J Am Chem Soc 118:7221
51. Albrecht M, Kotila S (1996) Chem Commun 2309
52. Albrecht M (1997) Chem Eur J 3:1466
53. Albrecht M, Kotila S (1996) Angew Chem Int Ed Engl 35:1208
54. Albrecht M, Röttele H, Burger P (1996) Chem Eur J 2:1264
55. Albrecht M, Schneider M, Röttele H (1999) Angew Chem Int Ed Engl 38:557
56. Albrecht M, Schneider M (2002) Eur J Inorg Chem 1301
57. Albrecht M, Schneider M, Fröhlich R (1998) New J Chem 753
58. Albrecht M, Kamptmann S, Fröhlich R (2003) Polyhedron 22:643
59. Albrecht M, Janser I, Meyer S, Weis P, Fröhlich R (2003) Chem Commun 2854
60. Albrecht M, Blau O (1997) Synthesis 213
61. Albrecht M, Blau O (1997) Chem Commun 345
62. Albrecht M, Blau O, Fröhlich R (1999) Chem Eur J 5:48
63. Albrecht M, Blau O, Zauner J (1999) Eur J Org Chem 3165
64. Albrecht M, Blau O, Fröhlich R (2002) Proc Natl Acad Sci USA 99:4876
65. Albrecht M, Witt K, Weis P, Wegelius E, Rissanen K, Fröhlich R (2002) Inorg Chim Acta 341:25
66. Albrecht M, Witt K, Röttele H, Fröhlich R (2001) Chem Commun 1330
67. Das AK, Rueda A, Falvello LR, Peng S-M, Bhattacharya S (1999) Inorg Chem 38:4365
68. Al-Mandhary MRA, Steel PJ (2002) Inorg Chem Commun 5:954
69. Hasenknopf B, Lehn J-M, Boumediene N, Dupont-Gervais A, Van Dorsselaer A, Kneisel BO, Fenske D (1997) J Am Chem Soc 119:10956
70. Mamula O, Von Zelewsky A, Bernardinelli G (1998) Angew Chem Int Ed Engl 37:290
71. Baxter PNW, Lehn J-M, Rissanen K (1997) Chem Commun 1323
72. Baum G, Constable EC, Fenske D, Housecroft CE, Kulke T (1999) Chem Commun 195
73. Jones PL, Byrom KJ, Jeffery JC, McCleverty JA, Ward MD (1997) Chem Commun 1361
74. Hasenknopf B, Lehn J-M, Boumediene N, Leize E, Van Dorsselaer A (1998) Angew Chem Int Ed Engl 37:3265
75. Dietrich-Buchecker CO, Sauvage J-P, Kintzinger J-P, Maltèse P, Pascard C, Guilhem J (1992) New J Chem 16:931

76. Dietrich-Buchecker CO, Rapenne G, Sauvage J-P (1997) Chem Commun 2053
77. Rapenne G, Dietrich-Buchecker CO, Sauvage J-P (1999) J Am Chem Soc 121:994
78. Schubert US, Eschbaumer C (2002) Angew Chem Int Ed Engl 41:2892
79. Eisenbach CD, Schubert US, Baker GR, Newkome GR (1995) J Chem Soc Chem Commun 69
80. Lavalette A, Hamblin J, Marsh A, Haddleton DM, Hannon MJ (2002) Chem Commun 3040
81. Horn CJ, Blake AJ, Champness NR, Lippolis V, Schröder M (2003) Chem Commun 1488
82. Houjou H, Shimizu Y, Koshizaki N, Kanesato M (2003) Adv Mater 15:1458

Topics in Current Chemistry (2004) 248: 141–200
DOI 10.1007/978-3-540-44505-0

Hydrogen-Bond-Mediated Template Synthesis of Rotaxanes, Catenanes, and Knotanes

Christoph A. Schalley (✉) · Torsten Weilandt · Jens Brüggemann · Fritz Vögtle

Kekulé-Institut für Organische Chemie und Biochemie der Universität, Gerhard-Domagk-Strasse 1, 53121 Bonn, Germany
c.schalley@uni-bonn.de

Abstract The synthesis of rotaxanes, catenanes, and trefoil knotanes can only be efficiently accomplished, when a suitable template effect is used that provides a means either to thread a string-like molecule through a macrocycle or to wrap an open macrocycle around a string. This overview is devoted to the template effects for the synthesis of mechanically bound species which are mediated by hydrogen bonding. Three major classes can be distinguished by the charges involved: (i) cation-mediated template effects utilizing the formation of crown ether complexes with secondary ammonium ions, (ii) recognition of neutral amides by oligolactam macrocycles, and (iii) threading through hydrogen bonding to anions. The basic idea which we put forward in this chapter is that none of the template effects found so far and discussed here is "traceless" in the sense that they would not leave functional groups behind in the final rotaxane, catenane, or knotane structure. Consequently, the variability of the structures available through these template syntheses is limited in this respect. Also, the properties of the mechanically bound species are intimately interconnected with the template effect used for their synthesis. However, what appears here as a drawback may also be an advantage, when the functional groups used for templating can be utilized to control the properties of the final product which hopefully can in future be exploited for the development of new materials. We aim at providing sufficient examples in order to strengthen this point in throughout the article. Among the properties and functions discussed are the realization of molecular machines, topological chirality, and the stability of rotaxanes as determined from deslipping experiments.

Keywords Catenanes · Hydrogen bonding · Knots · Mechanical bond · Molecular machines · Molecular recognition · Rotaxanes · Template synthesis · Topology

1 Introduction

1.1 The Art of Template Design

It is almost impossible to give a concise definition of the term "template" which accounts for all chemistry to which it has been related [1–6]. Templates span the whole range from biochemistry with its complex apparatus for DNA replication [7] to biomineralization and the formation of structured inorganic materials [8] to the covalently templated synthesis of macrocycles [9] to the preparation of supramolecular catalysts [10] or host-guest complexes [11] and to metal-organic reactions for the synthesis of small ring systems [12]. Nevertheless, all these have in common that a template must serve different purposes. (i) It organizes reaction partners in such a way that a desired product is formed that would not form as easily in the absence of the template. Thus, a template controls reactivity through organizing the reaction partners in a favorable arrangement producing form. (ii) In order to achieve this, the template needs to bind to the reaction partners. Molecular recognition is thus a necessary prerequisite for template syntheses and the binding sites of the components must be complementary to each other. (iii) The control of reactivity and the recognition of the reaction partners imply that information is stored in the template and transferred to the product of the reaction. The third important aspect is thus information transfer. In fulfilling these three tasks, templates help to tame molecular complexity and provide access to otherwise inaccessible products.

There are different ways to categorize templates (Fig. 1). One could, for example, try to distinguish template effects according to the (non-)covalent interactions involved. This classification has the disadvantage to be ambiguous

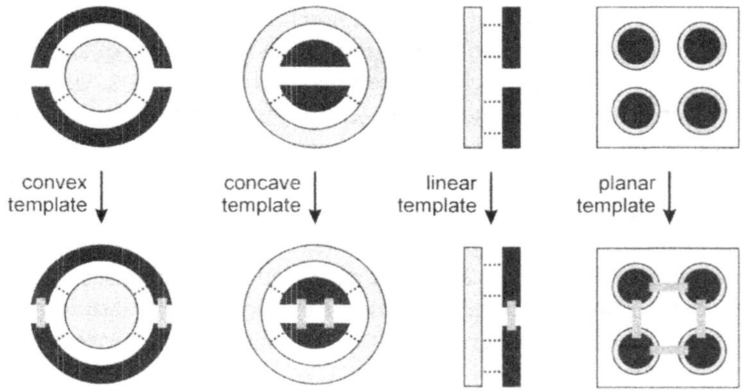

Fig. 1 Schematic classification of templates according to their topology. The template is shown in *gray*, the reaction partners in *black*. *Dotted lines* represent non-covalent bonds, *gray sticks* are new covalent bonds formed between the reaction partners in the templation step

for templates which operate via several different types of forces. The template effect could, for example, involve hydrogen bonds and π-donor π-acceptor interactions making it difficult to decide to which category it should be assigned. Maybe, a better way is to classify the template according to its topography. The early template syntheses of crown ethers utilized a central metal ion around which a macrocycle formed with a certain size selectivity from wrapped-around open-chain precursors [13]. Such templates could be called convex, because of the convex surface of the cation mediating the template effect. In contrast, a receptor which binds two reaction partners inside a cavity and brings them to reaction is concave. This is true for many templates leading to mechanically interlocked species. One of the most prominent natural templates, i.e. single-stranded DNA, could be called a linear template according to this classification and a surface on which molecules self-assemble into an ordered array [14] may be coined a planar template. A beautiful example of a planar template is a recently reported electropolymerization using a liquid crystalline phase as the template [15].

Another important point is to distinguish between a reactant, a template, and a catalyst [16]. A strict definition would imply that the template must be removable after successfully mediating a certain reaction. Instead, a reactant becomes part of the product in one or another form. However, we will see that definitions become blurred, when discussing the synthesis of rotaxanes, catenanes, and knots. Many of the methods for the preparation of interlocked molecules are based on macrocycles which bind part of the other component in a pseudorotaxane fashion. This part is then reacted so that the desired species, e.g., a catenane or a rotaxane, is formed. The macrocyclic template thus finally becomes part of the product and might consequently be considered as a reactant rather than a template. Nevertheless, these methods are widely accepted as template syntheses in the chemical literature, although the use of removable transition metal ions for the synthesis of mechanically interlocked molecules [17–20] is the only true template synthesis in the strict sense of above.

It is similarly difficult to separate templates from catalysts: On one hand, there are templates which do not promote catalytic reactions, because the template is bound too strongly to generate turnover. They have to be removed from the product by separation techniques and need to be used in stoichiometric amounts. On the other hand, there are catalysts which do not organize the reactants in space but rather change their intrinsic reactivity, e.g., general acid or base catalysis. Thus, they cannot be regarded as templates. These are the clear-cut cases. However, mixed forms exist, where a template is bound reversibly to the product or where a catalyst organizes the reactants with respect to their geometry.

All definitions of "template" necessarily suffer from this continuous transition from reactants to templates and to catalysts and we should keep in mind that a distinction between these categories is not always straightforward. We therefore put forward a more abstract definition of a template which is based on its ability to organize the reactants appropriately through covalent or non-covalent bonding interactions. In our definition, a template is not a defined particle, i.e., an atom or molecule, but rather the sum of all connections between the species reacting with each other which are involved in geometrically controlling the reactivity in the desired way. It is the array of interactions and their spatial arrangement that count.

This brings us to one of the central ideas of this account. Once we accept that templates may not be "traceless", but leave behind a memory of their action in that functional groups remain in the final product once needed for templation, it immediately becomes clear that the range of possible structures is limited. The functional groups left in the product may well change or even dominate its properties and any attempt to change that must severely affect the template synthesis. However, what appears to be a disadvantage can be transformed into a virtue, if one thinks backwards just like in retrosynthesis. Although many templates are still found by coincidence, one might thus think of a rational template design starting with the desired final structure. Properties and functions of a molecule are based on its structure and the presence of suitable functional groups in a well-defined spatial arrangement [21]. The structure is generated through templation. Therefore, one may think of designing template effects which result in products bearing just those functional groups needed for the desired properties and functions to emerge. This chapter intends to illustrate the "art of template design" and the intimate relation of template and function with a focus on mechanically interlocked molecules that are produced mediated through hydrogen bonding.

1.2 Molecules with Mechanical Bonds

The research of mechanically interlocked molecules [22–29] has been inspired much by the scientists' sense for aesthetic molecules [30]. The generation of ever new topologies, of which a small selection is shown in Fig. 2, has lead to more and more complex species. Applying the mathematical discipline of topology to these species has paved the road to the chemical realization of topological chirality as a new form of chirality beyond its classical Euclidean counterpart. However, the focus has shifted away from the beauty of topologically interesting molecules towards their functional properties. This will form the second focus of this account, which is closely related to the first one discussed above. The generation of switches, shuttles, and other molecular devices is currently under intense examination by a number of different groups [31–40].

Templation as it is discussed in this article is strongly dependent on molecular recognition. Useful geometries for the syntheses of rotaxanes, catenanes, and knotanes require a step, in which a thread-like molecule is threaded into a macrocyclic one or vice versa. Thus, the range of possible precursor structures is rather narrow. The molecular recognition employed for templating consequently not only requires binding, but binding in the correct structural arrangement. In order to account for the importance of molecular recognition, we include in this review the non-covalently bound pseudorotaxane precursors which upon stopper attachment can be converted into rotaxanes or which by ring closure yield catenanes. These pseudorotaxanes are not held together by mechanical bonds, but provide profound insight into the interactions on which the template effects are based.

In this article, we include those syntheses of mechanically bound molecules that involve hydrogen bonding [41–43] as the major supramolecular interac-

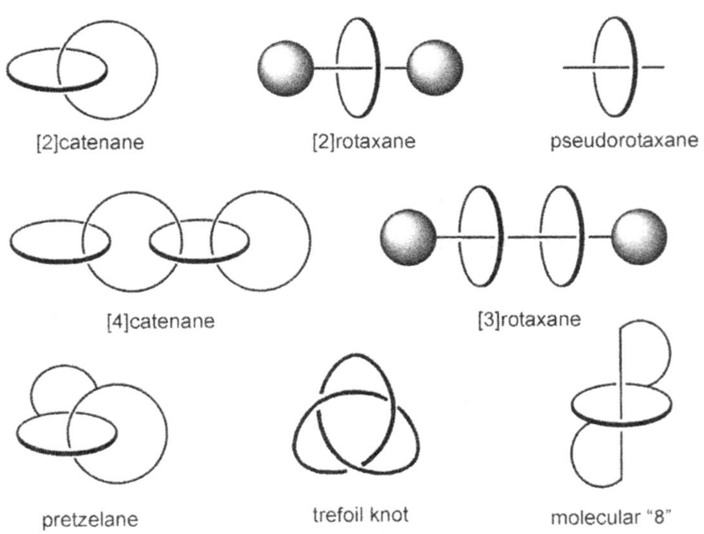

Fig. 2 Different mechanically interlocked molecules. *Figures in brackets* count the number of separate, mechanically bound subunits. Note that the pretzelane, the trefoil knot, and the molecular figure eight are comprised of only one subunit, while the others bear mechanical bonds between several independent and not covalently linked building blocks

tion between the reactants in the templation step. Other template syntheses such as the Cu(I)-mediated generation of catenates and rotaxates or those based on π-donor π-acceptor interactions will form the scope of other reviews in this series. However, there are cases at the borderline between these categories. For example, C-H···O [44] or C-H···π hydrogen bonds play a certain role in catenanes and rotaxanes synthesized through π-donor π-acceptor interactions [45–48]. Nevertheless, we can mention them here only briefly and concentrate in the following on template syntheses of mechanically bound species based on (a) the interaction of crown ethers with secondary ammonium ions, (b) the hydrogen bonding between tetralactam macrocycles and secondary amides, and (c) the binding of anions inside macrocycles. The article is thus roughly organized with respect to the charge of the species under study.

2 Cationic Templates: Secondary Alkyl Ammonium Ion/Crown Ether Complexes

2.1 Pseudorotaxanes through Ammonium Ion/Crown Ether Complexes

The binding of NH_4^+ and primary ammonium ions to crown ethers has been long known and well studied [49, 50], while studies of the binding of secondary ammonium ions to smaller crown ethers are relatively scarce. This might be

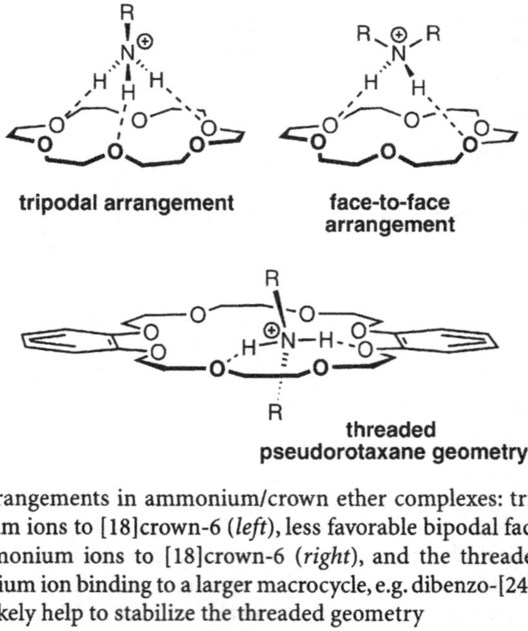

Fig. 3 Possible arrangements in ammonium/crown ether complexes: tripodal binding of primary ammonium ions to [18]crown-6 (*left*), less favorable bipodal face-to-face binding of secondary ammonium ions to [18]crown-6 (*right*), and the threaded geometry of a secondary ammonium ion binding to a larger macrocycle, e.g. dibenzo-[24]crown-8. C-H···O hydrogen bonds likely help to stabilize the threaded geometry

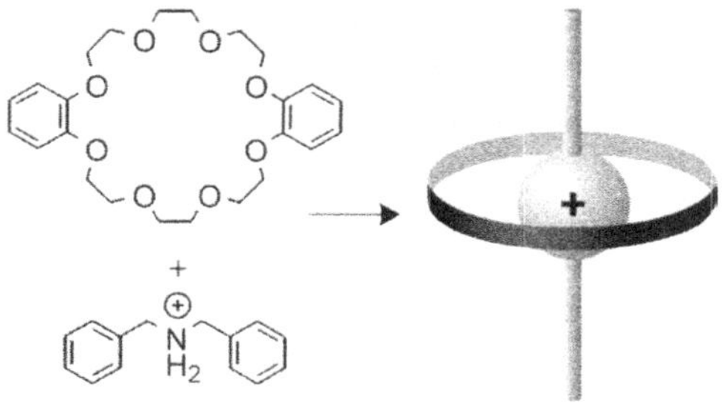

Fig. 4 Complex formation from dibenzyl ammonium ion and dibenzo-[24]crown-8 **1**

due to the fact that NH_4^+ and primary ammonium ions can bind to the crowns in a favorable tripodal fashion (Fig. 3). This is not possible for secondary ammonium ions, which necessarily bind in a face-to-face manner. In 1995, Stoddart et al. reported the formation of complexes between secondary ammonium ions and dibenzo-[24]crown-8 (1) (Fig. 4) [51, 52]. The larger crown allows the formation of a threaded arrangement of the ammonium ion inside its cavity in which the binding interactions are maximized through the formation of additional C-H···O hydrogen bonds [53]. Such a geometry is found in the crystal as well as in solution and the term "pseudorotaxane" has been coined to specify the threaded nature of the guest cation inside the crown.

A number of studies followed which reported variants of the early pseudorotaxanes with crown ethers of different sizes up to tribenzo-[27]crown-9 and ammonium ions with various side chains [54–57]. These include threading of secondary ammonium ions through macrobicyclic hosts [58] and the self-assembly of C_{60} dimers through pseudorotaxane formation [59]. Quite a large selection of binding constants is available (Table 1) which as expected for hydrogen-bound species depend to a large extent on the solvent, in which complex formation occurs. For the example in Fig. 3, no binding is observed in highly competitive dimethylsulfoxide (DMSO), while the binding constant is ca. $K=2\times10^4\ M^{-1}$ in chloroform. In solvents with low dielectric constants, ion pairing needs to be considered, and a model taking it into account explicitly has been developed by Jones and Gibson [60].

However, not only the solvent significantly affects the strength of interaction in the pseudorotaxanes. From a series of binding constant measurements of a large number of different secondary ammonium ions to dibenzo-[24]crown-8, Busch et al. were able to deduce substitution patterns of the ammonium ions which favor the formation of pseudorotaxanes [61]. While the chain length of aliphatic side chains do not alter the binding strength of the pseudorotaxanes much, branching within the chain, in particular in the α- or to a lesser extent in the β-position to the nitrogen atom lowers the binding energy significantly. Another change is observed, when aromatic rings are positioned at different distances relative to the nitrogen atom. The binding constants increase from benzyl to phenylethyl to phenylpropyl side chains, probably due to increasingly

Table 1 Binding constants of secondary ammonium ions to different crown ethers including solvent and temperature as the major parameters which change the binding strengths

Macrocycle	Ammonium ion		Solvent	T (°C)	K_a M^{-1}	Ref
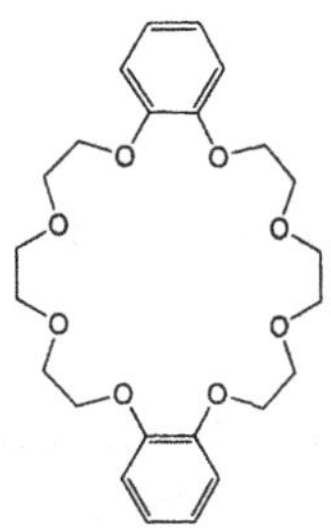	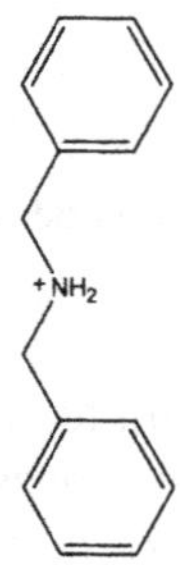		$CDCl_3$	25	27000	51
			CD_3CN	25	420	51
			CD_3COCD_3	25	360	51
			CD_3SOCD_3	25	a)	51
			$CDCl_3/CD_3CN = 1$	25	1700	52
			CD_3CN	31	200	52
			$CDCl_3/CD_3CN = 1$	31	1110	52
	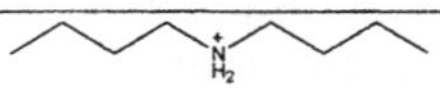		CD_3CN	25	50	51
	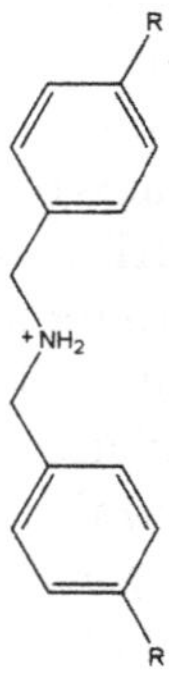	R= MeO	CD_3CN	31	130	52
			$CDCl_3/CD_3CN = 1$	31	990	52
		Me	CD_3CN	31	170	52
			$CDCl_3/CD_3CN = 1$	31	960	52
			$CDCl_3/CD_3CN = 3$	27	1900	91
		Cl	CD_3CN	31	470	52
			$CDCl_3/CD_3CN = 1$	31	1890	52
		Br	CD_3CN	31	460	52
			$CDCl_3/CD_3CN = 1$	31	2010	52
		CO_2H	CD_3CN	31	510	52
			$CDCl_3/CD_3CN = 1$	31	2520	52

[a] No binding observed.
[b] Not determined due to insolubility of the salt.
[c] Not determined due to intermediate exchange rates of the guest hampering the evaluation of NMR experiments.

Table 1 (continued)

Macrocycle	Ammonium ion		Solvent	T (°C)	K_a M^{-1}	Ref
		NO_2	CD_3CN	31	1300	52
			$CDCl_3/CD_3CN = 1$	31	5940	52
		CHO	CD_3CN	27	1600	102
		COOMe	CD_3CN	27	110	90
		CH_2Br	CD_3CN		352	93
			$CDCl_3/CD_3CN = 1$		1778	93
		CH_2N_3	$CDCl_3$	25	4000	96
		i-Pr	$CDCl_3/CD_3CN = 3/1$	20	2870	85
		t-Bu	$CDCl_3/CD_3CN = 3/1$	40	2470	85
			$CDCl_3/CD_3CN = 3/1$	<50	a)	85
	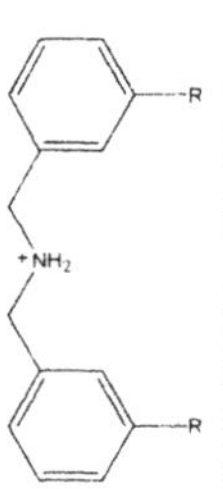	R= Me	CD_3CN	31	180	52
			$CDCl_3/CD_3CN = 1$	31	1100	52
		Cl	CD_3CN	31	580	52
			$CDCl_3/CD_3CN = 1$	31	2810	52
		Br	CD_3CN	31	570	52
			$CDCl_3/CD_3CN = 1$	31	2760	52
		CO_2H	CD_3CN	31	660	52
			$CDCl_3/CD_3CN = 1$	31	2850	52
		NO_2	CD_3CN	31	1350	52
			$CDCl_3/CD_3CN = 1$	31	6460	52
	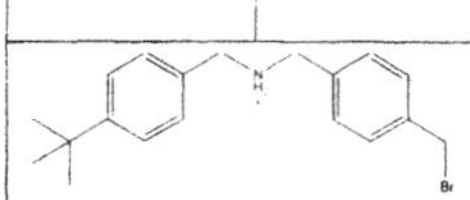		CD_3CN		400	93

Table 1 (continued)

Macrocycle	Ammonium ion		Solvent	T (°C)	K_a M^{-1}	Ref
	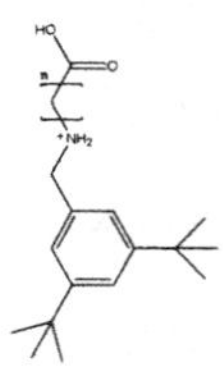	n = 1	CD_3CN	25	104	87
		n = 2	CD_3CN	25	70	87
		n = 3	CD_3CN	25	220	87
		n = 4	CD_3CN	25	250	87
	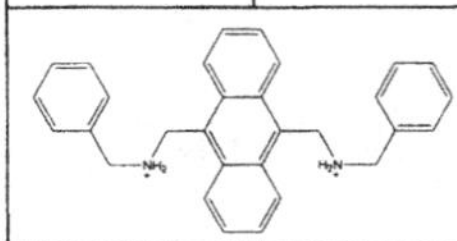		CD_3CN		2800	94
			$CDCl_3/CD_3NO_2$ = 1		6200	94
	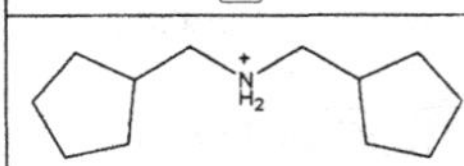		$CDCl_3/CD_3CN$ = 3/1	20	290	85
			$CDCl_3/CD_3CN$ = 3/1	40	110	85
	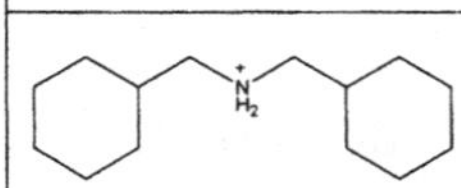		$CDCl_3/CD_3CN$ = 3/1	40	110	85
	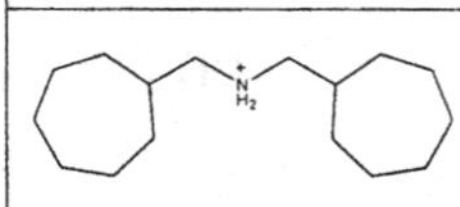		$CDCl_3/CD_3CN$ = 3/1	<50	a)	85
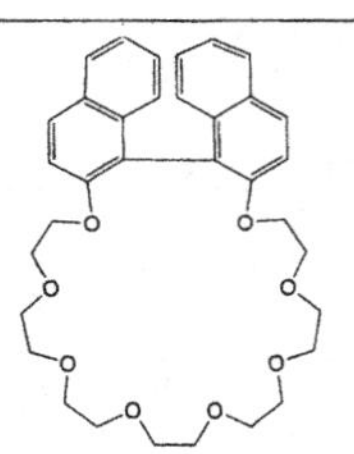	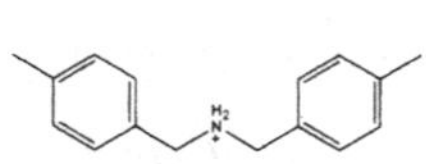		$CDCl_3/CD_3CN$ = 3	27	960	91
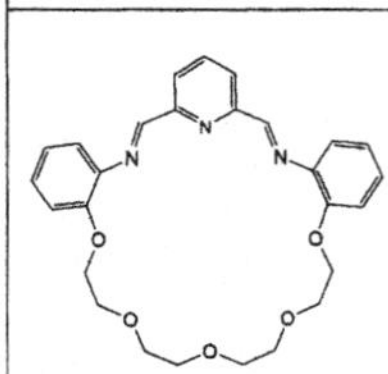	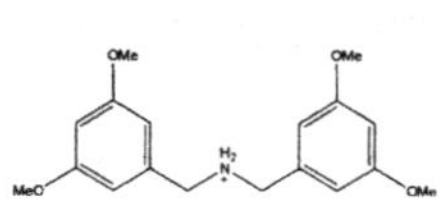		CD_3CN	25	300	103

Table 1 (continued)

Macrocycle	Ammonium ion	Solvent	T (°C)	K_a M^{-1}	Ref
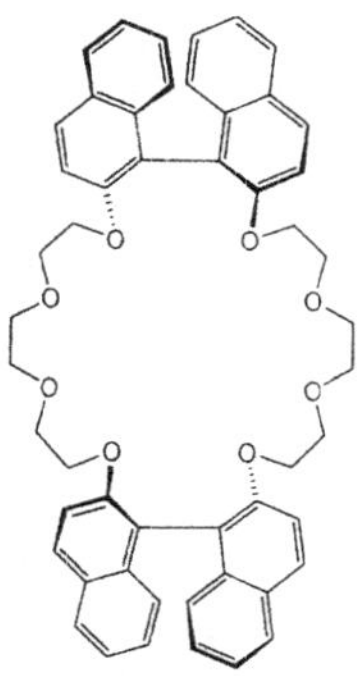	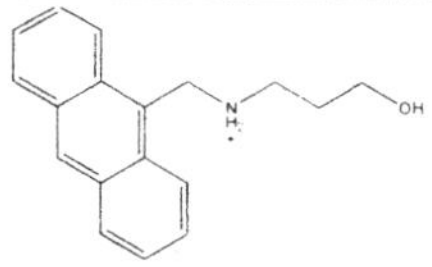	$CDCl_3$	22	730	12
	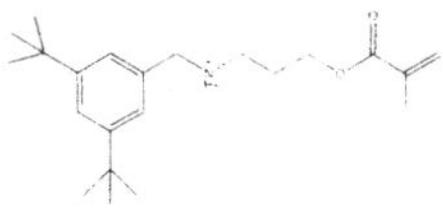	$CDCl_3$	10	3.3	12
		CD_2Cl_2	10	8.6	12
		CD_2Cl_2	-60	22.3	12
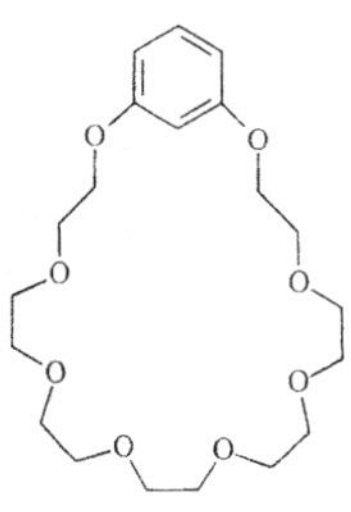	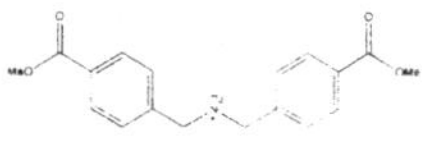	CD_3CN	27	50	90
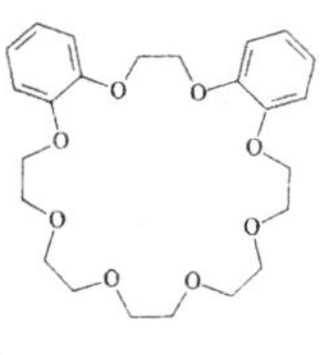	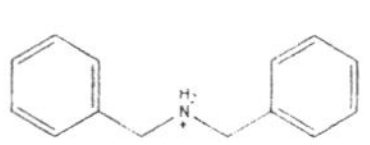	CD_3CN	25	360	54
		$CDCl_3$	25	22000	54
		CD_3COCD_3	25	310	54

Table 1 (continued)

Macrocycle	Ammonium ion		Solvent	T (°C)	K_a M^{-1}	Ref
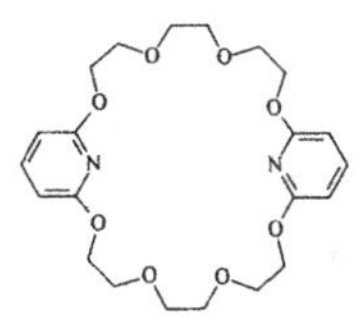	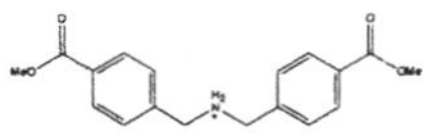		CD_3CN	25	1700	95
	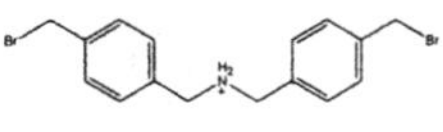		CD_3CN	25	1100	95
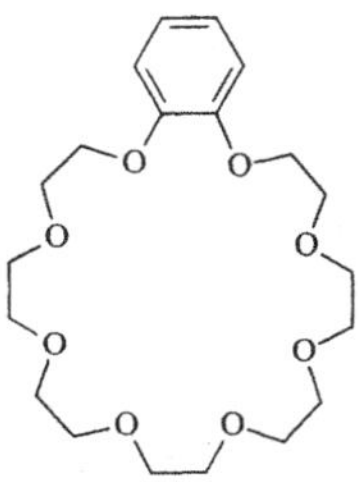		R= Me	CD_3COCD_3	25	b)	61
		n-Pr	CD_3COCD_3	25	149	61
		i-Pr	CD_3COCD_3	25	17.1	61
		i-Bu	CD_3COCD_3	25	43.5	61
		t-Bu	CD_3COCD_3	25	a)	61
		n-Bu	CD_3COCD_3	25	127	61
		Pen	CD_3COCD_3	25	c)	61
		Hex	CD_3COCD_3	25	177	61
		Cyhex	CD_3COCD_3	25	16.9	61
		Hep	CD_3COCD_3	25	161	61
		Oc	CD_3COCD_3	25	151	61
		Non	CD_3COCD_3	25	120	61
		Ocdec	CD_3COCD_3	25	178	61
		Benz	CD_3COCD_3	25	94	61
		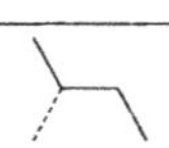	CD_3COCD_3	25	12.9	61
		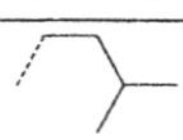	CD_3COCD_3	25	163	61
		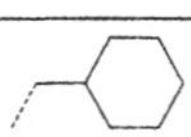	CD_3COCD_3	25	a)	61
		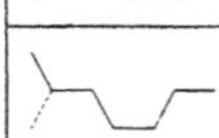	CD_3COCD_3	25	27	61
		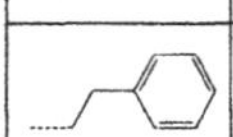	CD_3COCD_3	25	194	61
		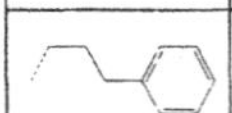	CD_3COCD_3	25	281.5	61

Table 1 (continued)

Macrocycle	Ammonium ion	Solvent	T (°C)	K_a M^{-1}	Ref
		CD_3COCD_3	25	92	61
		CD_3COCD_3	25	82	61
		CD_3COCD_3	25	c)	61
		CD_3COCD_3	25	99.8	61
	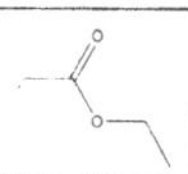	CD_3COCD_3	25	c)	61
		CD_3COCD_3	25	280	61
		CD_3COCD_3	25	223	61
		CD_3COCD_3	25	221	61

facile interactions of the phenyl rings of the ammonium ions with the aromatic moieties of the crown ether host.

In marked contrast to dibenzo-[24]crown-8, its derivative tetrabenzo-[24]crown-8 does not form pseudorotaxanes in solution [62]. In the crystal structure, however, the hexafluorophosphate salt forms pseudorotaxanes in an arrangement which is highly ordered through C-H···F hydrogen bonds. These results as well as the formation of a non-threaded complex from a bis-ammonium salt and bis(*m*-phenylene)-[32]crown-10 in solution and in the crystal point to the pivotal role of small structural changes. They indicate a careful evaluation of the geometry of the complexes to be necessary before postulating pseudorotaxane formation [63].

2.2
Multivalency and Oligo-/Polypseudorotaxanes

The ammonium/crown binding motif can also be employed to generate larger species with a higher number of subunits. If, for example, bis-*p*-phenylene-[34]-crown-10 is mixed with dibenzyl ammonium ions, two guests can be threaded through the somewhat larger cavity of the crown ether [64]. When mixed with

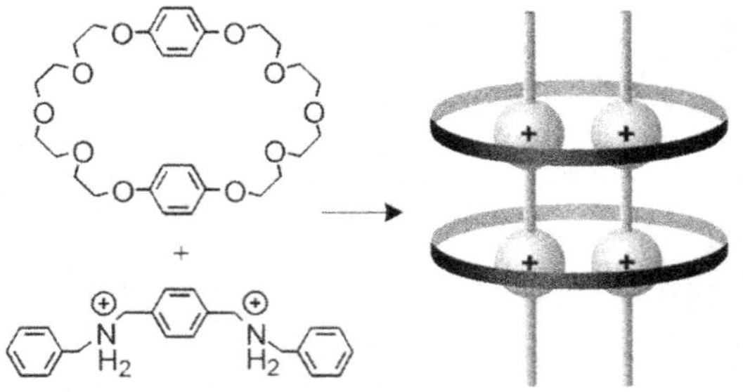

Fig. 5 Doubly docked pseudorotaxanes as formed from the precursors *on the left*

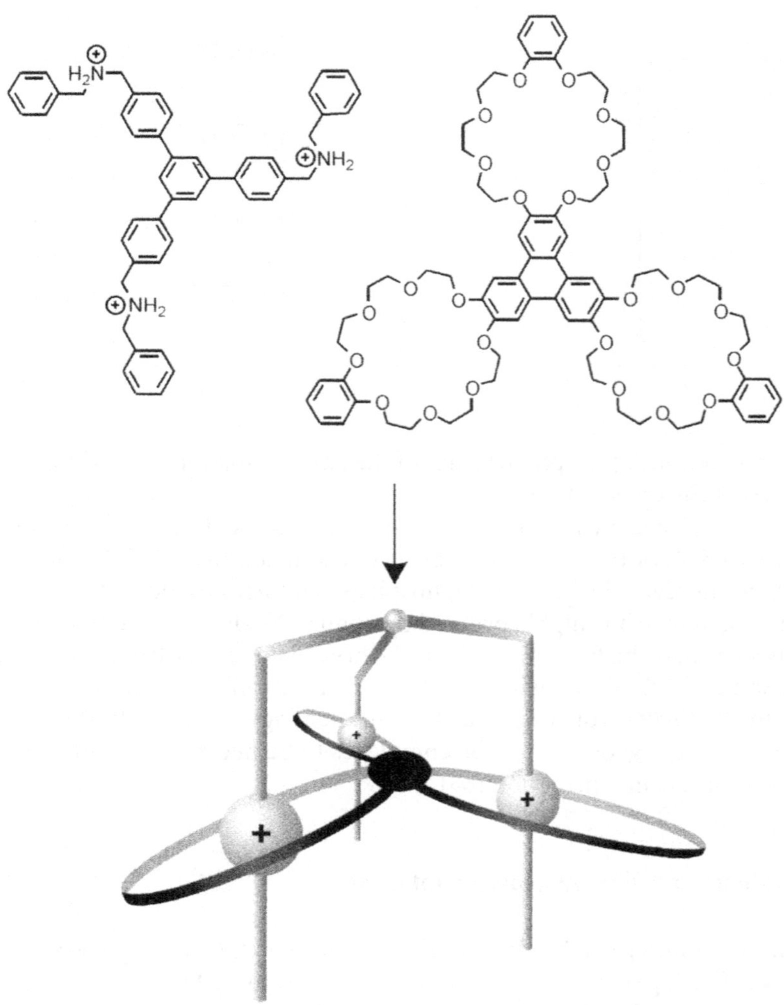

Fig. 6 A triply threaded superbundle

a bidentate ammonium ion, 2:2 complexes are formed and doubly docked pseudorotaxanes can be synthesized (Fig. 5) [65]. This approach can be extended to even larger species with larger crowns and longer tri- and tetradentate ammonium guests [66, 67] and evidence for their formation comes from X-ray crystal structure analysis, NMR experiments, and mass spectrometry. Multivalent assemblies have been reported [68] recently, in which a triply "crowned" triphenylene interacts through three ammonium crown interactions with a trifurcated trication (Fig. 6). Evidence for the formation of 1:1 complexes comes from X-ray crystallography, NMR experiments, and mass spectrometry. The photochemical and electrochemical features of the complex differ from those of the components. For example, fluorescence quenching is observed upon complex formation. Furthermore, threading/dethreading can be controlled by addition of acids and bases, respectively.

Supramolecular, i.e., non-covalently bound oligo- and polymers of pseudorotaxanes utilizing the ammonium/crown binding motif have been reported [69–71]. One of the problems associated with this is illustrated in Fig. 7. A ditopic monomer can in principle form cyclic and open-chain oligomeric complexes of different sizes [72, 73]. However, as long as geometry and flexibility permit, entropy favors the formation of the smallest possible cyclic structure. Indeed, most of the studies dealing with this aspect report a cyclic dimer to be the major species in solution, as long as the concentration of the building blocks is not too high [74–76]. Quite interestingly, stilbene-containing "daisy chain" pseudorotaxanes as well as stilbene-containing doubly docked pseudorotaxanes related to those mentioned above (Fig. 5) can be subjected to photodimerization

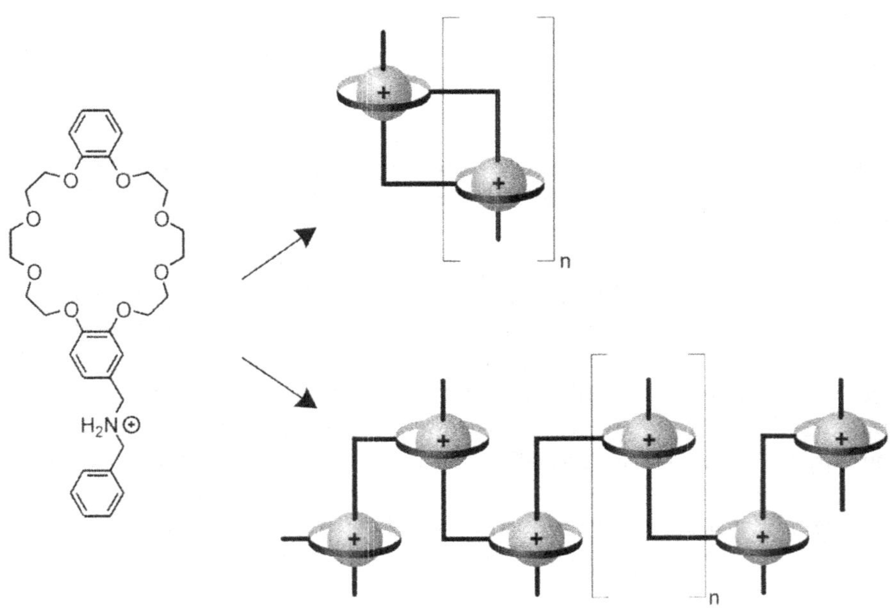

Fig. 7 Daisy chain pseudorotaxanes

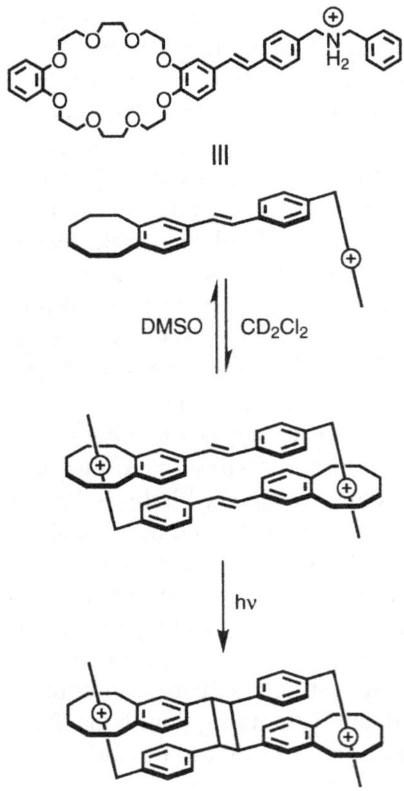

Fig. 8 The formation of daisy chain pseudorotaxanes bearing stilbene units in their axle components permits control over the stereochemistry of the photochemical dimerization in the solid state as well as in solution

experiments in the solid state and in solution (Fig. 8) which provide stereochemically pure products, although several different complexes can be formed [77, 78].

Another approach to polymeric pseudorotaxanes is the covalent synthesis of a polymer decorated with side-chain crown ethers, which can bind secondary ammonium ions along the polymer chain [79–81]. This concept has been realized utilizing polymethacrylates. The host-guest properties were studied using Scatchard and Hill plots. Not unexpectedly, the binding of the ammonium ion is anti-cooperative with a Hill coefficient of n_H=0.45. Consequently, the binding of additional ammonium ions occurs with lower and lower binding constants when more and more guests are already present in the polymer. Finally, it should be briefly mentioned that dendritic structures have been self-assembled employing the ammonium/crown recognition motif [82–84].

2.3 Rotaxane and Catenane Syntheses Involving Ammonium/Crown Ether Complexes

After identifying the ammonium/crown binding motif as a useful precursor for the synthesis of interlocked molecules, several approaches were developed for the attachment of stopper groups to yield stable rotaxanes which would not deslip, even when the ammonium group is deprotonated. One prerequisite is that the reactions used for stopper attachment to the axle of the pseudorotaxane is compatible with the binding of secondary ammonium ions in crown ethers. Consequently, in particular basic conditions must be avoided which deprotonate the ammonium ion and thus interfere with the hydrogen bonds formed between the two components of the pseudorotaxane. Another obvious prerequisite is an appropriate stopper size. This aspect was studied by Stoddart et al. who used ammonium ions with end groups of different sizes and studied the kinetics of the slipping reaction which leads to (pseudo)rotaxane formation [85, 86].

Several approaches to stopper attachment are summarized in Fig. 9, which are all based on the threading-followed-by-stoppering strategy. Amide bond formation was utilized (Fig. 9a,b) [87], one of which was carried out at the chloroform/water interface. The rotaxane yield could be increased significantly in this two-phase system, likely due to favorable arrangements of the components at the interface [88]. With isocyanates (Fig. 9c), ureas are readily formed connecting the second stopper to the axle while threaded through the crown ether [89]. Different crown ethers have been used instead of dibenzo-[24]crown-8, for example benzo-*m*-phenylene-[25]crown-8 [90] and a binaphthyl-incorporating chiral derivative of [24]crown-8 [91, 92]. A very useful approach is the reaction of pseudorotaxane bearing benzyl bromides at the axle ends with triphenyl phosphine. In a simple S_N1 reaction, bulky stoppers are thus attached (Fig. 9d) [93–95]. The triphenyl phosphonium stoppers can then be further exchanged with other stopper aldehydes in Wittig reactions opening a wide field of different accessible axle structures with a large variety of properties [96]. This reaction has also been used to generate the molecular necklaces [97] and the daisy chain rotaxanes shown in Fig. 10 [98]. Also, [1,3]-dipolar cycloadditions employing azide groups at the axle center piece and acetylene dicarboxylic esters as the dipolarophiles can be used for stopper attachment (Fig. 9e). With this strategy, not only [2]- but also [3]rotaxanes have been synthesized [99]. As a last example for the threading-followed-by-stoppering strategy, the coordination of a pyridine terminated pseudorotaxane to a metalloporphyrin should be mentioned here (Fig. 11) [100, 101].

Most covalent syntheses of organic compounds are kinetically controlled including the stopper attachments discussed above. In contrast, truly reversible reactions would provide access to rotaxanes under thermodynamic control, which is usually realized, when weak non-covalent interactions mediate what we could call supramolecular synthesis. Such a rotaxane synthesis would ben-

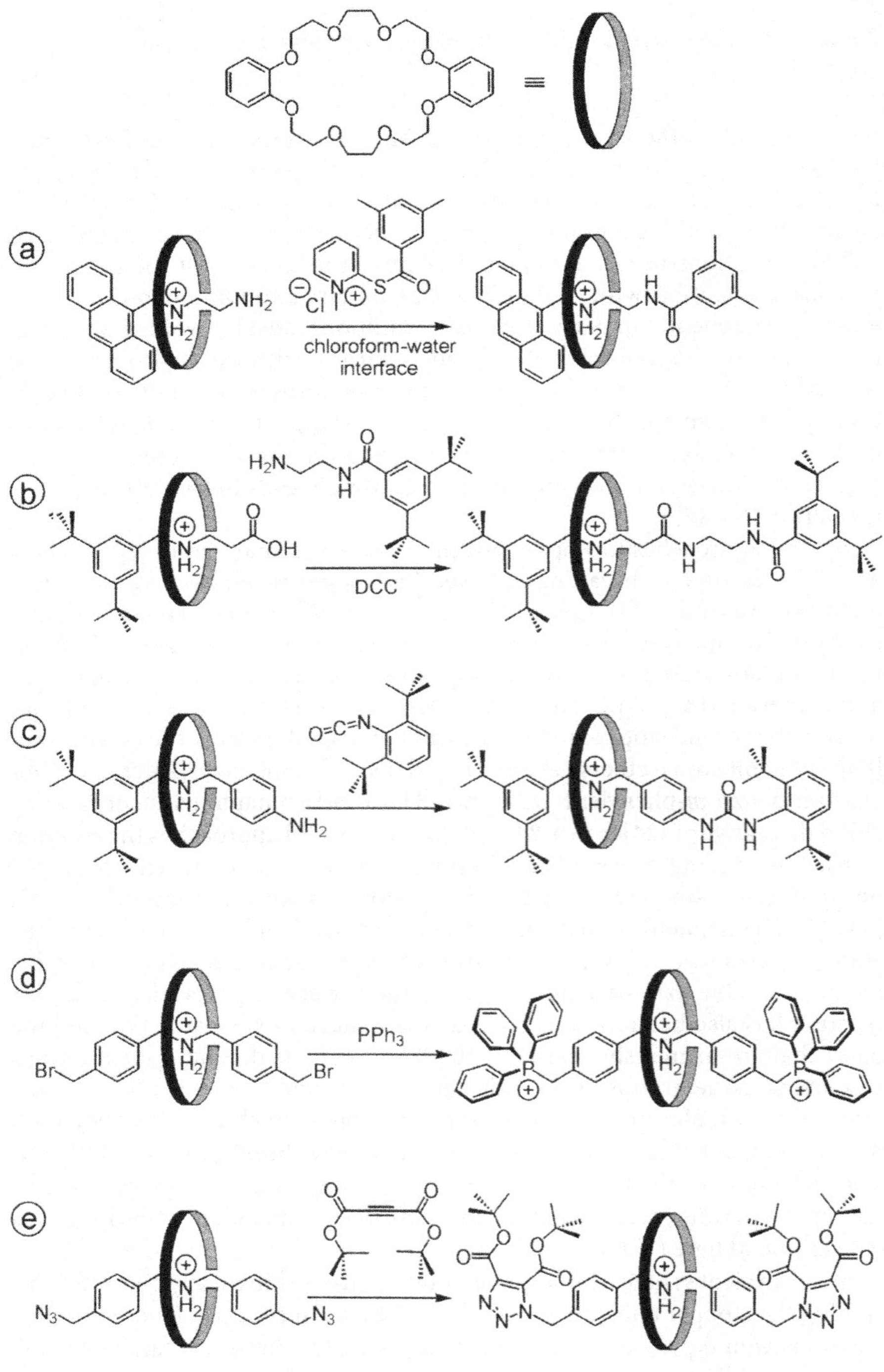

Fig. 9 Rotaxane syntheses employing the ammonium/crown recognition motif. These syntheses follow the threading-followed-by-stoppering strategy

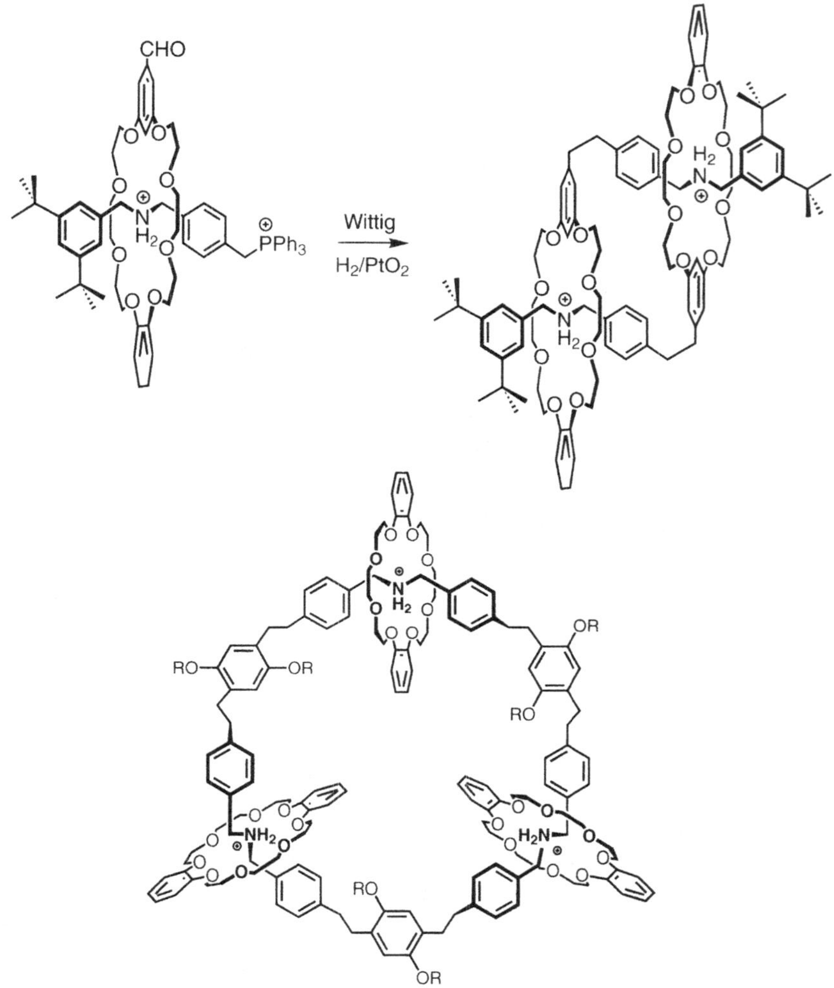

Fig. 10 Synthesis of hermaphrodite daisy chain rotaxanes (*top*) and of molecular necklaces (*bottom*) utilizing Wittig reactions for the exchange of the triphenyl phosphonium stoppers

efit from all the advantages of self-assembly, in particular from multiple error checking and error correction steps. There are several reversible reactions which can be and have been used for rotaxane syntheses under thermodynamic control. (i) Imine formation from aldehydes and amines was employed in two different ways. On one hand, stoppers with amino groups were attached to aldehyde-terminated axle center pieces [102]. On the other hand, the crown ether was modified so that an amino-terminated open-chain precursor was reacted with a dialdehyde. In this synthesis, the axle is already equipped with two stoppers and the wheel is clipped around the axle [103, 104]. In both cases, reduction of the C=N double bonds provides stable rotaxanes. (ii) Ring closure

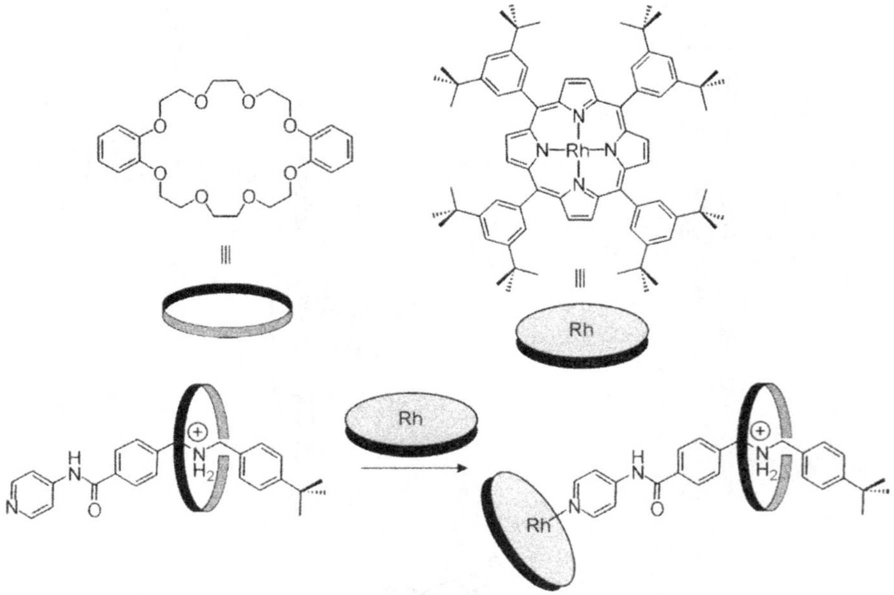

Fig. 11 A metalloporphyrin used as the stopper

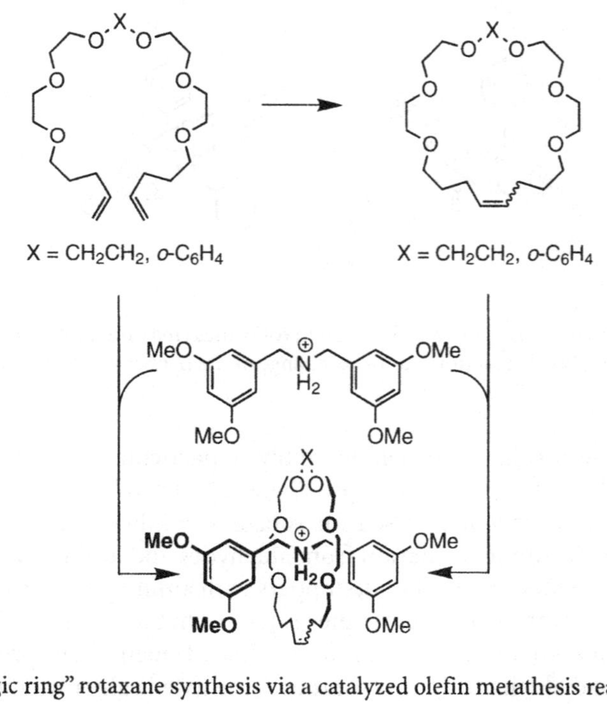

Fig. 12 "Magic ring" rotaxane synthesis via a catalyzed olefin metathesis reaction

via olefin metathesis (Fig. 12) is another alternative which starts from an open-chain crown derivative bearing two vinyl groups at its ends [105]. Complex formation with a secondary ammonium ion followed by ring closure provides the desired rotaxane which can be hydrogenated to yield a stable rotaxane. In what was coined a "magic ring" experiment, Grubbs et al. were able to demonstrate the reaction to be reversible. Starting with a closed macrocycle containing a double bond, the incorporation of the axle occurred with almost quantitative yields when a metathesis catalyst was added. A similar approach to [3]catenanes was recently reported [106]. (iii) The formation of disulfide bridges can be used to connect two pseudorotaxanes whose axle bears one stopper and is terminated at the opposite end by a thio group. In the presence of thiols, this reaction is reversible and the product distribution can be controlled by temperature, solvents, and concentrations. The synthesis of [2]- as well as [3]rotaxanes is possible [107, 108].

2.4
Controlling Mechanical Motion Through Protonation/Deprotonation

Coming back to the idea that the template leaves behind functional groups in the product as a trace of its existence, all rotaxanes and catenanes discussed so far have in common that they bear a secondary ammonium group and a crown ether within their structure. Since they cannot be split into their subunits due to catenation or due to the presence of sufficiently bulky stoppers, they can be expected to retain the intertwined molecular topology upon deprotonation. Consequently, deprotonation should change the properties of the rotaxanes and catenanes significantly making molecular switches possible.

However, several reports describe difficulties in deprotonating the ammonium group, when inserted into the crown ether, irrespective of the base (potassium carbonate, triethylamine, DBU) [109]. The unusually low acidity is also reflected in the much slower H/D exchange observed for the rotaxanes in comparison to the free secondary ammonium salts. When a second functional group is present in the axle which can bind and stabilize the wheel in the deprotonated state of the rotaxane, deprotonation becomes possible [110]. Such a rotaxane is shown in Fig. 13. In its protonated form the hydrogen bonds connecting the secondary ammonium station in the axle with the crown ether are stronger and outweigh the interactions with the paraquat station at the opposite end of the axle. When a base is added, the wheel is bound more tightly at the paraquat station, where π-donor π-acceptor interactions between the electron-poor paraquat moiety and the electron-rich aromatic rings of the wheel are efficiently competing with the amine station. The rotaxane in Fig. 13 can be regarded as a molecular switch that responds to an external stimulus with a shuttling motion of the wheel along the axle. The switching is fully reversible and the displacement of the wheel is quantitative.

Instead of a motion along the axle of a rotaxane, the almost quantitative, fully reversible de- and rethreading of pseudorotaxanes depending on the pro-

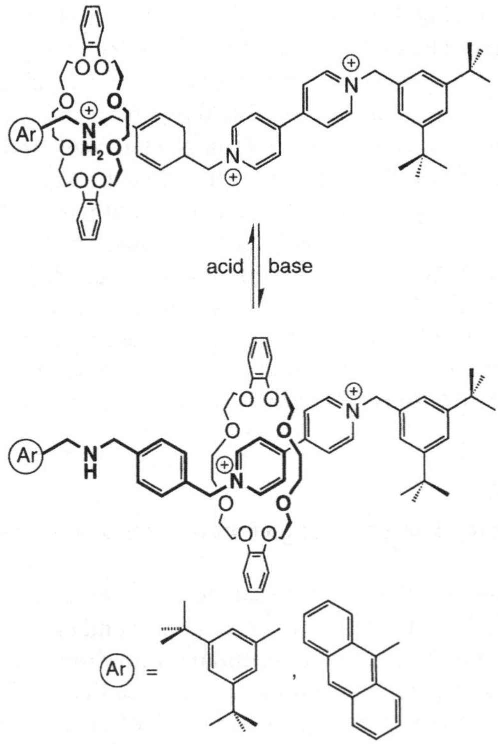

Fig. 13 An acid/base controllable molecular shuttle. In its protonated form, the wheel is located at the secondary ammonium station, while it moves to the paraquat end of the axle when deprotonated

tonation state can be utilized for molecular switches. This has been carried out with the components shown in Fig. 14 [111], which bear photoactive groups such as naphthylenes and anthracenes. While both the crown ethers and the anthracene-bearing axles are fluorescent, the fluorescence of the two components is almost completely quenched when the pseudorotaxane is formed. Notable exceptions are the complexes of dinaphtho-[30]crown-10, in which strong interactions between the two naphthyl groups of the macrocycle and the anthracene moiety of the axle are feasible. This leads to the disappearance of the fluorescence bands of the two components and the growth of a new, broad fluorescence band for the complex. Besides the rather simple dethreading/rethreading of these pseudorotaxanes, much more complex systems have been synthesized and examined, which employ multiple different recognition motifs. Depending on the protonation state of the axle, the number of wheels threaded onto the axle can then be controlled [112].

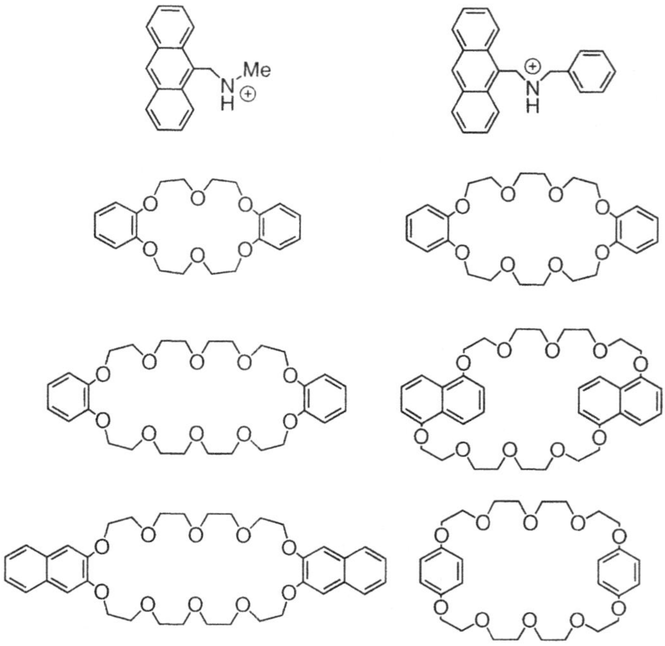

Fig. 14 Photoactive components utilized for pseudorotaxane formation. The generation of complexes changes the fluorescence properties of the complexes. While the fluorescence is usually quenched, a new fluorescence band appears in the emission spectra of complexes with dinaphtho-[30]crown-10 due to strong interactions of the aromatic units of axle and wheel

3 Neutral Templates: The Amide Template Effect

3.1 Tetralactam Macrocycles as Hosts for Carbonyl Compounds

Neutral, hydrogen-bonded templates for the synthesis of interlocked molecules are based on the ability of amides to form hydrogen bonds to each other [113–117]. Figure 15 depicts a tetralactam macrocycle that bears four convergent hydrogen-bond donors capable of binding carbonyl compounds inside the macrocycle cavity. Originally designed as a receptor for *p*-benzoquinone ($K(CDCl_3)$=1.200 M^{-1}) [118], other guest such as cyclic dipeptides [119] have been used as guests and their binding constants were determined. Glycine anhydride showed a remarkably strong binding with a binding constant of $K(CDCl_3)=10^6$ M^{-1}. Even binding of these dipeptides in water was possible.

Furthermore, the substituents at the diamide moieties of the wheel have a significant influence on the binding energies [120]. From the binding constants listed in Fig. 16, it becomes clear that electron-donating substituents decrease

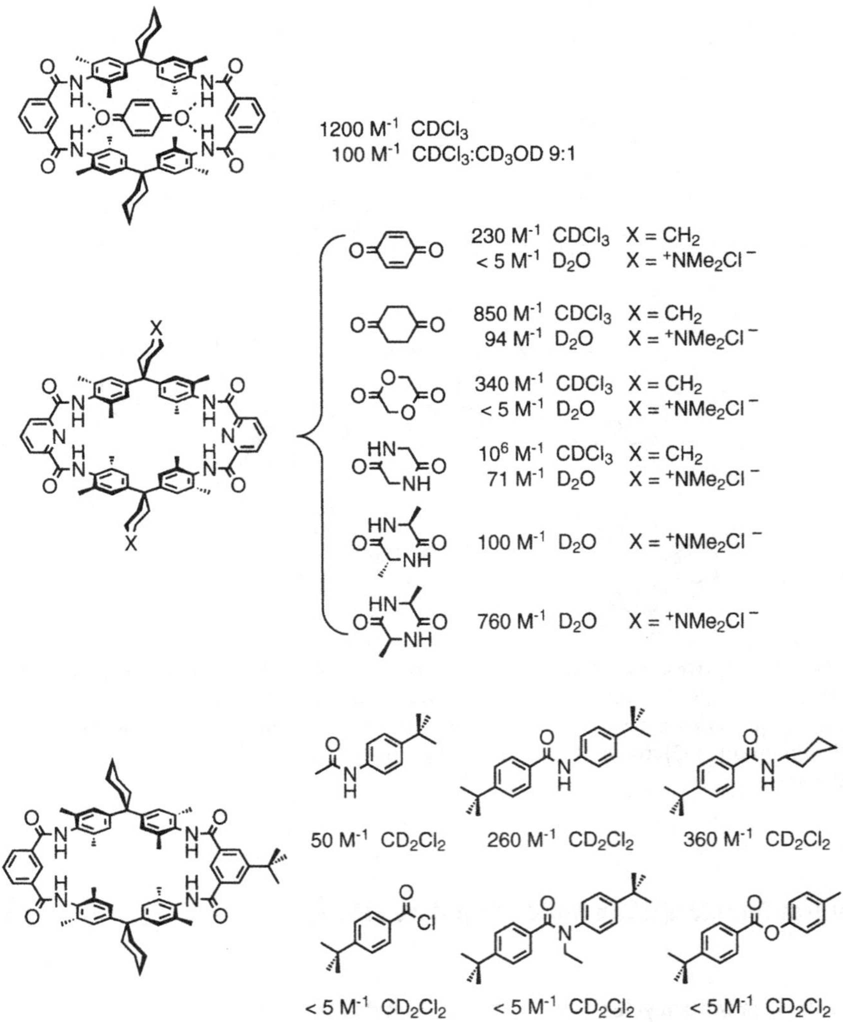

Fig. 15 *Hunter's* tetralactam macrocycle binding *p*-benzoquinone inside its cavity through four convergent amide NH hydrogen-bond donors together with a selection of binding constants of several derivatives to a series of guests. In addition to hydrogen bonding, edge-to-face interactions with the four aromatic rings may contribute to the binding strength

the binding constant, while electron-withdrawing groups increase binding significantly.

However, not only dipeptides and diamides bind to this macrocycle and it has indeed been utilized as sensor-active layer on quartz microbalance sensors for food control [121, 122], where simple carbonyl compounds such as acrylnitril, hexanone, or hexenal are lead compounds. Simple secondary amides can also act as guests with binding constants around K=200–300 M^{-1} in non-com-

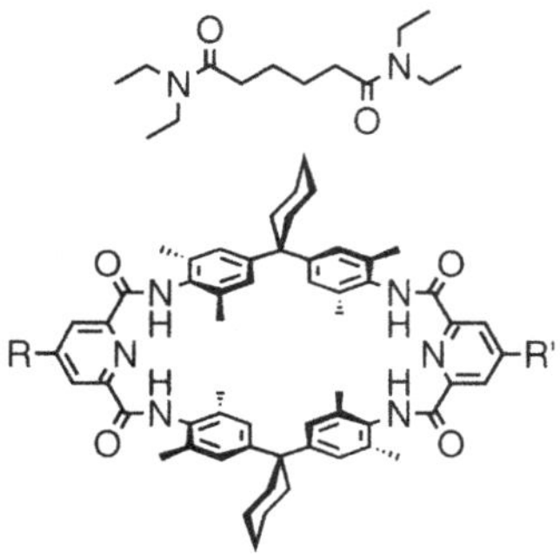

R = R' = H	780 M^{-}
R = H, R' = NMe_2	240 M^{-1}
R = H, R' = OMe	550 M^{-1}
R = H, R' = Cl	1360 M^{-1}
R = H, R' = NO_2	4730 M^{-1}
R = R' = NMe_2	70 M^{-1}
R = R' = OMe	410 M^{-1}
R = R' = Cl	2430 M^{-1}
R = R' = NO_2	24200 M^{-1}

Fig. 16 Binding constants in $CDCl_3$ of hexanoic diamide to differently substituted tetralactam macrocycles

petitive solvents such as dichloromethane [123]. In contrast, other monocarbonyl compounds such as esters, tertiary amides, or acid chlorides do not bind strongly to the macrocycle. This indicates that the hydrogen-bonding pattern for the binding of secondary amides involves more than a twofold hydrogen bridge [124] to the carbonyl oxygen which is also feasible with the other derivatives. Indeed, X-ray crystal structures of rotaxanes and catenanes [125] as well as theoretical calculations at the density functional level [126] confirm (i) that the macrocycle is flexible enough to invert one of the amide groups so that one amide carbonyl group is pointing into the cavity and (ii) that a secondary amide forms a third hydrogen bond with the inverted carbonyl oxygen (Fig. 17, top). In addition to the twofold hydrogen bond, an additional hydrogen bond is thus formed between the NH group of the guest and the wheel. This significantly strengthens the binding of secondary amides in comparison to all derivatives which are not capable of forming this additional hydrogen bond. According to theory, the energy difference between a merely twofold hydrogen-bonded arrangement (Fig. 17 right) and that with a triply hydrogen-bound connection (Fig. 17 center) between amide guest and macrocycle amounts to ca. 25 kJ/mol.

The hydrogen-bonding pattern is somewhat different and more complex for catenanes (Fig. 17 bottom). Here, the two macrocycles bear a total of eight amide groups. The most stable conformation found in the calculations is capable of forming a total of six hydrogen bonds, because each wheel serves as the host for one of the amide groups of the other. Each of the patterns is analogous to that mediating the binding between axle and wheel in rotaxane. A second arrangement which bears a total of five hydrogen bonds is less stable by ca. 23 kJ/mol in agreement with the crystal structures of analogous catenanes, in which the more stable conformer was observed [127]. Interestingly, the hydrogen-bonding pattern observed in the rotaxanes is already formed completely before the final ring closure of the second wheel (Fig. 17, bottom right) so that these calculations provide in-depth insight into the details of the template effect mediating catenane synthesis.

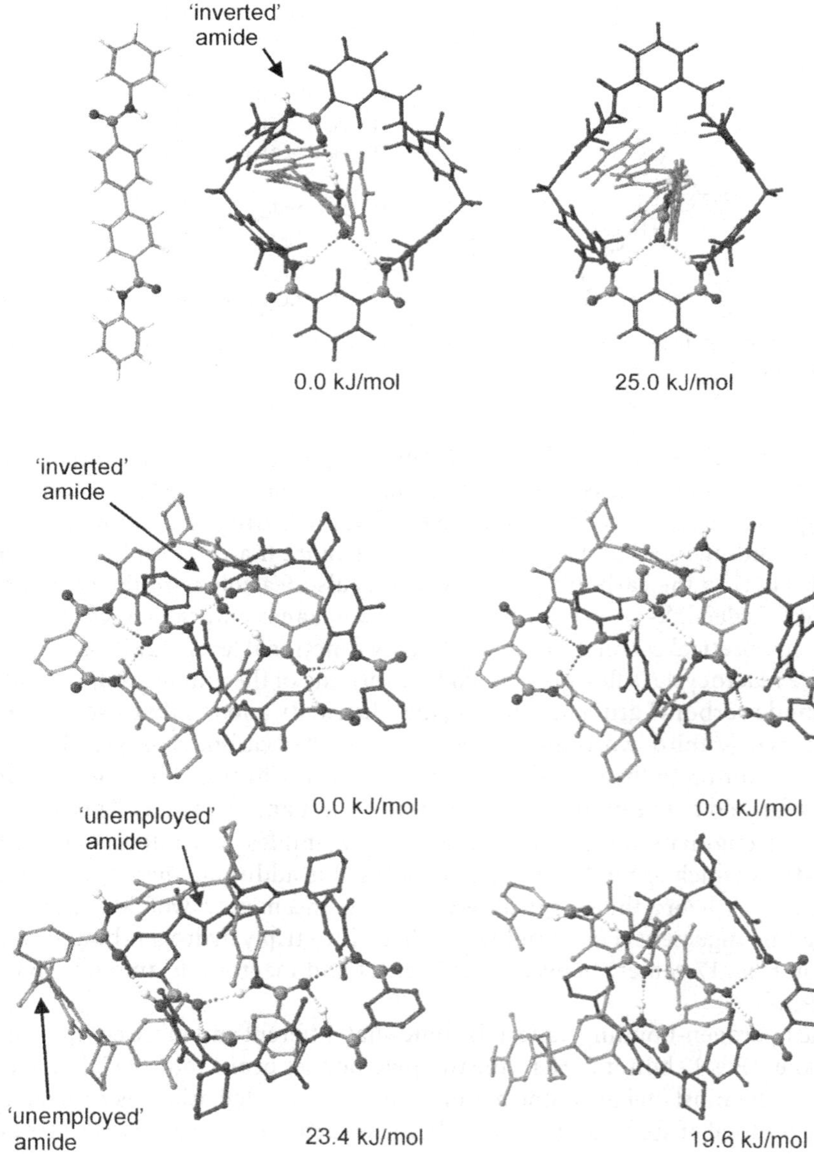

Fig. 17 *Top row*: structure of a fragment of a typical rotaxane axle as geometry-optimized at the B3LYP/DZP density functional level of theory (*left*). Binding of the axle fragment in the cavity through an arrangement which permits the formation of three hydrogen bonds (*center*). Analogous host-guest complex with only a two-fold hydrogen bond connecting the two components (*right*). The four structures *at the bottom* represent AM1-optimized conformers of a catenane (*left side*) and the corresponding precursor formed before the last ring-closure step. Energies give relative energies and can only be compared pairwise for directly related conformers. The amide groups involved in hydrogen bonding are shown in *ball-and-stick representation*. *Dotted lines* represent hydrogen bonds

3.2 Synthesis and Structural Variability of Amide Catenanes and Rotaxanes

Again, molecular recognition of a guest within a macrocycle provides a threaded arrangement which can be utilized for the construction of rotaxanes and catenanes. A large structural variety was synthesized during the last decade starting with the catenane which is formed as a side product with 34% yield during the synthesis of the unsubstituted macrocycle shown in Fig. 15 (top) [128]. NOESY NMR experiments confirmed the catenated structure of this compound and NMR experiments also provided insight into the hydrogen-bonding pattern which connects the two macrocycles in the catenane. These experiments also showed that one carbonyl group in each macrocycle is pointing into the cavity. Clear evidence for the catenated structure can be obtained from tandem mass spectrometric experiments [129, 130] in the positive [131] or negative [132] ionization mode. A comparison of tetralactam macrocycles, octalactam macrocycles, and catenanes consisting of two of the smaller macrocycles exhibits two completely different fragmentation pattern, when mass-isolated ions are subjected to collisional activation. All species likely undergo ring cleavage. Mono-macrocycles then loose water in the major, lowest-energy fragmentation channel. In marked contrast, the catenanes first expel one of the rings leaving the second, opened wheel as the major fragment which then consecutively fragments almost exactly like the corresponding tetralactam macrocycles.

Almost simultaneously with *Hunter's* report on the formation of a [2]catenane during the macrocyclization reaction, analogous catenanes with four methoxy substituents at the isophthalic acid subunits were prepared in one step [133]. These syntheses were followed by the step-wise preparation of catenanes mono-methoxy-functionalized in one or both rings [134]. These catenanes revealed a highly interesting type of isomerism. Due to restricted circumrotation which is hampered by the cyclohexyl side chains, *in-* and *out-*isomers can be distinguished (Fig. 18) in which the methoxy substituents either point into the cavity of the other wheel or are located at the outer periphery. For the catenanes carrying one substituent on each wheel, three isomers, i.e., the *in*/*in*-, *in*/*out*-, and *out*/*out*-arrangements have been prepared. Depending on the sequence

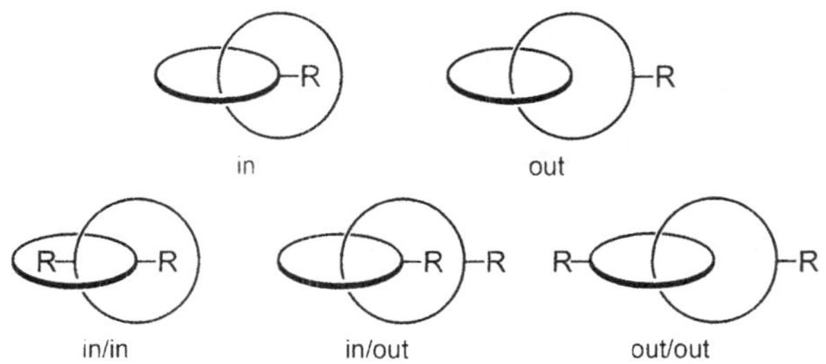

Fig. 18 Schematic representation of different isomeric catenanes which exist due to hampered circumrotation, when one or both of the wheels are mono-substituted

of substituted and unsubstituted isophthalic acid dichloride in the step-wise synthesis, specificity of formation of certain isomers is observed. Provided that the first macrocycle is present in the synthesis in its unsubstituted form, introduction of the substituted followed by the unsubstituted isophthalic acid dichloride in the formation of the second macrocycle gives rise to the *in*-isomer exclusively. Reversing the order provides access to the *out*-isomer. Finally, furane [135] and bipyrrole dicarboxylic acid dichloride [136] and flexible chain analogues [137] have been used replacing the isophthalic acid moiety in the catenane pointing to a broad applicability of this catenane synthesis which provides a large number of different structures.

At this point, a second type of amide catenane should be introduced (Fig. 19) [138]. It bears smaller wheels, but exhibits a similar hydrogen-bonding pattern involving again the combination of a twofold hydrogen bond binding one of the amide carbonyl groups of the other ring and an N-H···O hydrogen bond involving an inverted amide group [139]. Quite a large range of structures can again be accessed by substitution of the isophthalic acid dichloride moiety or by using different diamines (Fig. 19) [140]. Catenanes related to these have been prepared through olefin metathesis in a "magic ring" experiment [141]. In an elegant study, Takata et al. introduced a diene into one of the wheels, which was reacted with a dienophile in a *Diels-Alder* reaction [142]. After ozonolysis of the resulting double bond, the ring size was expanded by four atoms. By such an approach, the synthesis of catenanes becomes feasible which might otherwise be inaccessible. The *Diels-Alder* reaction has also been used to attach a C_{60} molecule to the wheel of a rotaxane in order to study the photoinduced intrarotaxane electron transfer between zinc porphyrin stoppers and the fullerene [143].

Both types of macrocycles have been employed not only for the synthesis of catenanes, but also for the preparation of rotaxanes. In 1995, we presented the first amide-based rotaxanes [144] that consisted of a tetralactam wheel and an axle with two tritylaniline stoppers attached to an isophthalic acid center piece. Again, a large variety of structures is accessible. The isophthalic acid axle center piece can be exchanged against the corresponding terephthalic acid, furan,

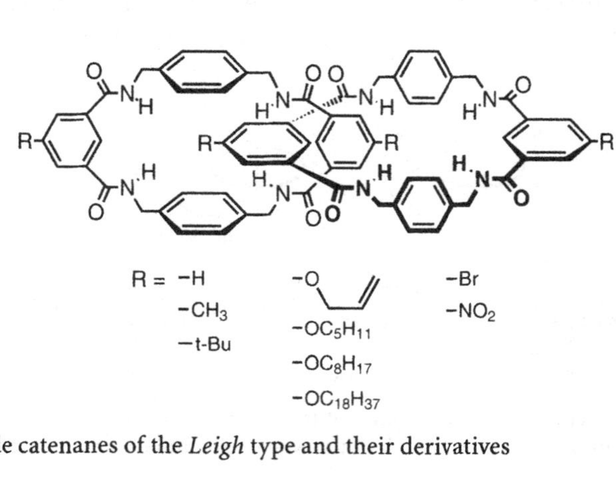

Fig. 19 Amide catenanes of the *Leigh* type and their derivatives

thiophene, and sulfonamide derivatives [145], and it is possible to use olefinic and aliphatic building blocks instead [146]. Also, the stoppers have been exchanged against porphyrins [147] and calixarenes [148]. While all rotaxanes mentioned so far have been synthesized from acid chloride center pieces and amine stopper groups, it is also possible to use isocyanates together with trityl aniline or trityl phenol stoppers providing access to urea and urethane groups in the axle [149, 150].

In order to get some idea of the stability of rotaxanes with different stoppers and macrocycles, the size complementarity of stopper and wheel cavity was examined through deslipping experiments [151]. This study showed that stoppers larger than trityl aniline are required, if the size of the macrocycle is only slightly increased. Also, the kinetics differed significantly depending on the solvent indicating that hydrogen bonding is an important factor for the stability of the rotaxane. Consequently, the choice of solvent may alter the properties of an amide rotaxane to quite an extent. Another interesting feature is the protection of the axle center piece through the wheel. When stilbenes are incorporated in the axle of a rotaxane, the catalytic hydrogenation is much slower than that of the free axle under the same experimental conditions [152].

The use of the smaller *Leigh*-type macrocycle for the generation was initially hampered by the fact that only the corresponding [2]catenane could be isolated, while the macrocycle, if formed was not separable from the other side products. Furthermore, solubility of amide-rich compounds is very limited due to the high enthalpy liberated upon the formation of intermolecular hydrogen bonds. Consequently, a synthetic detour using a stoppered axle as guest inside the cavity of the nascent macrocycle and forming a rotaxane whose axle is subsequently cleaved would help to circumvent these problems by tying up all hydrogen bonding sites intramolecularly. This approach was used by Leigh et al. demonstrating that on one hand rotaxane synthesis is possible with the smaller macrocycles and that on the other hand the macrocycle itself becomes accessible by this strategy [153]. In a follow-up study, axles bearing a glycylglycyl moiety in their center were shown to form through an analogous clipping synthesis [154]. Similarly, tertiary amide groups do not obstruct the synthesis of these so-called peptide rotaxanes [155]. The effects of size and shape have been examined [156, 157]. One of the most remarkable results is the finding that even ester groups which form weaker hydrogen bonds than amides can compete with a ca. 10000-fold stronger anion binding, when multiple interactions between axle and macrocycle precursor are formed in an ideal arrangement. Besides the threading strategy employed by Vögtle and the clipping approach used by Leigh, a magic rod rotaxane synthesis was described (Fig. 20) [158]. In this experiment, a complete axle fully equipped with stoppers and incorporating a double bond in its center piece is mixed with an already closed wheel. Due to the stopper size, slipping is not possible and no rotaxane is formed. Upon addition of a metathesis catalyst, however, the axle is temporarily cleaved in reversible olefin metathesis reactions and threading of the macrocycle on to the axle becomes possible. The threading is followed by reconstruction of the axle

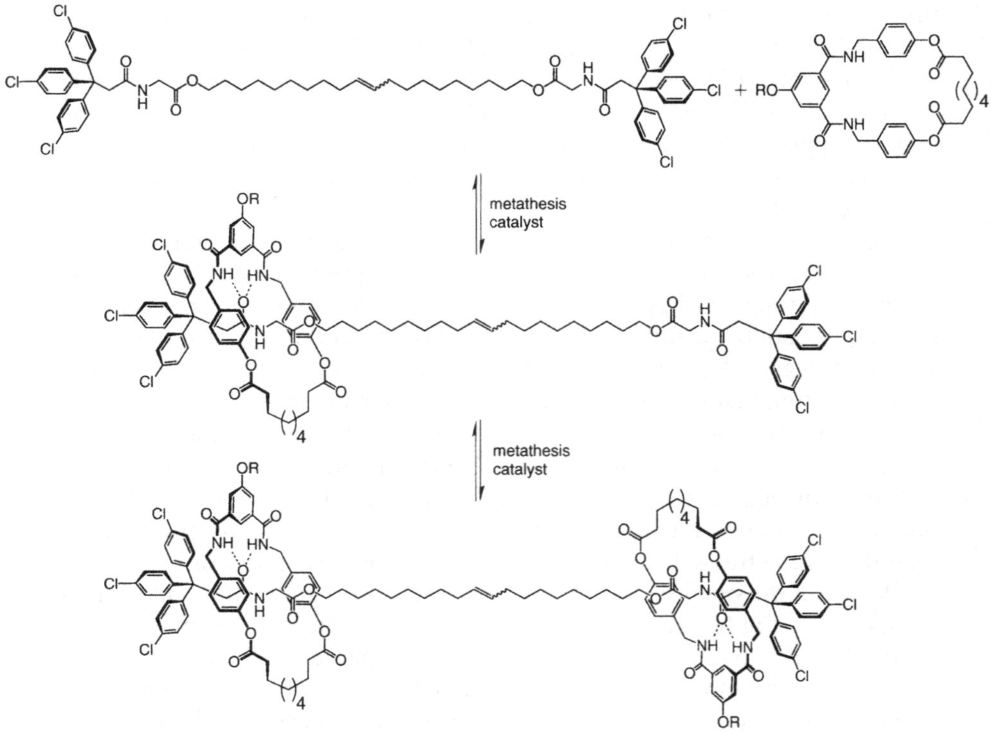

Fig. 20 The "magic rod" experiment for preparation of rotaxanes from intact axles and wheels through a reversible olefin metathesis reaction

in another metathesis step. Finally, even post-assembly chain extension to yield polyrotaxanes with different topologies is possible when rotaxane monomers with blocked isocyanates are subjected to a polymerization reaction [159].

Rotaxanes of the amide type have been self-assembled from macrocycles which bear the isophthaloyl diamide binding motif for axle recognition and metal complexes which provide the means for ring closure around the axle. Hunter et al. have used an isophthaloyl diamide connected to one metalloporphyrin and a pyridine moiety [160]. This molecule can dimerize through coordination of the pyridine to the metal center in the porphyrin and in the presence of an appropriate axle yields a rotaxane [161]. In an alternative strategy, Jeong et al. employ an open-chain analogue of the tetralactam macrocycle terminated at both ends with pyridine rings. After axle binding, the wheel can be closed through coordination of the pyridines to an osmium complex [162–165].

The structural variety of catenanes and rotaxanes of the amide type can be further extended by the synthesis of [n]catenanes (up to n=4) [166, 167] or [n]rotaxanes (up to n=5) [168, 169] that bear several catenated macrocycles or more than one macrocycle threaded onto the same axle, respectively. Also, introduction of a covalent bridge between the two components leads to new topologies, i.e., pretzelanes [170, 171] and [1]rotaxanes [172]. This was accomplished through replacement of one or more of the amide groups incorporated in the axle and wheel by sulfonamide groups [173], which can be deprotonated more readily and thus permit selective alkylation. Alternatively, substitution of the catenane macrocycles with OH groups at the isophthalic acid derivatives permits the reaction with an alkyl diiodide to yield the pretzelane. Both strategies have been employed not only to introduce an intramolecular bridge connecting the two components of the same molecule. Larger assemblies have been synthesized by connecting several rotaxanes or catenanes with each other [174–176]. Their topologies are schematically shown in Fig. 21. In conclusion, complex topologies have become accessible beyond a simple introduction of functional groups at the periphery of [2]catenanes and [2]rotaxanes.

3.3 Topological Chirality

Topology is a mathematical discipline dealing with certain well-defined geometrical transformations. While Euclidean geometry answers questions such as "How large is an object?", or "What is its three-dimensional shape?" the actual size or the details of the shape of an object do not matter in topology. All transformations which can be performed continuously *and* which can be reversed continuously are topologically allowed. These considerations can be applied to molecules. First a molecular graph needs to be defined – which can most easily be done by following the molecular backbone of each of the interlocked components.

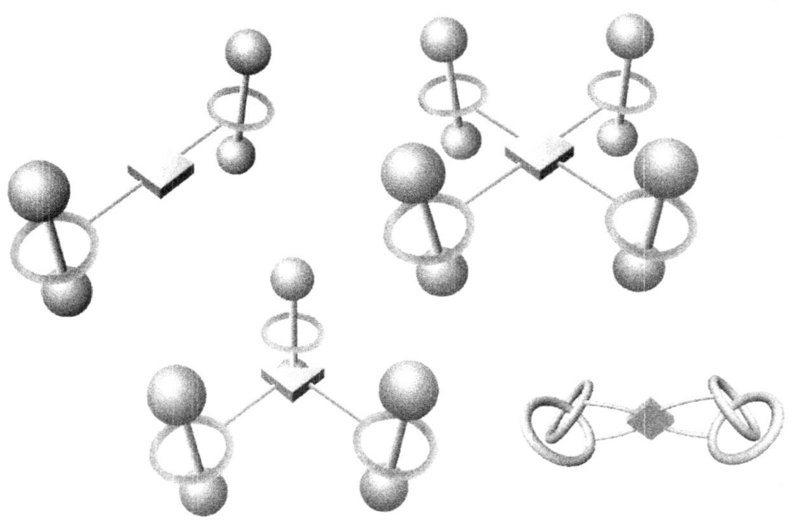

Fig. 21 Schematic representation of larger rotaxane and catenane assemblies which have been realized through a cross-linking alkylation of the corresponding monomers

This short introduction into topology may serve to help understand the phenomenon of topological chirality [177–179]. Without any stereogenic center involved, catenanes and rotaxanes can be made chiral [180, 181]. Just imagine that directionality is introduced in each of the subunits of the catenane and rotaxane in Fig. 22, for example by using a certain atom sequence which is not symmetrical. Then, each subunit is achiral, but the combination of two rings or a ring with an appropriate axle makes the catenane or rotaxane different from its mirror image thus generating two enantiomers. The knot is even inherently chiral without any modification concerning directionality [182]. It should be noted that rotaxanes cannot be called topologically chiral. "Cycloenantiomeric" is a better term, because topologically speaking, the rotaxane stoppers can be decreased in size until the mechanical bond is released. Consequently, the removal of the axle from the wheel is a topologically allowed transformation. Also, the argument that one could stretch the axle to infinite length thus preventing deslipping does not hold, since one needs to find only one topologically allowed transformation which leads to an achiral representation to destroy topological chirality. Nevertheless, topological chirality and cycloenantiomerism are closely related to each other.

Topological chirality thus is a very special form of chirality and one might well ask what consequences it has for the separability of the enantiomers and their chiroptical properties. In the beginning, chiral rotaxanes were made which were stoppered by chiral groups such as sugar derivatives [183–185]. However, the more interesting challenge was the preparation of topologically chiral catenanes [127, 186] and cycloenantiomeric rotaxanes [187–190]. There was not too much hope that a separation of the enantiomers of large and flexible molecules like rotaxanes or catenanes would be feasible, if they didn't contain chiral centers and thus differed only by marginal structural differences. However, HPLC

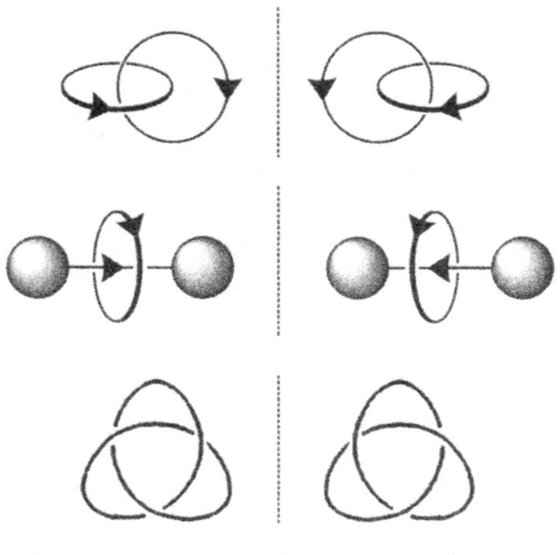

Fig. 22 Topological chirality does not require the presence of any stereogenic center. Just the topology of the compounds together with a certain directional atom sequence in the cases of rotaxanes and catenanes provides the basis for chirality

on a chiral stationary phase indeed allowed the separation of the enantiomers surprisingly well. One example is shown in Fig. 23 [191]. The rotaxanes shown bear a sulfonamide group in the wheel and another one in the axle. Both sulfonamides provide the directionality required for topological chirality. If one for example defines a path from the SO_2 group to the NH (the other way is of course possible, but it is important not to mix these definitions), one can draw arrows such as those shown in Fig. 22. The two enantiomers of this rotaxane could be separated after derivatization with a bridge connecting the two sul-

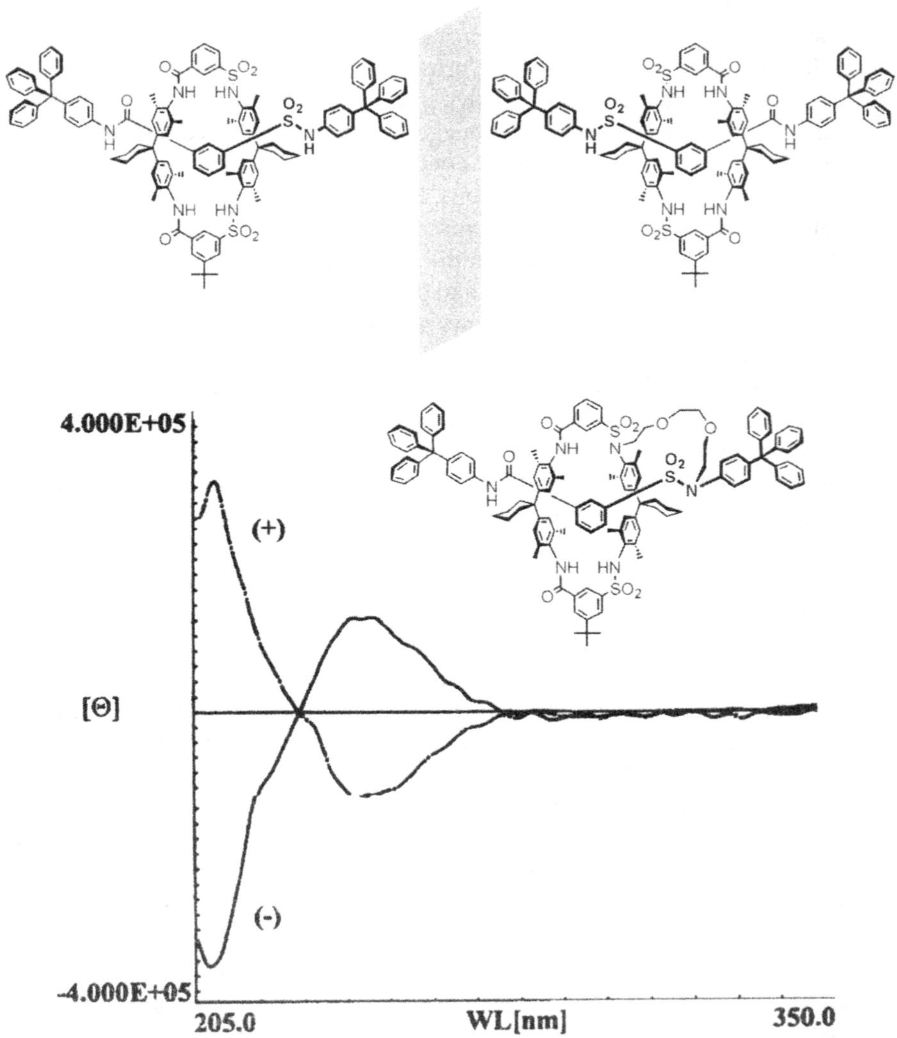

Fig. 23 *Top*: two enantiomers of a [2]rotaxane. *Bottom*: the *inset* shows the CD spectra of the two separated enantiomers of the corresponding [1]rotaxane after bridging the two sulfonamides

fonamides. CD spectroscopy of the two enantiomers yielded two CD spectra that are images of each other and thus provide evidence for non-racemic material with opposing optical rotational dispersions.

3.4 Molecular Machines Based on Amide Rotaxanes

As already discussed above, molecular motion in rotaxanes or catenanes requires the appropriate functional groups, even more so, when the motion is to be controlled by external stimuli such as pH, light, or electrons. The amide template effect is based on the presence and precise orientation of hydrogen-bond donors and exceptors and thus provides a means for motional control.

As a first step in this direction, the dynamic behavior of benzylic [2]catenanes was intensively studied. Experiments suggest the frequency of circumrotation to depend on a delicate balance of steric effects, the inter-macrocyclic array of hydrogen bonds, π-π stacking interactions, and other weak forces [192]. Tremendous effects have been found when comparing rotaxanes with 2,5-thiopheneoyl and 2,6-pyridinoyl building blocks in the wheel structure. The rate of rotation of the thiophene derivative is higher than that of the pyridine analogue by a factor of 3.2 million. Also, the solvent plays a pivotal role. When DMSO is used, a solvent that strongly competes with hydrogen bonds, the speed of rotation is significantly increased due to a weakening of the internal hydrogen bonds. The solvent effects can also be seen, when a catenane is examined that bears only one diamide moiety and a diester unit in each ring [193]. In non-competitive solvents the second wheel is hydrogen bonded to the diamide side, while the hydrogen bonds are weakened in competitive solvents causing a shift to the other side of the wheel. A concise theoretical treatment of the circumrotation in benzylic [2]catenanes [194–196] revealed the details of the mechanism of movement which involves several large amplitude motions of one ring along the other one. The different steps require rearrangements of the macrocycle in order to optimize steric and electrostatic interactions.

Not only amide catenanes have been investigated with respect to their molecular mobility. Also, pseudorotaxanes with photoswitchable threading/dethreading behavior [197] and molecular shuttles based on peptide [2]rotaxanes [198–200] and [3]rotaxanes [201, 202] have been developed. The ring rotation around the rotaxane axle can be induced by an external alternating electric field [203] or by electrochemical means [204].

Most work was, however, devoted to the generation of light-driven shuttles [205–208], one example of which is presented in Fig. 24 [209]. In its resting state, the rotaxane wheel is bound to the left diamide side by hydrogen bonding. A light-induced electron transfer from a donor present in solution to the stopper on the right generates an anion radical with better hydrogen-bond acceptor qualities and consequently, the wheel moves to this side surprisingly quickly. Back-conversion to the resting state by re-oxidation of the anion radical causes the wheel to shift back to the left.

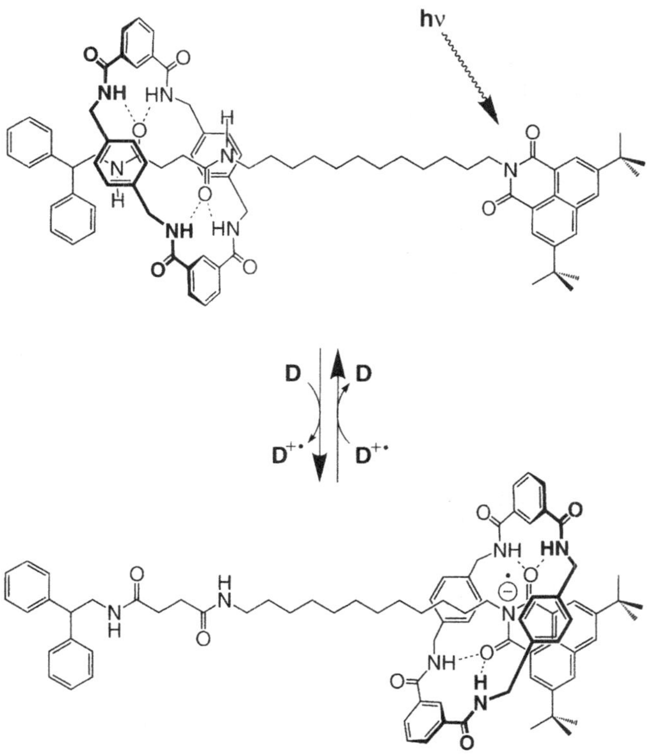

Fig. 24 Light-driven shuttling of a rotaxane due to changes in the quality of the internal hydrogen bonds. The donor molecule can for example be DABCO

Two more particular examples should be mentioned here. The first one is a peptide [2]rotaxane which bears two different stations on the axle. The wheel prefers the left station (Fig. 25) as long as the double bond remains in its E state. When switched by light to its Z form, the wheel moves to the other side of the axle. The right station is chiral and the wheel feels a strong influence of the chirality. The second example is based on a [3]catenane consisting of two small and one large wheel with three different stations. This system is particularly interesting, because it realizes unidirectional motion, which is a necessary prerequisite for the realization of molecular motors [210, 211]. Figure 26 provides a schematic representation of the mechanism. In each step, one of the small wheels blocks the movement of the other in one direction so that a net movement in only one direction results from a number of subsequent reaction steps that alter the binding properties along the stations of the large wheel and thus induce motion of the small ones.

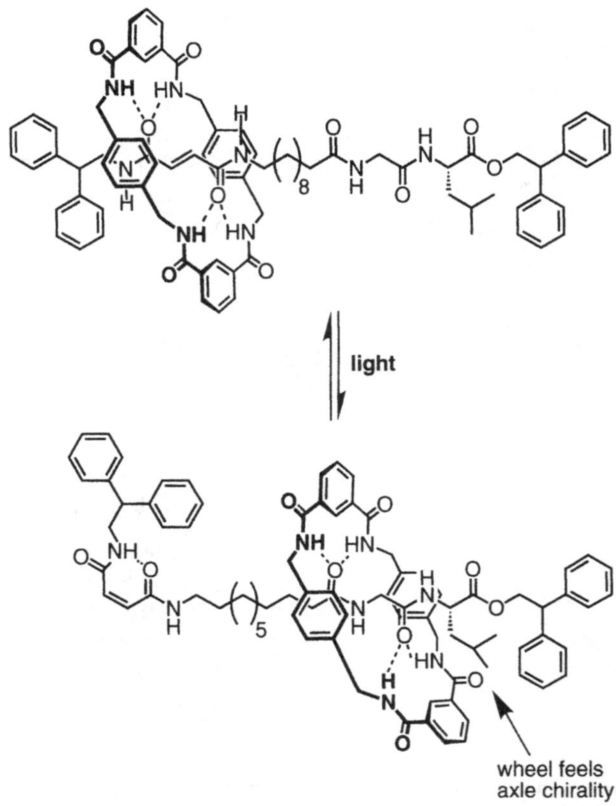

Fig. 25 Chiroptical switching in a peptide [2]rotaxane. E/Z isomerization by light of the double bond on the left side of the axle induces shuttling of the wheel to the other side, where it senses the chirality of the dipeptide station

3.5 Trefoil Dodecaamide Knotanes

It is a formidable challenge to design a template effect for the synthesis of a trefoil knot [212–216] or even more sophisticated topologies such as the Borromean link [217–219]. Exactly three crossing points need to be fixed during the synthesis by non-covalent bonds in order to generate a knot rather than a large non-intertwined macrocycle which would result from a precursor with only one crossing. Two crossings would give rise to a catenane after ring closure of the macrocycles. Similarly, more than three crossings would lead to different interlocked species. In view of these difficulties, it was a great surprise that a knot was formed as a byproduct in the synthesis of the pyridine containing macrocycle shown in Fig. 27. In this reaction three distinct reaction products were isolated: the expected tetralactam macrocycle, its larger octalactam relative, and the knot in up to 20% yield [220–222].

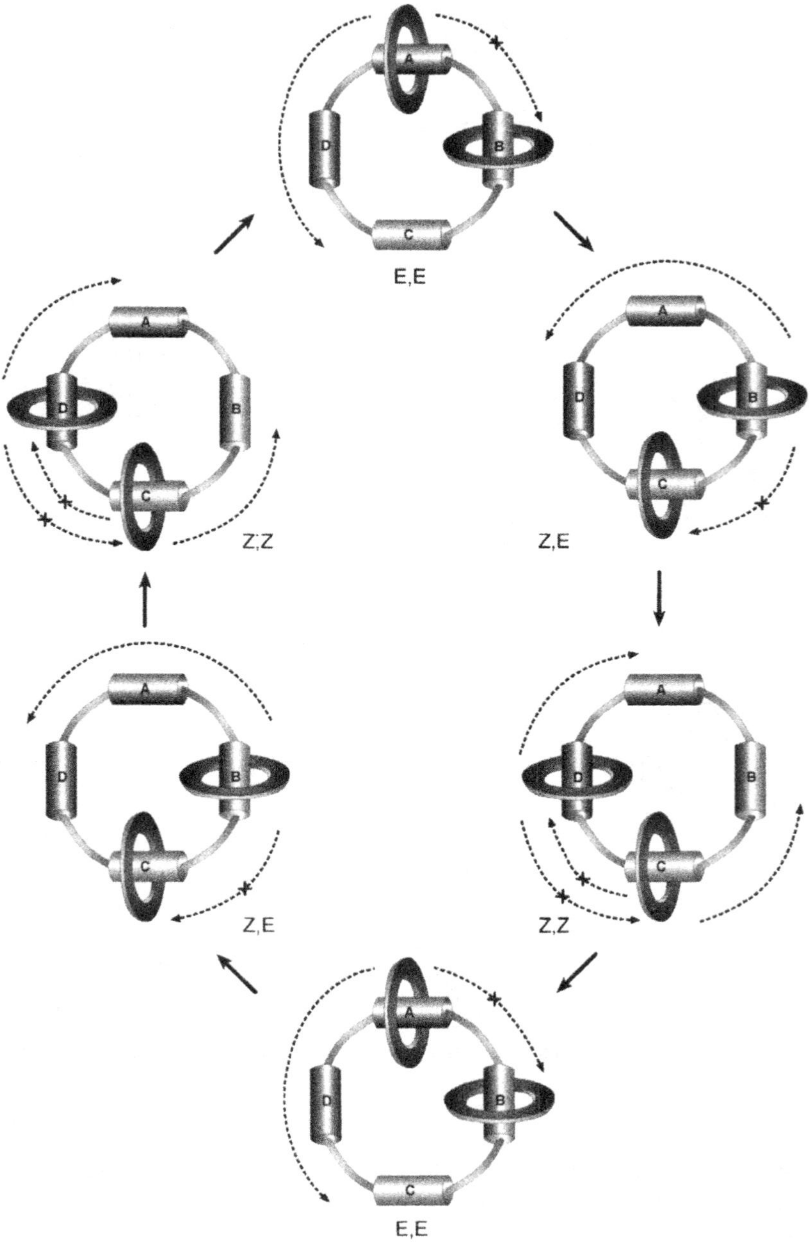

Fig. 26 Unidirectional motion along the large three-station wheel of a [3]catenane. In each step one of the two small wheels blocks the way back for the other moving one so that a net movement in one direction is realized

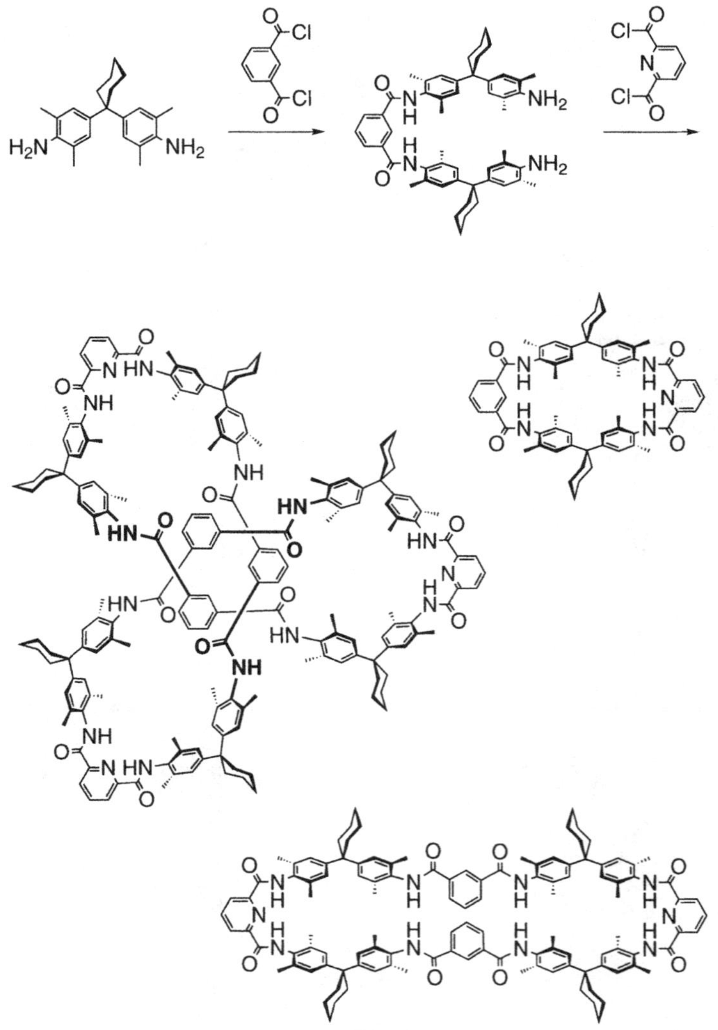

Fig. 27 A trefoil knot forms as a byproduct together with the expected macrocycle and its larger octalactam analogue

The characterization of the knotted structure was a non-trivial problem. The ^{1}H and ^{13}C NMR spectra even of the unsubstituted representative are too complex to permit a conclusive interpretation. Finally, after the single crystal X-ray structure showed that the molecule was indeed a knot, it became clear that this complexity comes from a dynamic process [223], which slowly interconverts three loops desymmetrized through the formation of intraknot hydrogen bonds. Two of the loops are different from the third one due to internal hydrogen bonding. In order to solve the problem of characterization, two experiments are most telling. The first one is X-ray crystallography, for which suitable crystals could be obtained and the structure of the knot in the solid state is shown in Fig. 28. The second method relies on the fact that the knot is inherently chiral due to its topology, while the corresponding catenane and simple macrocycles are not when generated from the same building blocks. Since the knot is formed as a racemate, the separation of the enantiomers is a prerequisite for obtaining CD spectra which then provide evidence for the knotted structure [224]. Although these separations are not trivial to carry out, enantiomerically pure samples could be obtained by HPLC on a chiral stationary phase. The UV- and CD-detected chromatograms of a representative HPLC run are depicted in Fig. 29 and show that a baseline separation of both knot enantiomers is possible. The two enantiomers give rise to signals in the CD spectra with opposite signs. Even the determination of the absolute configuration was possible by comparison of the experimental CD spectra with those obtained from a theoretical calculation. The separations are achieved more readily, when the knotane is functionalized with centrochiral groups, e.g., camphor derivatives, at the pyridine moieties [225].

Experiments and calculations [126] point to a template mechanism for knot formation involving the structure shown in Fig. 30 as the final intermediate. The hydrogen-bonding patterns resemble those found in the template synthesis of rotaxanes and catenanes. This intermediate structure is in line with the experimental finding that a *tert*-butyl substituted isophthalic acid unit does not yield detectable amounts of the knot, while the pyridine dicarboxylic acid building block can easily be substituted without drastic effects on the knot yield. The isophthalic acid units are buried inside a structure which does not permit large steric changes, while the pyridines are located at the periphery of that complex.

Several substituted derivatives of the knot have been synthesized. With the appropriate protective group strategy, the knot could be derivatized at the three

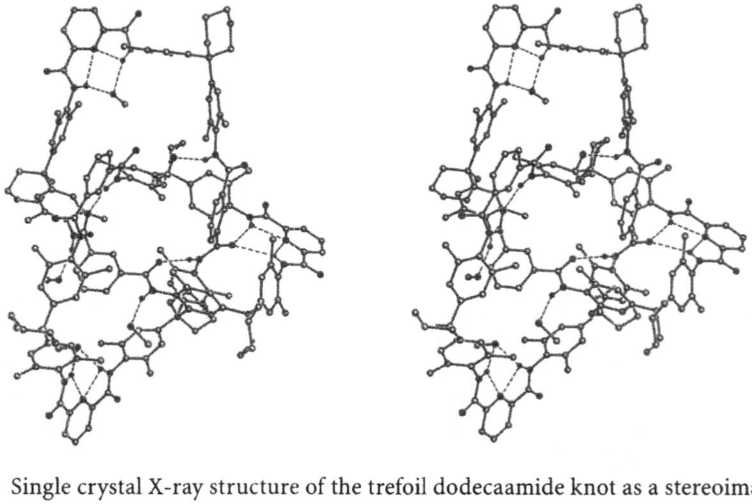

Fig. 28 Single crystal X-ray structure of the trefoil dodecaamide knot as a stereoimage for 3D-viewing

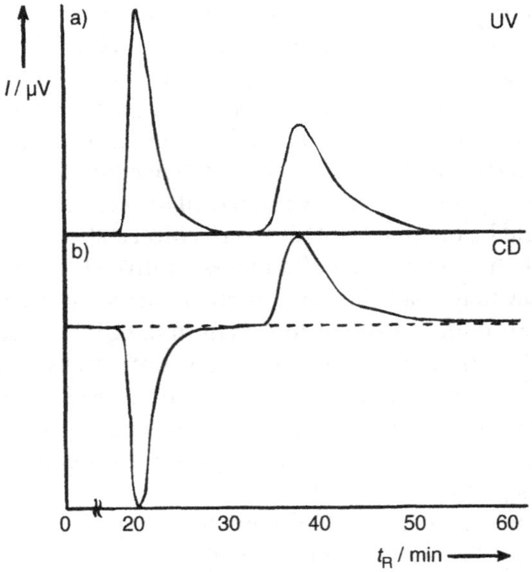

Fig. 29 UV chromatogram (*top trace*) and CD chromatogram (*bottom trace*) of the trefoil dodecaamide knot. Since the knot is the only possible chiral product, the CD trace, which shows two bands of equal integrations with opposite signs of the CD signal, provides evidence for knot formation and the successful separation of both enantiomers by chiral HPLC

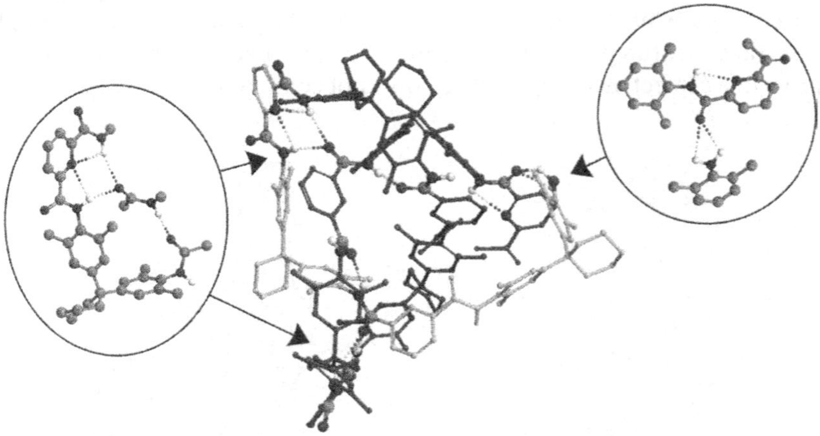

Fig. 30 Lowest-energy conformation of the immediate precursor for knot formation determined by an extensive Monte Carlo conformational search followed by reoptimization of the best conformer at the semiempirical AM1 level of theory. The intertwined structure is stabilized by hydrogen-bonding patterns as shown *in the two circles*. Note that hydrogen bonding also brings the amino and acid chloride termini closely together which form the last amide bond in order to complete the knot. The three intertwined loops are shown in different shades of *gray*. Amide groups involved in hydrogen bonding are shown as *ball-and-stick models*. Carbon-centered hydrogen atom are omitted for clarity

pyridines with a large variety of substituents. Even the synthesis of knots with three different substituents is possible, including dendrons of the Fréchet type [226]. Of great interest with respect to their topological chirality are the recently reported knotaxane, i.e., a rotaxane with two knots as the stoppers [227, 228] and larger assemblies of interconnected knotanes, which have been coined "knotanophanes", an expression that refers to cyclophanes built from knotanes [229].

4
New Topologies through Catenation along the Seam of Hydrogen-Bonded Calixarene Capsules

A complex hydrogen-bonding pattern, which is based on the formation of a hydrogen-bonded dimeric calixarene capsule, has recently been utilized to provide a doubly catenated capsule [230]. This approach represents an unconventional and elegant approach to new topologies. The hydrogen-bonding pattern involved is shown in Fig. 31. Two tetraurea-substituted calix[4]arenes can dimerize through a seam of 16 hydrogen bonds to form a capsule which is capable of guest encapsulation inside its cavity [231–233]. The urea groups interdigitate like the fingers of praying hands. The substituents R^1 and R^2 can be varied over a wide range. Typically, R^1 is an alkyl chain providing solubility in non-com-

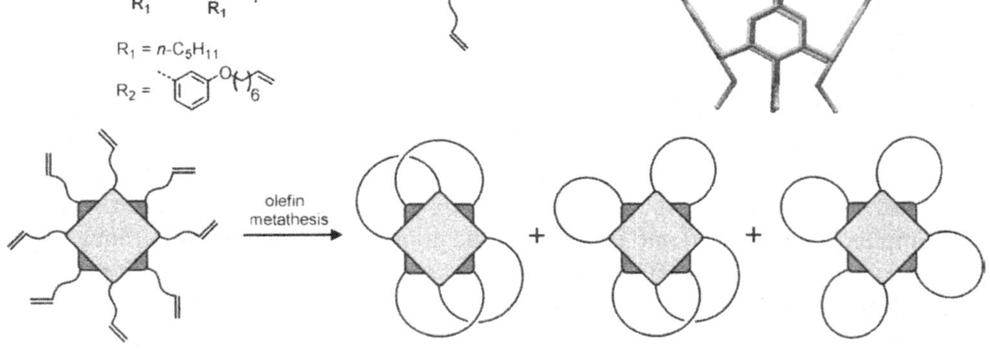

Fig. 31 *Top*: structural formula of the tetraurea calixarene monomer (*left*) and the hydrogen-bonding pattern of its dimer (*right*). Note that the cavity inside is capable of guest encapsulation. *Bottom*: olefin metathesis reaction catenates and/or bridges the two halves of the capsule leading to three different product

petitive organic solvents which are required for the self-assembly of the capsule through hydrogen-bonds.

In the present case, R^2 is a substituent with a terminal vinyl group. In a metathesis reaction with the dimer, three different products can be expected and have indeed been isolated: A doubly catenated capsule (5–12% yield), a doubly bridged mono-catenane (26–32% yield) and a tetrabridged dimer (10–15% yield) [234]. Such dimers held together not only by hydrogen bonds, but also through mechanical bonds provide a new type of container somewhere between *Cram's* covalently bound carcerands and the self-assembled capsules which merely dimerize through non-covalent forces. Even under conditions, which destroy the seam of hydrogen bonds, the two halves remain connected, but retain much of their flexibility to move relative to each other. This is reflected in the host-guest properties of the capsule. Compared to the dimer before metathesis reaction, the kinetics of the guest exchange remain virtually unchanged. Nevertheless, the mobility of an encapsulated tetraethylammonium ion appears to be more restricted.

5
Anionic Templates: The Role of Charged Hydrogen Bonds

5.1
Anion-Templated Synthesis of Rotaxanes and Their Structural Variability

The last section of this account is devoted to anion templates [235] for the synthesis of interlocked molecules [236]. Often, anion recognition [237–242] is more difficult to achieve with a controlled relative geometry of host and guest so that it may be more challenging to design a threaded host-guest complex based on anions. This may be the reason, why the anion templates for the formation of catenanes or rotaxanes are limited in number and have been used less often than those based on cation recognition.

The first anion template for rotaxane synthesis was discovered in an experiment which initially was intended to be a control experiment for analyzing the amide template effect. If hydrogen bonding of the tetralactam macrocycle in Fig. 32 to an axle amide group is indeed important for rotaxane formation, one would expect that the non-amidic axle center piece in Fig. 32 would not give rise to any rotaxane due to the absence of amide groups. However, the result was completely unexpected and surprising. The non-amidic center piece, although not appropriately functionalized, was consumed in rotaxane formation proceeding with 80% to 95% yield [243]. As the mechanism, initial formation of a "wheeled" nucleophile through hydrogen bonding of the phenolate stopper to the macrocycle amide groups was suggested. In the subsequent step, the stopper reacts with the electrophilic semiaxle through the wheel and the wheel is trapped on the axle which finally contains only ether oxygens as functional groups. Consequently, the interactions between axle and wheel in the rotaxane are likely of minor importance and strength. The "wheeled" nucleophile can be utilized for the synthesis of rotaxanes with differently functionalized axles. Ester, carbonate, and acetal rotaxanes have been prepared [244]. The nucleophile was employed for the introduction of *N*-tosylamide, thioester, and phosphate bonds into the axle with yields between 41% and 88% [245]. The second stopper has also been attached via a Michael addition reaction [246] so that the structural variety of rotaxanes is again quite large. A tetralactam macrocycle equipped with a biquinoline moiety could be used for rotaxane synthesis, when

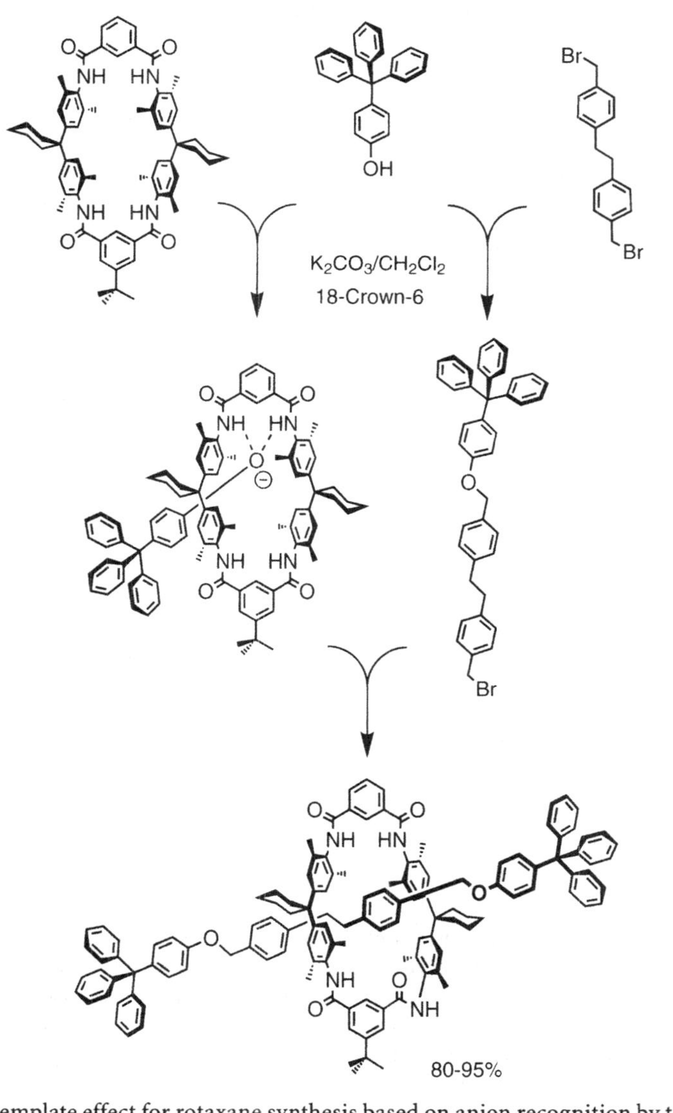

Fig. 32 A template effect for rotaxane synthesis based on anion recognition by tetralactam macrocycles

the stoppers were chosen large enough to prevent deslippage. Addition of Cu(I) ions provided a basis for dimerization of the wheel or the rotaxanes [247]. These complexes promise to have some potential for the generation of larger, more complex structures through self-assembly in the future.

The first problems with the new anion-templated rotaxane synthesis were encountered when the preparation of rotaxanes with smaller 3,5-di-*tert*-butyl phenol stoppers was attempted [248]. These rotaxanes were intended for the investigation of their deslipping behavior (see below) which should provide insight into the size complementarity of the stopper and the wheel cavity. For this purpose, smaller stoppers are needed and a yield of 2% to 5% which was ob-

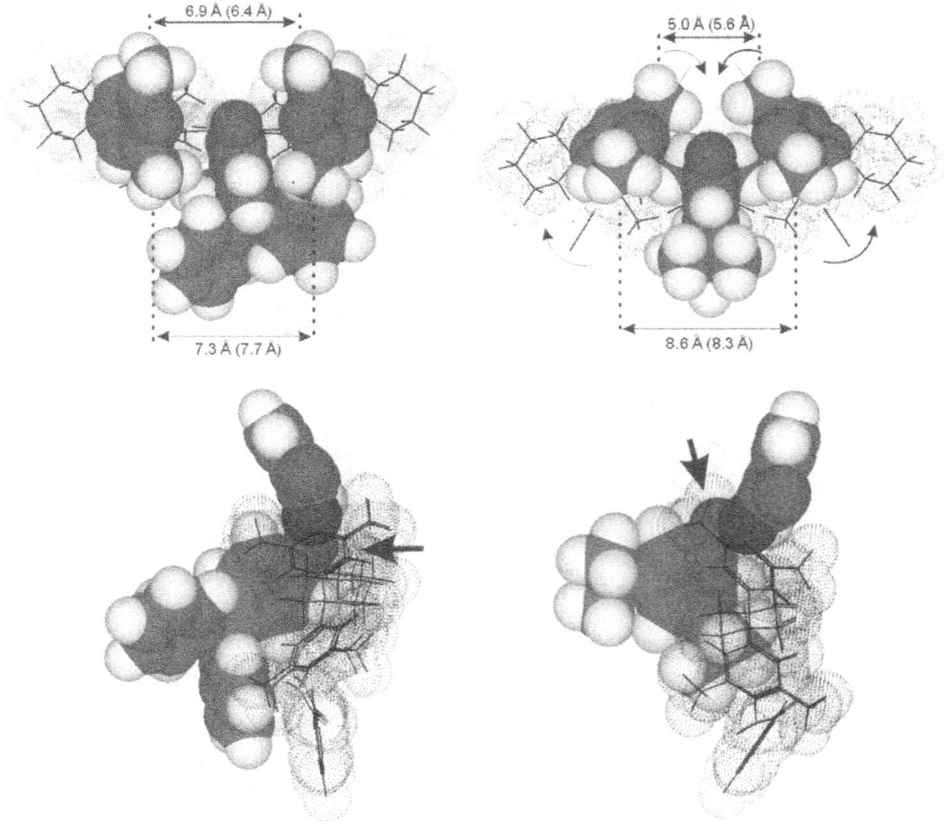

Fig. 33 Minimal energy conformations of stopper wheel complexes with tritylphenolate (*left*) and 3,5-di-*tert*-butylphenolate (*right*). *Top row*: view into the wheel cavity which shows the open attack path for the semi-axle electrophile *on the left* and the closed path *on the right*. Distances between each pair of two opposite methyl groups of the wheel are given (average for a family of favorable conformations *in parenthesis*). *Bottom row*: the different conformations of the stopper wheel complexes present the reactive phenolate oxygen towards the side indicated by the *arrows*. In the case of the 3,5-di-*tert*-butyl phenolate, the electrophile approaches from the wrong side preventing rotaxane formation

tained for some of the rotaxanes was thus quite disappointing. An analysis of the different stopper/wheel complexes by molecular modeling provided insight into the reasons why the yield decreases so much depending on the details of the stopper structure (Fig. 33). With the trityl phenolate stopper, the complex is open for the attack of the electrophile from the side of the wheel opposite to the stopper. The reactive free electron pair of the phenolate (which is not involved in hydrogen bonding) is presented to that side so that a reaction with the incoming electrophile will yield the rotaxane by trapping the wheel on the axle. In contrast, the smaller 3,5-di-*tert*-butyl phenolate stopper inserts itself deeper into the cavity of the wheel which leads to a conformation closed for attack from the opposite side. Instead, the electrophile can more favorably attack from the same side so that initially a weakly hydrogen-bonded wheel/axle complex is formed. Since the axle and wheel are not bound to each other by a mechanical bond, the complex easily dissociates into free axle and free wheel and the rotaxane yields are drastically reduced. A solution to this problem is shown in Fig. 34. By spatially separating the phenolate site needed for recognition from the

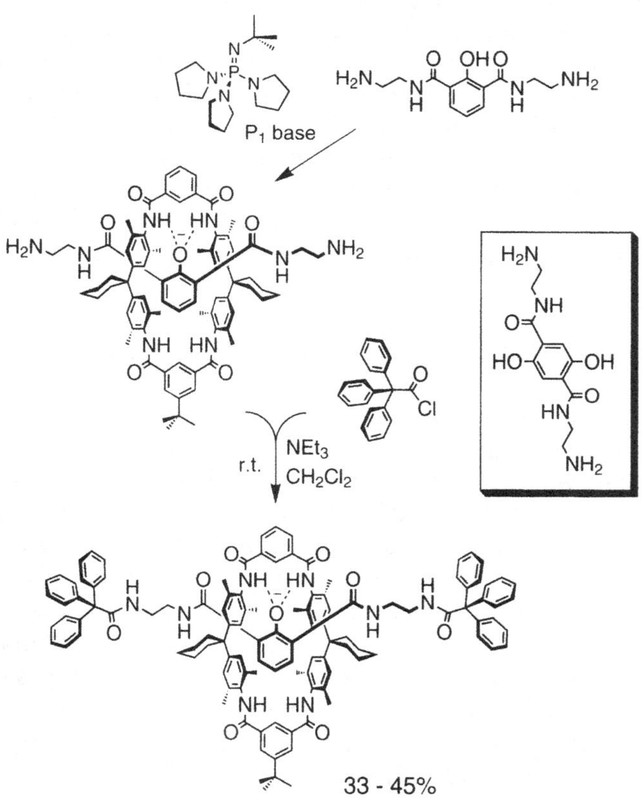

Fig. 34 A novel template effect for the syntheses of rotaxanes with functionalized centerpieces. The stoppers are attached at a position remote from the wheel cavity so that steric problems can be avoided. Instead of the phenolate axle center piece, the hydroquinone derivative on the right can also be used

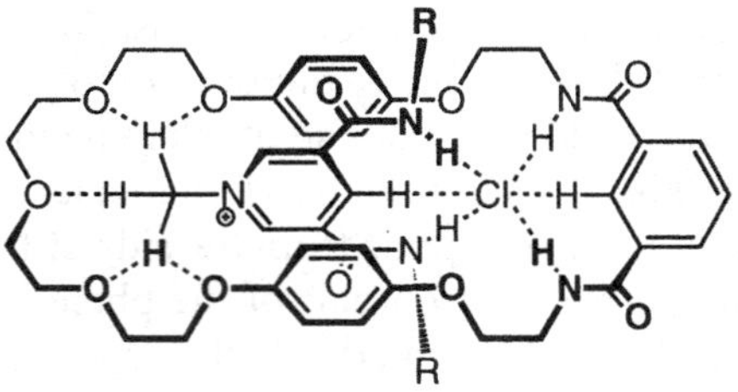

Fig. 35 An anion template utilizing recognition of the anion through hydrogen bonding to amide groups. Note that the *N*-methyl group provides a means to control the threaded geometry with C-H···O hydrogen bonds to the crown ether half of the macrocycle

stopper attachment sites, steric problems are avoided and rotaxanes with a OH group in their axle center piece can be generated with yields up to 45% [249].

A different anion template effect was recently developed by Beer et al. who use the crown ether/bislactam hybrid macrocycle in Fig. 35 for binding an *N*-methyl pyridinium dicarboxylic amide derivative as a thread [250–252]. A chloride ion is tetrahedrally surrounded by amide NH groups, two coming from the wheel and the other two incorporated in the thread. The perpendicular geometry is locked in by additional C-H···O hydrogen bonds between the *N*-methyl group and the crown ether part of the macrocycle. These pseudorotaxanes again provide a suitable spatial arrangement for the synthesis of rotaxanes. Starting with an open-chain analogue of the macrocycle and an axle already equipped with large stoppers, olefin metathesis was used to close the macrocycle after binding to the axle mediated through the anion template [253].

5.2 Deslippage as a Precision Tool for the Measurement of Steric Size

The ether rotaxanes with tetralactam macrocycles as the wheels, which are easily accessible through anion templation, only undergo very weak interactions between axle and wheel. No strong hydrogen bonds are formed as was the case for the amide rotaxanes. Stacking interactions likely do not play a significant role, because (i) the distances to the aromatic rings between axle and wheel are ca. 5 Å [126] and (ii) the organic solvents used are not a particularly supportive environment for them. Consequently, ether rotaxanes are well suited for a study of their deslipping behavior and the effects of small structural changes on it. As a first step, rotaxanes with alkyl chains of different lengths in the axle center piece were studied with respect to their deslipping kinetics [254]. These experiments intended to test the validity of the simple model that a rotaxane is just a thin filament threaded through the cavity of a macrocycle equipped with two bulky stopper groups that prevent deslipping. If the stopper groups are the thickest part, this model would predict that no large differences in the deslipping kinetics should be observed irrespective of the length of the axle. This is indeed the case. The next step was to test, whether the deslipping reaction would provide a basis to determine steric size in a precise way. Provided the absence of strong non-covalent forces between wheel and axle, the use of differently substituted stoppers with the same wheel should have an effect on the deslipping kinetics. The smallest steric difference which can be incorporated in such a structure is the exchange of hydrogen atoms at the stopper periphery against deuterium. Since the deuterated stoppers are expected to be somewhat smaller mainly due to a lower vibrational amplitude in the C-D bond stretching vibrations as compared to the C-H bond, one would predict that the deuterated rotaxanes deslip faster than their unlabeled analogues. The experiment is unambiguous: The observed kinetic isotope effect for the rotaxanes shown in Fig. 36 is inverse and amounts to ca. 0.9, i.e., the deuterated rotaxane deslips faster by ca. 10% as compared to the unlabeled one [255]. These experiments demonstrate the sensitivity of the deslipping process against extremely small changes in steric size. Another study dealing with the problem of steric size employs the deslipping reaction to measure the bulkiness of dendritic stoppers [256]. The fact that a G1 Frechét-type dendron deslips more quickly than a di-*tert*-butylphenol stoppered rotaxane suggests that the flexibility of the dendron permits a branch-by-branch deslippage which is not as easily accomplished with the less flexible di-*tert*-butylphenol stopper group.

The situation changes significantly, when rotaxanes are examined which bear functional groups that exert either a significant change in the structure of

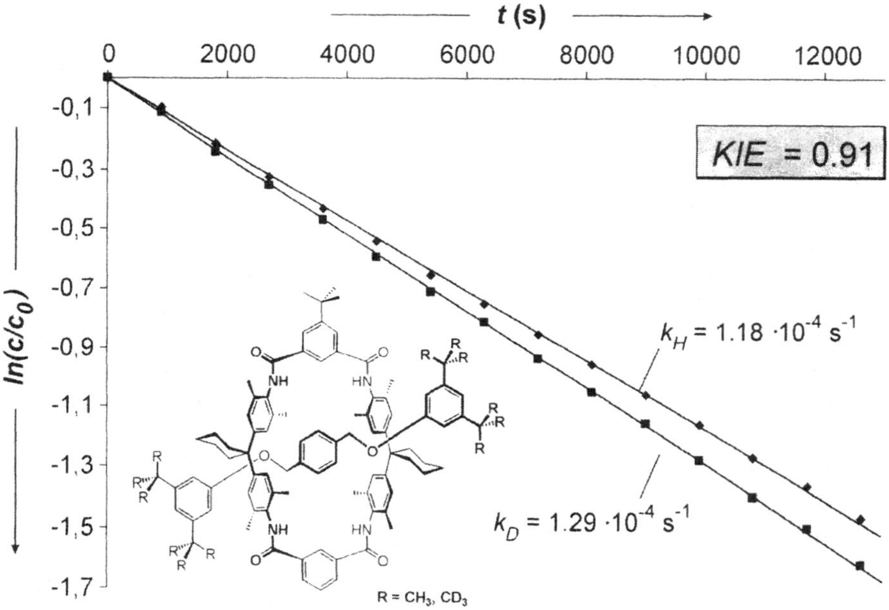

Fig. 36 Determination of the kinetic isotope effect for the deuterated rotaxane and its unlabeled analogue. The *KIE* is defined as k_H/k_D and is inverse in this case because the deuterated stoppers deslip faster than the non-deuterated ones by ca. 10%

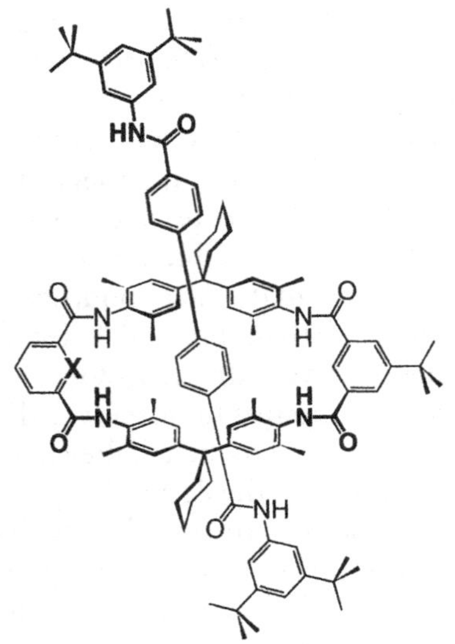

	$\Delta H^{\ddagger}$ [kJ mol^{-1}]	$\Delta S^{\ddagger}$ [J K^{-1} mol^{-1}]	$t_{1/2}$ (333 K) [h]
X = CH	78	-49	61
X = N	130	-40	> 10^6

Fig. 37 Exchanging the isophthaloyl unit by the corresponding pyridine derivative causes a significant increase in the stability of the rotaxane by a factor of more than 10,000

one of the components or which are capable of forming non-covalent bonds between axle and wheel. One example is shown in Fig. 37. Replacing not more than a C-H by an isoelectronic nitrogen in one of the isophthaloyl units of the wheel causes an increase in the half life by four magnitudes [257]. This effect is almost completely due to a change in activation enthalpy and can be traced back to the formation of intramolecular hydrogen bonds to the pyridine nitrogen which contract the wheel significantly. Another example is shown in Fig. 38. Two ester rotaxanes with structures identical with the exception of the orientation of the ester groups show dramatic differences in their deslipping behavior [258]. These differences can be traced back to (i) hydrogen bonding between axle and wheel as indicated by the large solvent effect encountered for one of the rotaxanes and (ii) to the rotational barriers at the stopper/ester junction.

These results may become important with respect to the following idea: Natural molecular machines such as the ATP synthase enzyme complex [259, 260] and many others represent extremely efficient devices. One reason, why nature uses very complex structures for implementing function may be that a complex system can be fine tuned and thus optimized by an accordingly larger number of subtle adjustments in structure. A detailed understanding of effects such as those reported here may then become an important tool for the design of artificial analogues. In particular, the fine tuning of the properties of such

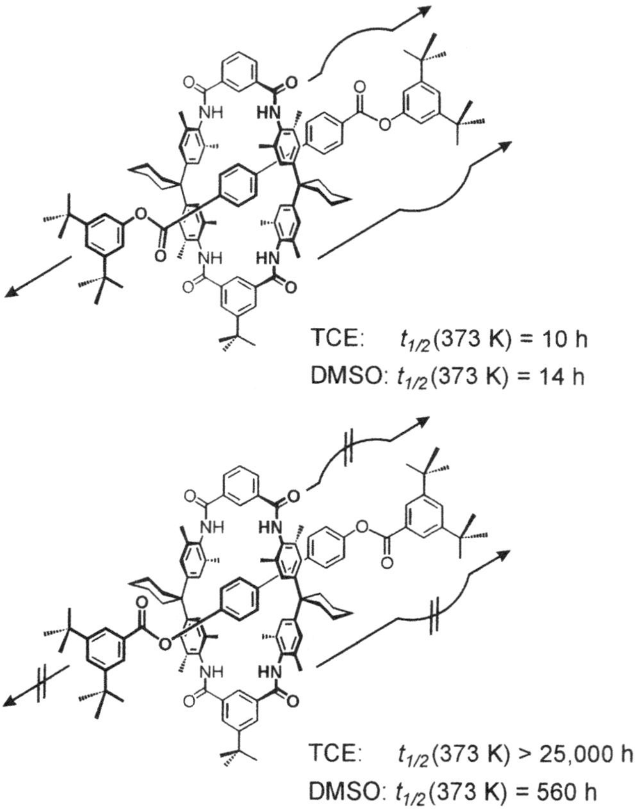

Fig. 38 Two ester rotaxanes which are identical with the sole exception of the orientation of the ester bond. A cooperative interplay of hydrogen bonds between the amide groups in the wheel and the carbonyl groups of the axle and the rotational barriers of the bonds connecting the stoppers to the ester groups leads to large changes in the deslippage for only one of the rotaxanes depending on the solvent. In unpolar solvents, one rotaxane is stable, while the other one deslips with a half-life of a few hours

machines will probably rely on and make use of these and other subtle structural variations in order to improve their performance and optimize their efficiency without causing large structural changes.

5.3 Anion-Induced Mechanical Motion

Molecular devices and machines whose function relies on the presence of anions are scarce. One example [261] employs a peptide [2]rotaxane which bears a phenol in the axle as the second station. Upon deprotonation to the phenolate, the wheel shifts to the anion due to the much stronger hydrogen bonds formed with the phenolate oxygen. A similar strategy is realized in the rotaxanes shown

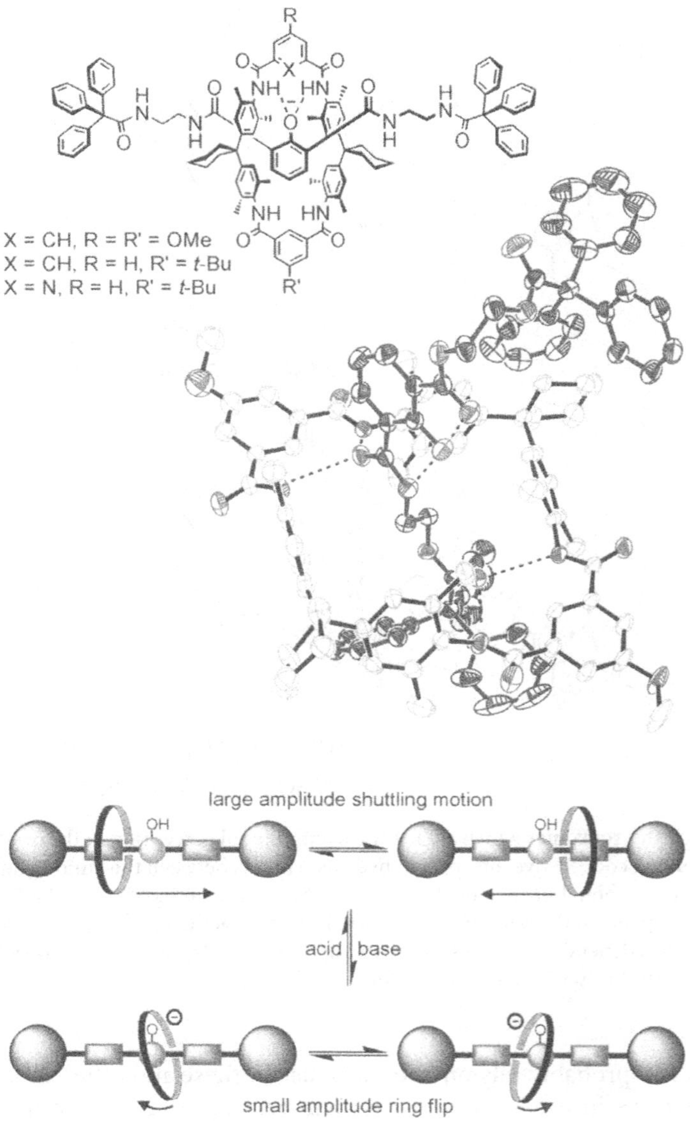

Fig. 39 A series of structurally different acid/base-controlled molecular shuttles. *Top*: structural formulae. *Center*: Single crystal x-ray structure of the methoxy-substituted rotaxane. *Bottom*: schematic drawing of the switching behavior and the dynamics. Deprotonation reduces the large amplitude shuttling motion observed for the protonated rotaxane in variable temperature NMR experiments to a low-distance ring flip during which one side of the wheel remains bound to the phenolate

in Fig. 39. These rotaxanes were – as described above – synthesized through anion templating [249, 262]. The OH group is left as a reminder of the template in these structures and serves as the control unit [263]. In the neutral state, hydrogen bonding occurs between two carbonyl groups of one of the spacers carrying the stoppers. The crystal structure is in line with NMR chemical shifts providing evidence for the wheel shifted towards one of the stoppers. Variable temperature NMR and exchange spectroscopy reveal that the wheel is shuttling back and forth between the two degenerate stations at the two ends of the axle. Consequently, the neutral rotaxane performs a long-distance shuttling motion accompanied by rotation of the wheel around the axle. Upon deprotonation, strong hydrogen bonds to the central phenolate cause the wheel to shift to the axle center. However, the wheel forms not only the two hydrogen bonds to the phenolate, but an additional pair to one of the carbonyl groups of the axle center piece so that the wheel is fixed in an unsymmetric position at the center. Dynamic NMR and exchange spectroscopy again show the wheel to move. It rotates around the axle and undergoes a small-amplitude ring flip. Consequently, protonation and deprotonation do not only control the position of the wheel along the axle, but also have a major influence on the dynamic properties and determine the amplitude of the shuttling motion. This example with regard to the initial discussion of templates also illustrates that the same functional groups can be used for templation and for controlling the properties of the interlocked species.

6 Epilogue

This account hopefully made clear that on one hand the rational design of template effects for the synthesis of interlocked molecules is still a challenge. On the other hand, it intended to shed light on the potential of using the template effects not only for producing the interlocked molecules, but also for introducing the functional groups required for the implementation of desired properties and functions. Although the vast number of articles on host-guest chemistry and template syntheses for such compounds suggests that this chemistry is well understood, most of the template effects used today have been found by coincidence rather than design. Many surprising findings have led to the state-of-the-art of the field and it is by no means trivial to find, develop, and optimize new template effects. A profound understanding of the non-covalent interactions mediating the template synthesis will help much in the design efforts and is thus of utmost importance for the whole field of catenane, rotaxane, and knot chemistry. This is particularly true, since rotaxanes and catenanes have entered a stage in their development, in which they have undergone the transition from beautiful structures with ever-new topologies to those which exert functions. Many so-called molecular devices have been synthesized and studied with respect to a controlled motion of the two components relative to each other, some of which are described above. In order to control these devices, functional groups are required which can be addressed by external stimuli such as light, electrons, or chemical reactions. Time has come to revert the process of template development by starting with the desired function followed by the identification of the required functional groups, with which it can be realized in a molecular system, and finally by designing the appropriate template effect for the synthesis of the device. These arguments also point to the necessity to continue with intense research efforts in the template area, because many different templates will become necessary in order to satisfy the requirements of a growing variety of properties and functions to be fulfilled.

Acknowledgements We are grateful to the many co-workers who have skillfully contributed significantly to the scientific adventure here. Their names are incorporated in the list of references. C.A.S. thanks the Deutsche Forschungsgemeinschaft for support with a Heisenberg fellowship and the Fond der Chemischen Industrie for a Dozentenstipendium. We thank Gabriele Silva for help with the preparation of the manuscript. Our work has been supported for a long time by the Fonds der Chemischen Industrie and the Deutsche Forschungsgemeinschaft, recently in particular through the Collaborative Research Initiative (SFB) 624. Part of our catenane work has been funded in the CERC3 framework of the European Union.

References

1. Busch DH, Stephensen NA (1990) Coord Chem Rev 100:119
2. Cacciapaglia R, Mandolini L (1993) Chem Soc Rev 22:221
3. Gerbeleu NV, Arion VB, Burgess J (1999) Template synthesis of macrocyclic compounds. Wiley-VCH, Weinheim, Germany
4. Hubin TJ, Kolchinski AG, Vance AL, Busch DH (1999) Adv Supramol Chem 5:237
5. Diederich F, Stang PJ (eds) (2000) Templated organic synthesis. Wiley-VCH, Weinheim, Germany
6. Hubin TJ, Busch DH (2000) Coord Chem Rev 200–202:5
7. See, for example: Voet D, Voet JG (1990) Biochemistry. Wiley, Chichester, UK
8. van Bommel KJC, Friggeri A, Shinkai S (2003) Angew Chem 115:1010; Angew Chem Int Ed 42:980
9. Gibb BC (2003) Chem Eur J 9:5181
10. Sanders JKM (2000) Pure Appl Chem 72:2265
11. Sherman J (2003) Chem Commun 1617
12. Dötz KH (1984) Angew Chem 96:573; Angew Chem Int Ed 23:587
13. Pedersen CJ (1967) J Am Chem Soc 89:7019
14. Safarowsky C, Merz L, Rang A, Broekmann P, Hermann BA, Schalley CA (2004) Angew Chem 116:1311; Angew Chem Int Ed 43:1291
15. Hulvat JF, Stupp SI (2003) Angew Chem 115:802; Angew Chem Int Ed 42:778
16. Anderson S, Anderson HL (2000) Templates in organic synthesis: definitions and roles. In: Diederich F, Stang PJ (eds) Templated organic synthesis. Wiley-VCH, Weinheim, Germany, p 1
17. Sauvage J-P (1990) Acc Chem Res 23:319
18. Chambron J-C, Dietrich-Buchecker CO, Heitz V, Nierengarten J-F, Sauvage J-P, Pascard C, Guilhem J (1995) Pure Appl Chem 67:233

19. Leigh DA, Lusby PJ, Teat SJ, Wilson AJ, Wong JKY (2001) Angew Chem 113:1586; Angew Chem Int Ed 40:1538
20. Mobian P, Kern J-M, Sauvage J-P (2003) J Am Chem Soc 125:2016
21. Schalley CA, Lützen A, Albrecht M (2004) Chem Eur J 10:1072
22. Dietrich-Buchecker CO, Sauvage J-P (1987) Chem Rev 87:795
23. Amabilino D, Stoddart J (1995) Chem Rev 95:2725
24. Glink P, Stoddart J (1998) Pure Appl Chem 70:419
25. Nepogodiev SA, Stoddart JF (1998) Chem Rev 98:1959
26. Sauvage J-P, Dietrich-Buchecker C (eds) (1999) Molecular catenanes, rotaxanes, and knots. Wiley-VCH, Weinheim, Germany
27. Raehm L, Hamilton D, Sanders J (2002) Synlett 11:1743
28. Rang A, Schalley CA (2004) Catenanes and interlinked molecules. In: Atwood JL, Steed JW (eds) Encyclopedia of supramolecular chemistry. Dekker, New York, p 206
29. Linnartz P, Schalley CA (2004) Rotaxanes and pseudorotaxanes. In: Atwood JL, Steed JW (eds) Encyclopedia of supramolecular chemistry. Dekker, New York, p 1194
30. Schill G (1971) Catenanes, rotaxanes and knots. Academic Press, New York, USA
31. Drexler KE (1994) Annu Rev Biophys Biomol Struct 23:377
32. Asfari Z, Vicens J (2000) J Incl Phenom Macrocyc Chem 36:103
33. Balzani V, Credi A, Raymo FM, Stoddart JF (2000) Angew Chem 112:3484; Angew Chem Int Ed 39:3348
34. Pease AR, Jeppesen JO, Stoddart JF, Luo Y, Collier CP, Heath JR (2001) Acc Chem Res 34:433
35. Ballardini R, Balzani V, Credi A, Gandolfi MT, Venturi M (2001) Acc Chem Res 34:445
36. Schalley CA, Beizai K, Vögtle F (2001) Acc Chem Res 34:465
37. Collin J-P, Dietrich-Buchecker C, Gaviña P, Jimenez-Molero MC, Sauvage J-P (2001) Acc Chem Res 34:477
38. Schalley CA (2002) Angew Chem 114:1583; Angew Chem Int Ed 41:1513
39. Balzani V, Venturi M, Credi A. (2003) Molecular devices and machines: a journey into the nanoworld. Wiley-VCH, Weinheim, Germany
40. Felder T, Schalley CA (2004) Artificial molecular rotary motors based on rotaxanes. In: Schmuck C, Wennemers H (eds) Highlights in bioorganic chemistry. Wiley-VCH, Weinheim, Germany, p 526
41. Etter MC (1990) Acc Chem Res 23:120
42. Aakeröy CB, Seddon KR (1993) Chem Soc Rev 22:397
43. Gillard RE, Raymo FM, Stoddart JF (1997) Chem Eur J 3:1933
44. Desiraju G (1991) Acc Chem Res 24:290
45. Houk KN, Menzer S, Newton SP, Raymo FM, Stoddart JF, Williams DJ (1999) J Am Chem Soc 121:1479
46. Hansen JG, Feeder N, Hamilton DG, Gunter MJ, Becher J, Sanders JKM (1999) Org Lett 2:449
47. Loeb SJ, Wisner JA (1998) Angew Chem 110:3010; Angew Chem Int Ed 37:2838
48. Raymo FM, Bartberger MD, Houk KN, Stoddart JF (2001) J Am Chem Soc 123:9264
49. Sutherland IO (1986) Chem Soc Rev 15:63
50. Gokel GW (1991) Crown ethers and cryptands. The Royal Society of Chemistry, Cambridge, UK
51. Ashton PR, Campbell PJ, Chrystal EJT, Glink PT, Menzer S, Philp D, Spencer N, Stoddart JF, Tasker PA, Williams DJ (1995) Angew Chem 107:1997; Angew Chem Int Ed 34:1865
52. Ashton PR, Fyfe MCT, Hickingbottom SK, Stoddart JF, White AJP, Williams DJ (1998) J Chem Soc Perkin Trans 2 2117
53. Review: Glink PT, Schiavo C, Stoddart JF, Williams DJ (1996) Chem Commun 1483

54. Ashton PR, Chrystal EJT, Glink PT, Menzer S, Schiavo C, Spencer N, Stoddart JF, Tasker PA, White AJP, Williams DJ (1996) Chem Eur J 2:709
55. Bryant WS, Guzei IA, Rheingold AL, Merola JS, Gibson HW (1998) J Org Chem 63:7634
56. Ashton PR, Bartsch RA, Cantrill SJ, Hane RE Jr, Hickingbottom SK, Lowe JN, Preece JA, Stoddart JF, Talanov VS, Wang Z-H (1999) Tetrahedron Lett 40:3661
57. Cantrill SJ, Fyfe MCT, Heiss AM, Stoddart JF, White AJP, Williams DJ (2000) Org Lett 2:61
58. Huang F, Zakharow LN, Rheingold AL, Jones JW, Gibson HW (2003) Chem Commun 2122
59. Diederich F, Echogoyen L, Gómez-López M, Kessinger R, Stoddart JF (1999) J Chem Soc Perkin Trans 2 1577
60. Jones JW, Gibson HW (2003) J Am Chem Soc 125:7001
61. Clifford T, Abushamleh A, Busch DH (2002) Proc Natl Acad Sci USA 99:4830
62. Ashton PR, Cantrill SJ, Preece JA, Stoddart JF, Wang Z-H, White AJP, Williams DJ (1999) Org Lett 1:1917
63. Bryant WS, Guzei IA, Rheingold AL, Gibson HW (1999) Org Lett 1:47
64. Ashton PR, Chrystal EJT, Glink PT, Menzer S, Schiavo C, Stoddart JF, Tasker PA, Williams DJ (1995) Angew Chem 107:2001; Angew Chem Int Ed 34:1869
65. Ashton PR, Fyfe MCT, Martinez-Diaz M-V, Menzer S, Schiavo C, Stoddart JF, White AJP, Williams DJ (1998) Chem Eur J 4:1523
66. Ashton PR, Fyfe MCT, Glink PT, Menzer S, Stoddart JF, White AJP, Williams DJ (1997) J Am Chem Soc 119:12514
67. Ashton PR, Becher J, Fyfe MCT, Nielsen MB, Stoddart JF, White AJP, Williams DJ (2001) Tetrahedron 57:947
68. Balzani V, Clemente-León M, Credi A, Lowe JN, Badjić JD, Stoddart JF, Williams DJ (2003) Chem Eur J 9:5348
69. Review: Fyfe MCT, Stoddart JF (1999) Coord Chem Rev 183:139
70. Ashton PR, Glink PT, Martínez-Díaz M-V, Stoddart JF, White AJP, Williams DJ (1996) Angew Chem 108:2058; Angew Chem Int Ed 35:1930
71. Ashton PR, Fyfe MCT, Hickingbottom SK, Menzer S, Stoddart JF, White AJP, Williams DJ (1998) Chem Eur J 4:577
72. Ashton PR, Baxter I, Cantrill SJ, Fyfe MCT, Glink PT, Stoddart JF, White AJP, Williams DJ (1998) Angew Chem 110:1344; Angew Chem Int Ed 37:1294
73. Cantrill SJ, Youn GJ, Stoddart JF (2001) J Org Chem 66:6857
74. Yamaguchi N, Gibson HW (1999) Chem Commun 789
75. Yamaguchi N, Gibson HW (1999) Angew Chem Int Ed 111:195; Angew Chem Int Ed 38:143
76. Gibson HW, Yamaguchi N, Jones JW (2003) J Am Chem Soc 125:3522
77. Amirsakis DG, Garcia-Garibay MA, Rowan SJ, Stoddart JF, White AJP, Williams DJ (2001) Angew Chem 113:4386; Angew Chem Int Ed 40:4256
78. Amirsakis DG, Elizarov AM, Garcia-Garibay MA, Glink PT, Stoddart JF, White AJP, Williams DJ (2003) Angew Chem 115:1158; Angew Chem Int Ed 42:1126
79. Review: Gibson HW, Marand H (1993) Adv Mater 5:11
80. Yamaguchi N, Gibson HW (1999) Macromol Chem Phys 201:815
81. Takata T, Kawasaki H, Asai S, Kihara N, Furusho Y (1999) Chem Lett 111
82. Yamaguchi N, Hamilton LM, Gibson HW (1998) Angew Chem 110:3463; Angew Chem Int Ed 37:3275
83. Elizarov AM, Chang T, Chiu S-H, Stoddart JF (2002) Org Lett 4:3565
84. Gibson HW, Yamaguchi N, Hamilton LM, Jones JW (2002) J Am Chem Soc 124:4653
85. Ashton PR, Baxter I, Fyfe MCT, Raymo FM, Spencer N, Stoddart JF, White AJP, Williams DJ (1998) J Am Chem Soc 120:2297

86. Ashton PR, Fyfe MCT, Schiavo C, Stoddart JF, White AJP, Williams DJ (1998) Tetrahedron Lett 39:5455
87. Zehnder DW, Smithrud DB (2001) Org Lett 3:2485
88. Kolchinski AG, Busch DH, Alcock NW (1995) J Chem Soc, Chem Commun 1289
89. Cantrill SJ, Fulton DA, Fyfe MCT, Stoddart JF, White AJP, Williams DJ (1999) Tetrahedron Lett 40:3669
90. Cantrill SJ, Fulton DA, Heiss AM, Pease AR, Stoddart JF, White AJP, Williams DJ (2000) Chem Eur J 6:2274
91. Cantrill SJ, Fyfe MCT, Heiss AM, Stoddart JF, White AJP, Williams DJ (1999) Chem Commun 1251
92. For a similar example of rotaxanes with chiral crown ether wheels, see: Tachibana Y, Kihara N, Ohga Y, Takata T (2000) Chem Lett 806
93. Rowan SJ, Cantrill SJ, Stoddart JF (1999) Org Lett 1:129
94. Chang T, Heiss AM, Cantrill SJ, Fyfe MCT, Pease AR, Rowan SJ, Stoddart JF, White AJP, Williams DJ (2000) Org Lett 2:2943
95. Chang T, Heiss AM, Cantrill SJ, Fyfe MCT, Pease AR, Rowan SJ, Stoddart JF, White AJP, Williams DJ (2000) Org Lett 2:2947
96. Rowan SJ, Stoddart JF (2000) J Am Chem Soc 122:164
97. Chiu S-H, Rowan SJ, Cantrill SJ, Ridvan L, Ashton PR, Garell RL, Stoddart JF (2002) Tetrahedron 58:807
98. Chiu S-H, Rowan SJ, Cantrill SJ, Stoddart JF, White AJP, Williams DJ (2002) Chem Commun 2948
99. Ashton PR, Glink PT, Stoddart JF, Tasker PA, White AJP, Williams DJ (1996) Chem Eur J 2:729
100. Asakawa M, Ikeda T, Yui N, Shimizu T (2002) Chem Lett 174
101. Ikeda T, Asakawa M, Goto M, Nagawa Y, Shimizu T (2003) Eur J Org Chem 3744
102. Cantrill SJ, Rowan SJ, Stoddart JF (1999) Org Lett 1:1363
103. Glink PT, Oliva AI, Stoddart JF, White AJP, Williams DJ (2001) Angew Chem 113:1922; Angew Chem Int Ed 40:1870
104. Horn M, Ihringer J, Glink PT, Stoddart JF (2003) Chem Eur J 9:4046
105. Kilbinger AFM, Cantrill SJ, Waltman AW, Day MW, Grubbs RH (2003) Angew Chem 115:3403; Angew Chem Int Ed 42:3281
106. Iwamoto H, Itoh K, Nagamiya H, Fukazawa Y (2003) Tetrahedron Lett 44:5773
107. Kolchinski AG, Alcock NW, Roesner RA, Busch DH (1998) Chem Commun 1437
108. Furusho Y, Oku T, Hasegawa T, Tsuboi A, Kihara N, Takata T (2003) Chem Eur J 9:2895
109. Kihara N, Tachibana Y, Kawasaki H, Takata T (2000) Chem Lett 506
110. Ashton PR, Ballardini R, Balzani V, Baxter I, Credi A, Fyfe MCT, Gandolfi MT, Gómez-López M, Martínez-Díaz M-V, Piersanti A, Spencer N, Stoddart JF, Venturi M, White AJP, Williams DJ (1998) J Am Chem Soc 120:11932
111. Ashton PR, Ballardini R, Balzani V, Gómez-López M, Lawrence SE, Martínez-Díaz MV, Montalti M, Piersanti A, Prodi L, Stoddart JF, Williams DJ (1997) J Am Chem Soc 119:10641
112. Ashton PR, Ballardini R, Balzani V, Fyfe MCT, Gandolfi MT, Martínez-Díaz MV, Morosini M, Schiavo C, Shibata K, Stoddart JF, White AJP, Williams DJ (1998) Chem Eur J 4:2332
113. Vögtle F, Jäger R, Händel M, Ottens-Hildebrandt S (1996) Pure Appl Chem 68:225
114. Vögtle F, Dünnwald T, Schmidt T (1996) Acc Chem Res 29:451
115. Jäger R, Vögtle F (1997) Angew Chem 109:966; Angew Chem Int Ed 36:931
116. Dünnwald T, Parham AH, Vögtle F (1997) Synthesis 339
117. Kogej M, Ghosh P, Schalley CA (2004) How to thread a string into the eye of a molecular needle: template-directed synthesis of mechanically interlocked molecules. In:

Harmata M (ed) Strategies and tactics in organic synthesis, vol 4. Elsevier, Amsterdam, p. 171
118. Hunter CA (1991) J Chem Soc Chem Commun 749
119. Allott C, Adams H, Bernad PL Jr, Hunter CA, Rotger C, Thomas JA (1998) Chem Commun 2449
120. Chang S-Y, Kim HS, Chang K-J, Jeong K-S (2003) Org Lett 6:181
121. Herrmann U, Jonischkeit T, Bargon J, Hahn U, Li Q-Y, Schalley CA, Vogel E, Vögtle F (2002) Anal Bioanal Chem 372:611
122. Bargon J, Braschoß S, Flörke J, Herrmann U, Klein L, Lörgen JW, Lopez M, Maric S, Parham AH, Piacenza P, Schäfgen H, Schalley CA, Silva G, Schwierz H, Vögtle F, Windscheif G (2003) Sensors Actuators B 96:6
123. Seel C, Parham AH, Safarowsky O, Hübner GM, Vögtle F (1999) J Org Chem 64:7236
124. For a theoretical treatment on these twofold hydrogen bonds, see: Reckien W, Peyerimhoff S (2003) J Phys Chem 107:9634
125. Reuter C, Seel C, Nieger M, Vögtle F (2000) Helv Chim Acta 83:630
126. Schalley CA, Reckien W, Peyerimhoff S, Baytekin B, Vögtle F (2004) Chem Eur J 10: 4777
127. Mohry A, Vögtle F, Nieger M, Hupfer H (1999) Chirality 12:76
128. Hunter CA (1992) J Am Chem Soc 114:5303
129. Schalley CA (2000) Int J Mass Spectrom 194:11
130. Schalley CA (2001) Mass Spectrom Rev 20:253
131. Schalley CA, Hoernschemeyer J, Li X-Y, Silva G, Weis P (2003) Int J Mass Spectrom 228:373
132. Schalley CA, Ghosh P, Engeser M (2004) Int J Mass Spectrom 232/233:249
133. Vögtle F, Meier S, Hoss R (1992) Angew Chem 104:1628; Angew Chem Int Ed 31:1619
134. Ottens-Hildebrandt S, Meier S, Schmidt W, Vögtle F (1994) Angew Chem 106:1818; Angew Chem Int Ed 33:1767
135. Ottens-Hildebrandt S, Nieger M, Rissanen K, Rouvinen J, Meier S, Harder G, Vögtle F (1995) J Chem Soc Chem Commun 777
136. Andrievsky A, Ahuis F, Sessler JL, Vögtle F, Gudat D, Moini M (1998) J Am Chem Soc 120:9712
137. Baumann S, Jäger R, Ahuis F, Kray B, Vögtle F (1996) Liebigs Ann 761
138. Johnston AG, Leigh DA, Pritchard RJ, Deegan MD (1995) Angew Chem 107:1324; Angew Chem Int Ed 34:1209
139. Fustin C-A, Leigh DA, Rudolf P, Timpel D, Zerbetto F (2000) ChemPhysChem 2:97
140. Johnston AG, Leigh DA, Nezhat L, Smart JP, Deegan MD (1995) Angew Chem 107:1327; Angew Chem Int Ed 34:1212
141. Kidd TJ, Leigh DA, Wilson AJ (1999) J Am Chem Soc 121:1599
142. Watanabe N, Kihara N, Takata T (2001) Org Lett 3:3519
143. Watanabe N, Kihara N, Furusho Y, Takata T, Araki Y, Ito O (2003) Angew Chem 115:705; Angew Chem Int Ed 42:681
144. Vögtle F, Händel M, Meier S, Ottens-Hildebrandt S, Ott F, Schmidt T (1994) Liebigs Ann 739
145. Vögtle F, Jäger R, Händel M, Ottens-Hildebrandt S, Schmidt W (1996) Synthesis 3:353
146. Jäger R, Baumann S, Fischer M, Safarowsky O, Nieger M, Vögtle F (1997) Liebigs Ann 2269
147. Vögtle F, Ahuis F, Baumann S, Sessler JL (1996) Liebigs Ann 921
148. Fischer C, Nieger M, Mogck O, Böhmer V, Ungaro R, Vögtle F (1997) Eur J Org Chem 155
149. Braun O, Vögtle F (1997) Synlett 10:1184
150. Braun O, Hünten A, Vögtle F (1999) J Prakt Chem 341:542

151. Heim C, Affeld A, Nieger M, Vögtle F (1999) Helv Chim Acta 82:746
152. Parham AH, Windisch B, Vögtle F (1999) Eur J Org Chem 1233
153. Johnston AG, Leigh DA, Murphy A, Smart JP, Deegan MD (1996) J Am Chem Soc 118:10662
154. Leigh DA, Murphy A, Smart JP, Slawin AMZ (1997) Angew Chem 109:752; Angew Chem Int Ed 36:728
155. Clegg W, Gimenez-Saiz C, Leigh DA, Murphy A, Slawin AMZ, Teat SJ (1999) J Am Chem Soc 121:4124
156. Gatti FG, Leigh DA, Nepogodiev SA, Slawin AMZ, Teat SJ, Wong JKY (2001) J Am Chem Soc 123:5983
157. Brancato G, Coutrot F, Leigh DA, Murphy A, Wong JKY, Zerbetto F (2002) Proc Natl Acad Sci USA 99:4967
158. Hannam JS, Kidd TJ, Leigh DA, Wilson AJ (2003) Org Lett 5:1907
159. Kidd TJ, Loontjens TJA, Leigh DA, Wong JKY (2003) Angew Chem 115:3501; Angew Chem Int Ed 42:3379
160. Hunter CA, Sarson LD (1994) Angew Chem 106:2424; Angew Chem Int Ed 33:2313
161. Hunter C, Low C, Packer M, Spey S, Vinter J, Vysotsky M, Zonta C (2001) Angew Chem 113:2750
162. Jeong K-S, Choi JS, Chang S-Y, Chang H-Y (2000) Angew Chem 112:1758; Angew Chem Int Ed 39:1692
163. Chang S-Y, Choi JS, Jeong K-S (2001) Chem Eur J 7:2687
164. Chang S-Y, Jang H-Y, Jeong K-S (2003) Chem Eur J 9:1535
165. Chang S-Y, Jeong K-S (2003) J Org Chem 68:4014
166. Safarowsky O, Vogel E, Vögtle F (2000) Eur J Org Chem 499
167. Schwanke F, Safarowsky O, Heim C, Silva G, Vögtle F (2000) Helv Chim Acta 83:3279
168. Vögtle F, Dünnwald T, Händel M, Jäger R, Meier S, Harder G (1996) Chem Eur J 2:640
169. Parham AH, Schmieder R, Vögtle F (1999) Synlett 12:1887
170. Jäger R, Schmidt T, Karbach D, Vögtle F (1996) Synlett 8:723
171. Li Q-Y, Vogel E, Parham AH, Nieger M, Bolte M, Fröhlich R, Saarenketo P, Rissanen K, Vögtle F (2001) Eur J Org Chem 4041
172. Jäger R, Händel M, Harren J, Rissanen K, Vögtle F (1996) Liebigs Ann 1201
173. Ottens-Hildebrandt S, Schmidt T, Harren J, Vögtle F (1995) Liebigs Ann 1855
174. Dünnwald T, Jäger R, Vögtle F (1997) Chem Eur J 3:2043
175. Mohry A, Schwierz H, Vögtle F (1999) Synthesis 10:1753
176. Osswald F, Vogel E, Safarowsky O, Schwanke F, Vögtle F (2001) Adv Synth Catal 343:303
177. Chambron J-C, Dietrich-Buchecker C, Sauvage J-P (1993) Top Curr Chem 165:131
178. Mislow K (1999) Top Stereochem 1
179. Lukin O, Godt A, Vögtle F (2004) Chem Eur J 10:1878
180. Vögtle F, Safarowsky O, Heim C, Affeld A, Braun O, Mohry A (1999) Pure Appl Chem 71:247
181. Reuter C, Schmieder R, Vögtle F (2000) Pure Appl Chem 72:2233
182. Meskers SCJ, Dekkers HPJM, Rapenne G, Sauvage J-P (2000) Chem Eur J 6:2129
183. Archut A, Müller WM, Baumann S, Habel M, Vögtle F (1997) Liebigs Ann 495
184. Schmidt T, Schmieder R, Müller WM, Kiupel B, Vögtle F (1998) Eur J Org Chem 2003
185. Kauffmann C, Müller WM, Vögtle F, Weinman S, Abramson S, Fuchs B (1999) Synthesis 5:849
186. Reuter C, Pawlitzki G, Wörsdörfer U, Plevoets M, Mohry A, Kubota T, Okamoto Y, Vögtle F (2000) Eur J Org Chem 3059
187. Reuter C, Seel C, Nieger M, Vögtle F (2000) Helv Chim Acta 83:630
188. Reuter C, Mohry A, Sobanski A, Vögtle F (2000) Chem Eur J 6:1674

189. Schmieder R, Hübner G, Seel C, Vögtle F (1999) Angew Chem 111:3741; Angew Chem Int Ed 38:3528
190. Reuter C, Wienand W, Schmuck C, Vögtle F (2001) Chem Eur J 7:1728
191. Yamamoto C, Okamoto Y, Schmidt T, Jäger R, Vögtle F (1997) J Am Chem Soc 119: 10547
192. Leigh DA, Murphy A, Smart JP, Deleuze MS, Zerbetto F (1998) J Am Chem Soc 120: 6458
193. Leigh DA, Moody K, Smart JP, Watson KJ, Slawin AMZ (1996) Angew Chem 108:326; Angew Chem Int Ed 35:306
194. Leigh DA, Troisi A, Zerbetto F (2001) Chem Eur J 7:1450
195. Deleuze MS, Leigh DA, Zerbetto F (1999) J Am Chem Soc 121:2346
196. For a theoretical study of rotaxanes, see: Leigh DA, Troisi A, Zerbetto F (2000) Angew Chem 112:358; Angew Chem Int Ed 39:350
197. Jeong K-S, Chang K-J, An Y-J (2003) Chem Commun 1450
198. Da Ros T, Guldi DM, Morales AF, Leigh DA, Prato M, Turco R (2003) Org Lett 5:689
199. Lane AS, Leigh DA, Murphy A (1997) J Am Chem Soc 119:11092
200. Bottari G, Dehez F, Leigh DA, Nash PJ, Perez EM, Wong JKY, Zerbetto F (2003) Angew Chem 115:6066; Angew Chem Int Ed 42:5886
201. Zhao X, Jiang X-K, Shi M, Yu Y-H, Xia W, Li Z-T (2001) J Org Chem 66:7035
202. Chen L, Zhao X, Zhao C-X, Jiang X-K, Li Z-T (2003) J Org Chem 68:2704
203. Bermudez V, Capron N, Gase T, Gatti FG, Kajzar F, Leigh DA, Zerbetto F, Zhang S (2000) Nature 406:608
204. Altieri A, Gatti FG, Kay ER, Leigh DA, Martel D, Paolucci F, Slawin AMZ, Wong JKY (2003) J Am Chem Soc 125:8644
205. Review: Brouwer AM, Fazio SM, Frochot C, Gatti FG, Leigh DA, Wong JKY, Wurpel GWH (2003) Pure Appl Chem 75:1055
206. Wurpel GWH, Brouwer AM, van Stokkum IHM, Farran A, Leigh DA (2001) J Am Chem Soc 123:11327
207. Altieri A, Bottari G, Dehez F, Leigh DA, Wong JKY, Zerbetto F (2003) Angew Chem 115:2398; Angew Chem Int Ed 42:2296
208. Gatti FG, Léon S, Wong JKY, Bottari G, Altieri A, Farran Morales MA, Teat SJ, Frochot C, Leigh DA, Brouwer AM, Zerbetto F (2003) Proc Natl Acad Sci USA 100:10
209. Brouwer AM, Frochot C, Gatti FG, Leigh DA, Mottier L, Paolucci F, Roffia S, Wurpel GWH (2001) Science 291:2124
210. Feringa BL (2001) Acc Chem Res 34:504
211. Kelly TR (2001) Acc Chem Res 34:514
212. Ashton PR, Matthews OA, Menzer S, Raymo FM, Spenzer N, Stoddart JF, Williams DJ (1997) Liebigs Ann/Recueil 2485
213. Rapenne G, Dietrich-Buchecker C, Sauvage J-P (1999) J Am Chem Soc 121:994
214. Meskers SCJ, Dekkers HPJM, Rapenne G, Sauvage J-P (2000) Chem Eur J 6:2129
215. Adams H, Ashworth E, Breault GA, Guo J, Hunter CA, Mayers PC (2001) Nature 411:763
216. Woods CR, Benaglia M, Toyota S, Hardcastle K, Siegel JS (2001) Angew Chem 113:771; Angew Chem Int Ed 40:749
217. Loren JC, Yoshizawa M, Haldimann RF, Linden A, Siegel JS (2003) Angew Chem 115:5880; Angew Chem Int Ed 42:5702
218. Chichak KS, Catrill SJ, Pease AR, Chiu S-H, Cave GWV, Atwood JL, Stoddart JF (2004) Science 304:1308
219. Schalley CA (2004) Angew Chem 116:4499; Angew Chem Int Ed 43:4399
220. Review: Lukin O, Vögtle F (2004) Angew Chem (in press)

221. Recker J, Vögtle F (2001) J Incl Phenom Macrocyclic Chem 41:3
222. Safarowsky O, Nieger M, Fröhlich R, Vögtle F (2000) Angew Chem 112:1699; Angew Chem Int Ed 39:1616
223. Lukin O, Müller WM, Müller U, Kaufmann A, Schmidt C, Leszczynski J, Vögtle F (2003) Chem Eur J 9:3507
224. Vögtle F, Hünten A, Vogel E, Buschbeck S, Safarowsky O, Recker J, Parham AH, Knott M, Müller WM, Müller U, Okamoto Y, Kubota T, Lindner W, Francotte E, Grimme S (2001) Angew Chem 113:2534; Angew Chem Int Ed 40:2468
225. Lukin O, Yoneva A, Vögtle F (2004) Eur J Org Chem 1236
226. Recker J, Müller WM, Müller U, Kubota T, Okamoto Y, Nieger M, Vögtle F (2002) Chem Eur J 8:4434
227. Lukin O, Kubota T, Okamoto Y, Schelhase F, Yoneva A, Müller WM, Müller U, Vögtle F (2003) Angew Chem 115:4681; Angew Chem Int Ed 42:4542
228. For a topologically chiral dumbbell, which may act as the axle of a knotaxane, see: Lukin O, Recker J, Böhmer A, Müller WM, Kubota T, Okamoto Y, Nieger M, Vögtle F (2003) Angew Chem 115:458; Angew Chem Int Ed 42:442
229. Lukin O, Kubota T, Okamoto Y, Kaufmann A, Vögtle F (2004) Chem Eur J 10:2804
230. Vysotsky MO, Bolte M, Thondorf I, Böhmer V (2003) Chem Eur J 9:3375
231. Shimizu KD, Rebek J Jr (1995) Proc Natl Acad Sci USA 92:12403
232. Mogck O, Böhmer V, Vogt W (1996) Tetrahedron 52:8489
233. Schalley CA, Castellano RK, Brody MS, Rudkevich DM, Siuzdak G, Rebek J Jr (1999) J Am Chem Soc 121:4568
234. For a singly bridged tetraurea calixarene capsule, see: Brody MS, Schalley CA, Rudkevich DM, Rebek J Jr (1999) Angew Chem 111:1738; Angew Chem Int Ed 38:1640
235. Vilar R (2003) Angew Chem 115:1498; Angew Chem Int Ed 42:1460
236. Seel C, Vögtle F (2000) Chem Eur J 6:21
237. Schmidtchen FP, Berger M (1997) Chem Rev 97:1609
238. Beer PD (1998) Acc Chem Res 31:71
239. Gale PA, Sessler JL, Král V (1998) Chem Commun 1
240. Beer PD, Gale PA (2001) Angew Chem 113:502; Angew Chem Int Ed 40:487
241. Lavigne JJ, Anslyn EV (2001) Angew Chem 113:3212; Angew Chem Int Ed 40:3119
242. Sessler JL, Davis JM (2001) Acc Chem Res 34:989
243. Hübner GM, Gläser J, Seel C, Vögtle F (1999) Angew Chem 111:395; Angew Chem Int Ed 38:383
244. Reuter C, Wienand W, Hübner GM, Seel C, Vögtle F (1999) Chem Eur J 5:2692
245. Hübner GM, Reuter C, Seel C, Vögtle F (2000) Synthesis 1:103
246. Reuter C, Vögtle F (2000) Org Lett 2:593
247. Li X-Y, Illigen J, Nieger M, Michel S, Schalley CA (2003) Chem Eur J 9:1332
248. Schalley CA, Silva G, Nising CF, Linnartz P (2002) Helv Chim Acta 85:1578
249. Ghosh P, Mermagen O, Schalley CA (2002) Chem Commun 2628
250. Wisner JA, Beer PD, Drew MGB (2001) Angew Chem 113:3718; Angew Chem Int Ed 40:3606
251. Wisner JA, Beer PD, Berry NG, Tomapatanaget B (2002) Proc Natl Acad Sci USA 99:4983
252. Tomapatanaget B, Tuntulani T, Wisner JA, Beer PD (2004) Tetrahedron Lett 45:663
253. Wisner JA, Beer PD, Drew MGB, Sambrook MR (2002) J Am Chem Soc 124:12469
254. Linnartz P, Schalley CA (2004) Supramol Chem 16:263
255. Felder T, Schalley CA (2003) Angew Chem 115:2360; Angew Chem Int Ed 42:2258
256. Hübner GM, Nachtsheim G, Li Q-Y, Seel C, Vögtle F (2000) Angew Chem 112:1315; Angew Chem Int Ed 39:1269

257. Affeld A, Hübner GM, Seel C, Schalley CA (2001) Eur J Org Chem 2877
258. Linnartz P, Bitter S, Schalley CA (2003) Eur J Org Chem 4819
259. Boyer PD (1998) Angew Chem 110:2424; Angew Chem Int Ed 37:2296
260. Walker JE (1998) Angew Chem 110:2438; Angew Chem Int Ed 37:2308
261. Keaveney CM, Leigh DA (2004) Angew Chem 116:1242; Angew Chem Int Ed 43:1222
262. Schalley CA (2004) J Phys Org Chem (in press)
263. Ghosh P, Kogej M, Schalley CA, Haase D, Saak W, Lützen A, Federwisch G, Gschwind R (2004) J Am Chem Soc (submitted)

Topics in Current Chemistry (2004) 248: 201–221
DOI 10.1007/978-3-540-44505-0

Template-Controlled Synthesis in the Solid-State

Leonard R. MacGillivray (✉) · Giannis S. Papaefstathiou · Tomislav Friščić · Dushyant B. Varshney · Tamara D. Hamilton

Department of Chemistry, University of Iowa, Iowa City, Iowa 52245-1294, USA
len-macgillivray@uiowa.edu

Abstract The application of molecular templates to direct reactivity in the solid state is described. Specifically, molecules that function as linear templates are demonstrated to provide control of the [2+2] photodimerization in the solid state such that it is possible to use the templates to conduct target-oriented molecular syntheses, or molecular syntheses by design. The degree of organization provided by the solid state is also shown to provide chemists a reliable means to capture and transform ordered complexes formed by way of hydrogen-bonded-driven self-assembly into complex covalent molecular and polymeric structures.

Keywords Linear templates · Photodimerization · Hydrogen-bonding · Preorganization · Molecular recognition · Self-assembly

List of Abbreviations

C=C	Carbon-carbon double
ht	Head-to-tail
hh	Head-to-head
4,4′-bpe	*trans*-1,2-Bis(4-pyridyl)ethylene
4,4′-tpcb	*rctt*-Tetrakis(4-pyridyl)cyclobutane
bpp-34-crown-10	Bisparaphenylene-34-crown-10
amm-stilb	Bis(dialkylammonium)-substituted stilbene
5-OMe-res	5-Methoxyresorcinol
1,8-nap	1,8-Napthalenedicarboxylic acid
4-Cl-sbz	4-Chlorostilbazole
4-Cl-dpcb	*rctt*-1,2-Bis(4-pyridyl)-3,4-bis(4-chlorophenyl)cyclobutane
2,2′-bpe	*trans*-1,2-Bis(2-pyridyl)ethylene
2,2′-tpcb	*rctt*-Tetrakis(4-pyridyl)cyclobutane
4-Cl-res	4-Chlororesorcinol
2,4-bpe	*trans*-1-(2-Pyridyl)-2-(4-pyridyl)ethylene
2,4-tpcb-hh	*rctt*-1,3-Bis(2-pyridyl)-2,4-bis(4-pyridyl)cyclobutane-hh
3,4-bpe	*trans*-1-(3-Pyridyl)-2-(4-pyridyl)ethylene
3,4-tpcb-hh	*rctt*-1,3-Bis(3-pyridyl)-2,4-bis(4-pyridyl)cyclobutane-hh
4,4′-bipyeth	1,2-Bis(4-pyridyl)ethane
4-Et-res	4-Ethylresorcinol
4,6-di-Cl-res	4,6-Dichlororesorcinol
1,4-bpeb	1,4-Bis(2-(4-pyridyl)ethenyl)benzene
4-benz-res	4-Benzylresorcinol

1 Introduction

The organized environment of the solid state is an intriguing medium within which to control the formation of covalent bonds [1]. Being virtually frozen in an environment with atom-to-atom separations on the order of angstroms, reactants in the solid state can, in principle, adopt geometries that may be difficult, or impossible, to achieve in the liquid phase, meaning that access to molecules less available or unavailable in solution can be achieved. That a toxic organic solvent is not required for a reaction to occur in the solid state also means reactions in the solid state are inherently environmentally-friendly, or green [2]. When combined with the fact that a reaction in the solid state results in a change in composition and, therefore, can affect a bulk physical property (e.g., color) of a solid [3], it becomes apparent that controlling the formation of covalent bonds in the solid state can bear relevance to both chemical synthesis and materials science.

Unfortunately, however, whereas the fluid environment of the liquid phase is routinely used to conduct target-oriented syntheses (e.g., natural products) [4], it has remained difficult, owing to subtle structure demands of close packing (i.e., the driving force that provides the cohesive energy for crystallization) [5], to control reactivity in the solid state [6]. Such lack of control has meant

Chemical Synthesis

	reaction homology	target-oriented synthesis
liquid phase	√	√
solid state	?	X

Fig. 1 Scheme illustrating the role of reaction homology and target-oriented synthesis in liquid phase and solid-state chemistry

that reactions involving members of a homologous series of compounds – an idea central to solution-phase synthesis – has been achieved with limited success [7] and a general ability to construct molecules of prescribed size, shape, and functionality – a signature of the liquid phase – has not been realized (Fig. 1). The process of controlling reactivity in the solid state has, moreover, largely evolved as being based on trial and error, relying, in most cases, on weak interactions between molecules to organize reactants in a suitable arrangement for reaction.

2
Overview

It is with these ideas in mind that we will provide here a review of work from our laboratory, and the laboratories of others, performed primarily during the past five years that focuses upon utilizing molecules, in the form of linear reaction templates [8], to control reactivity in the solid state. Specifically, upon identifying the ability of a linear template to position two molecules, by way of hydrogen bonds, in an arrangement suitable for a [2+2] photodimerization [7] that is largely independent of structure effects of close packing (Fig. 2), we have demonstrated that linear templates provide a means to direct reactivity in the solid state with synthetic freedoms akin to the liquid phase such that it is now possible to conduct target-oriented solid-state syntheses, or molecular solid-state syntheses by design.

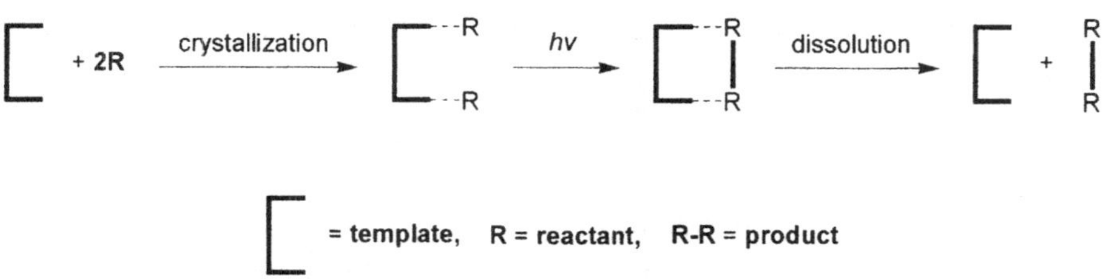

Fig. 2 Scheme showing the general synthesis behavior of a linear template

Our review will begin by outlining geometry criteria for a [2+2] photodimerization to occur in the solid state. We will then describe previous methods to control the [2+2] photoreaction and then present fundamentals of the template-controlled solid-state approach. We will also provide a brief account of the application of a similar method to control the formation of polymers, and then offer concluding remarks.

3 The Solid-State [2+2] Photodimerization

For a [2+2] photodimerization to occur in the solid state, two carbon-carbon double (C=C) bonds should conform to specific geometry criteria outlined by Schmidt [7]. Specifically, following crystal structure studies involving cinnamic acid, Schmidt has revealed that two C=C bonds should be aligned parallel and separated by <4.2 Å (Fig. 3). These criteria place the four *p*-orbitals of two C=C bonds in close enough proximity and proper orientation to undergo the photoreaction. Schmidt has also shown that the photoreaction is regioselective, the stereochemical relationship of reactants being preserved in the photoproducts. Thus, UV-irradiation of crystalline α-*trans*-cinnamic acid, a polymorph of cinnamic acid wherein the olefins are arranged head-to-tail (ht), produces α-truxillic acid while irradiation of β-*trans*-cinnamic acid, wherein the olefins are arranged head-to-head (hh), produces β-truxinic acid. These geometry criteria, which are based on a wealth of crystallographic data, make up the 'topochemical postulates' [6].

3.1 Approaches to Control the [2+2] Photodimerization in the Solid State

Past approaches to control the [2+2] photodimerization in the solid state have fallen into two general categories; specifically, methods that employ either (i) intramolecular substitution or (ii) cavities.

3.1.1 Intramolecular Substitution

In the first approach, substituents have been covalently attached to molecules to 'steer' reactants, upon crystallization, in a necessary arrangement for reaction.

Schmidt, for example, has revealed that Cl atoms when attached to an aromatic ring tend to steer a molecule, by way of Cl···Cl interactions, such that the molecule assembles head-to-head, in a β-structure, wherein neighboring olefins are photoactive [9]. Following this work, electrostatic, as well as donor-acceptor, interactions have been used for steering the synthesis of molecules and, in some cases, polymers [10–15] (Fig. 4).

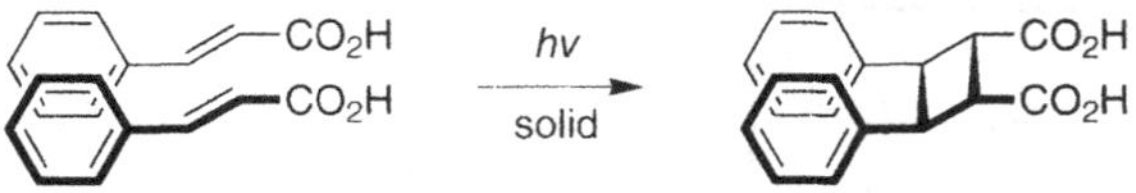

Fig. 3 Solid-state photoreaction of β-cinnamic acid to produce β-truxinic acid

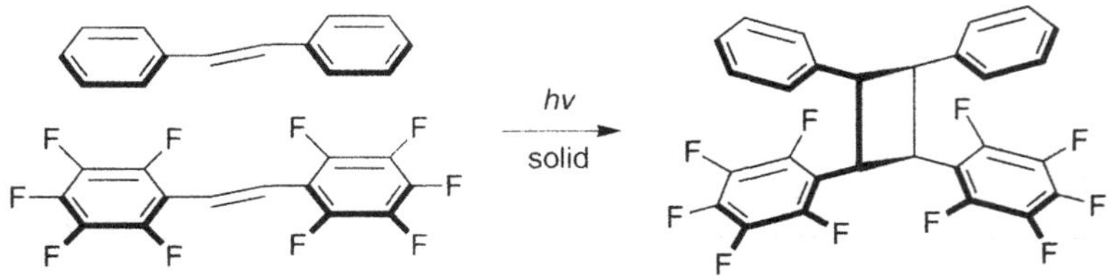

Fig. 4 Phenyl-perfluorophenyl interactions that orient two stilbenes in the solid state for a [2+2] photodimerization

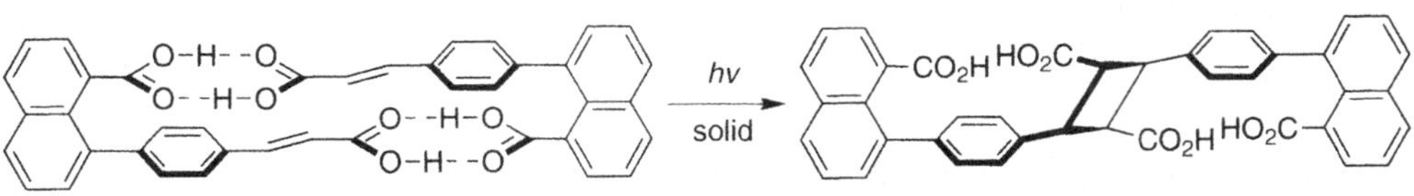

Fig. 5 'J'-shaped cinnamic acid that aligns two olefins within a hydrogen-bonded complex

In a notable success, a 'J'-shaped dicarboxylic acid has been shown by Feldman to self-assemble to form a dimer held together by four O-H···O hydrogen bonds wherein two C=C bonds conformed to the topochemical postulates (Fig. 5) [16]. The naphthalene unit served to preorganize the olefins at a separation distance suitable for the photoreaction. UV irradiation of the solid produced the expected photoproduct in quantitative yield.

3.1.2 Cavities

In the second approach, an auxiliary component has been employed to produce a cavity in the solid state that serves to host two molecules that undergo a [2+2] photoreaction.

Host cavities based on cyclodextrins [17] and saponite [18], for example, have been used to organize stilbenes by way of van der Waals and electrostatic forces, respectively. In the former, the host accommodated the guests within a tubular framework while, in the latter, the clay-like silicate accommodated the guests by swelling. In each case, the guest dimerized upon UV irradiation.

Extensive work involving auxiliary molecules that interact with reactants by way of hydrogen bonds and assemble to form solid-state cavities has also been described [19]. Specifically, tetraaryl diols have been shown by Toda to accommodate ketones within cavities wherein each host interacted with two guests by way of two O-H···O hydrogen bonds (Fig. 6). The guests packed within the

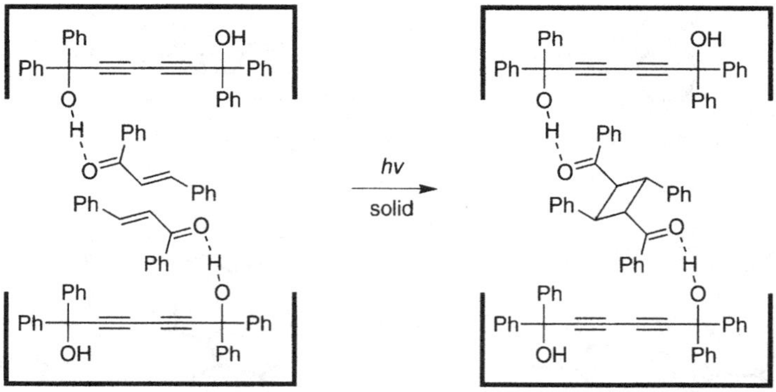

Fig. 6 Tetraaryl diol that forms cavities and interacts with olefins by way of hydrogen bonds

cavities such that two C=C bonds of two guests conformed to the topochemical postulates and photodimerized.

4 Solid-State Reactivity and Molecular Synthesis by Design

Although substituents and cavities have been successfully employed to direct the [2+2] photodimerization in the solid state, both methods have fallen short as general, or 'universal' [6], means to direct the solid-state reaction. In the former, the approach has typically involved steering forces that are relatively weak. This has made the method difficult to control since such forces are sensitive to, and therefore have difficulty competing with, subtle structure demands of close packing [5]. Numerous substituents have also been typically required to steer olefins for reaction which, in addition to yielding products with properties different than the parent reactant framework, can limit derivatization of the products. In the latter, the approach has required reactants that 'fit' within the cavities. This has limited the range of guests (i.e., sizes and shapes), and therefore products, accessible using the approach. Indeed, such lack of generality has meant that it is has remained difficult to synthesize molecules in the solid state [7] with degrees of synthetic freedom common to the liquid phase (e.g., reliability, functional groups). In effect, the products of [2+2] photodimerizations conducted in the solid state have been limited to each topochemical strategy, a fact that has thwarted efforts toward an ability to conduct target-oriented synthesis, or molecular solid-state synthesis by design [4].

5 Linear Templates

To develop a general means to control the [2+2] photodimerization in the solid state, and thereby provide a means to conduct molecular solid-state synthesis by design, recent efforts by us, and others, have focused on employing bifunctional molecules, in the form of linear templates [8], to direct the photoreaction (Fig. 2). It was anticipated that such molecules would offer a means to preorganize, according to principles of molecular recognition [20] and self-assembly [21], two molecules linearly within a discrete, solid-state complex for reaction. By enforcing the photoreaction to occur within a discrete complex using strong and directional noncovalent forces (e.g., hydrogen bonds), the approach could eliminate many vexatious problems of close packing that have made previous methods unreliable. Moreover, that the template would assemble along the periphery of the complex would mean that access to a variety of molecules may be achieved since the template could, in principle, adapt to changes to size (e.g., lengthening) and shape (e.g., bending) of the reactants. By making such changes systematically, an ability to conduct target-oriented solid-state synthesis could then be realized.

6 Linear Templates in the Solid State

The first study that illustrated that a linear template could be employed to direct a [2+2] photodimerization in the solid state was by Ito and Scheffer [22, 23]. Specifically, diamines (e.g., ethylenediamine) were revealed to undergo acid-base reactions with substituted cinnamic acids (e.g., *o*-methoxycinnamic acid), wherein each ammonium group formed an N^{+}-H···O^{-} hydrogen bond with a cinnamate (Fig. 7). Combinations of base and acid were photoactive, which was accounted for by the formation of three-component complexes wherein each base, in a *gauche* conformation, positioned two cinnamates for a [2+2] photo-

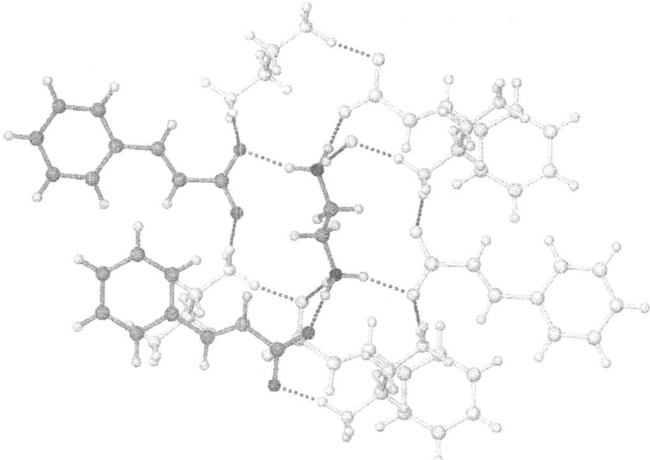

Fig. 7 Perspective of crystal structure of ethylenediammonium cinnamate. The dications and anions of a reactive complex are *highlighted in black*

reaction. It was also revealed that the ammonium ions could adopt an *anti* conformation that forced the cinnamates into an unreactive orientation while, in some cases, the cations participated in N^+-H···O^- forces with other components of the lattice to produce infinite assemblies that were photostable. Nevertheless, features of a linear template chemistry were demonstrated by the Ito and Scheffer approach.

Inspired by work of Ito and Scheffer, we turned to develop a linear template that could reliably organize two olefins for a [2+2] photoreaction. To achieve this goal, we anticipated that a rigid molecule with no more than two hydrogen bond donor groups separated at a distance consistent with the topochemical postulates [7] could serve as such a template. Specifically, from a structure study involving a bis(resorcinol)anthracene co-crystallized with anthraquinone [24], we anticipated that co-crystallization of resorcinol with *trans*-1,2-bis(4-pyridyl)ethylene (4,4′-bpe) would produce a discrete, four-component complex, 2(resorcinol)·2(4,4′-bpe), held together by four O-H···N hydrogen bonds wherein two molecules of 4,4′-bpe would be positioned for a [2+2] photodimerization (Fig. 8). Similar to Ito and Scheffer, the design would employ a bifunctional molecule to direct the photoreaction. In contrast to Ito and Scheffer, the rigidity of the aromatic ring and presence of only two hydrogen bond donor groups would largely ensure that 4,4′-bpe would be organized independent of structure effects of close packing in a solid.

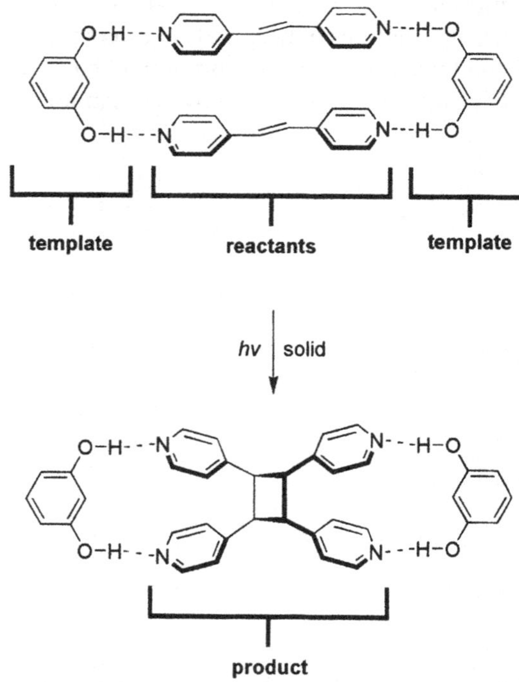

Fig. 8 Schematic of 2(resorcinol)·2(4,4′-bpe)

In line with our strategy, co-crystallization of resorcinol with 4,4′-bpe produced the discrete, four-component complex 2(resorcinol)·2(4,4′-bpe) (Fig. 9) [25]. In this arrangement, the four pyridyl units of the two olefins were stacked and twisted approximately orthogonal to the diol such that the two C=C bonds were oriented parallel and separated by 3.65 Å, an ideal position for a [2+2] photodimerization. UV-irradiation of the solid produced the anticipated photoproduct, *rctt*-tetrakis(4-pyridyl)cyclobutane (4,4′-tpcb), in 100% yield. The structure of the photoproduct was also determined by way of single-crystal X-ray analysis where 4,4′-tpcb was shown to assemble with resorcinol in a three-component complex, 2(resorcinol)·(4,4′-tpcb), held together by four O-H···N hydrogen bonds. The template-controlled reaction was also conducted on a gram scale where resorcinol and 4,4′-tpcb were separated by way of a liquid phase extraction.

Following our report, a flexible template based on a crown ether was shown to organize two C=C bonds within a discrete complex in the solid state for a [2+2] photoreaction [26]. Specifically, Garcia-Garibay and Stoddart demonstrated that reaction of a bisparaphenylene-34-crown-10 (bpp-34-crown-10) with a bis(dialkylammonium)-substituted stilbene (amm-stilb) produced the four-component complex 2(bpp-34-crown-10)·2(ammon-stilb) held together by eight N^+-H···O hydrogen bonds (Fig. 10). In this design, the cavity of the

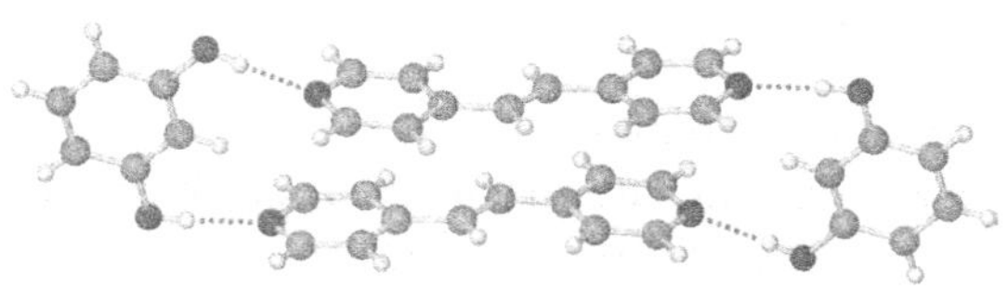

Fig. 9 Perspective of the four-component complex 2(resorcinol)·2(4,4′-bpe) (hydrogen bonds *dotted*)

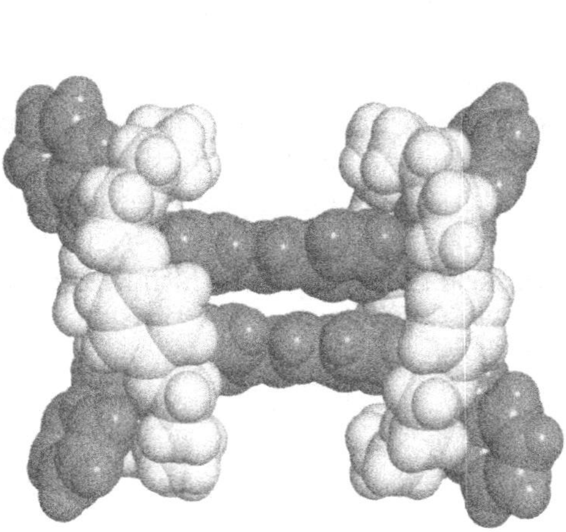

Fig. 10 A crown ether that directs a [2+2] photodimerization in the solid state: space-filling view of the four-component complex 2(bpp-34-crown-10)·2(ammon-stilb) (crown ether *light gray*; olefins *dark gray*)

crown ether was filled with the two reactants, the olefinic bonds being aligned parallel and separated by approximately 4.20 Å. UV irradiation of the solid resulted in a stereospecific dimerization of ammon-stilb to give a single diastereoisomer in approximate 80% yield, as confirmed by way of X-ray crystallography.

6.1 Fundamentals

To expand the scope of molecules accessible using linear templates, we anticipated that it would be necessary to determine the tolerance [27] of the self-assembly process to structural changes to the templates and reactants. Such modifications could, for example, introduce functionalities (e.g., lone pairs) that may interfere with the interaction between the template and reactants. In the ideal case, the template would interact preferentially with complementary sites of the reactants despite structural changes to the components of the complex. Moreover, with basic knowledge of structural changes tolerated by the assembly process, the designed synthesis of molecules that may be less available or completely inaccessible using conventional solution-based synthesis could be realized.

6.1.1 Modifying the Template

We have revealed that the assembly process is tolerant to substituents placed along the periphery of the template. In particular, derivatives of resorcinol with substituents largely unable to disrupt the hydrogen bonds between the templates and reactants (e.g., -R, -OR) have been shown to serve as templates. Thus, co-crystallization of 5-methoxyresorcinol (5-OMe-res) with 4,4′-bpe produced the four-component complex 2(5-OMe-res)·2(4,4′-bpe) with two C=C bonds aligned parallel and separated by 3.66 Å. UV-irradiation of the solid with broadband UV-radiation produced 4,4′-tpcb in quantitative yield [25].

In addition to resorcinol, we have shown that 1,8-naphthalenedicarboxylic acid (1,8-nap) serves as a template. Similar to 2(resorcinol)·2(4,4′-bpe), co-crystallization of 1,8-nap with 4,4′-bpe produced the four-component complex 2(1,8-nap)·2(4,4′-bpe) held together by four O-H···N hydrogen bonds (Fig. 11) [28]. The two carboxylic acid groups, each of which was twisted out of the plane

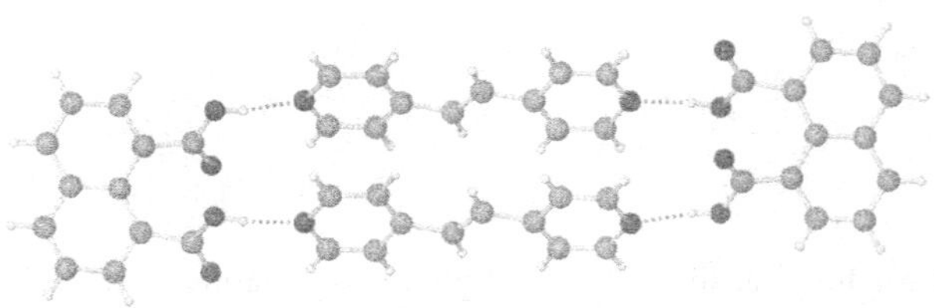

Fig. 11 Perspective of the four-component complex 2(1,8-nap)·2(4,4′-bpe)

of the naphthalene unit, positioned the reactants for reaction, the two C=C bonds being aligned parallel and separated by 3.73 Å. UV-irradiation of the solid using broadband UV radiation produced 4,4′-tpcb in quantitative yield.

6.1.2
Modifying the Reactants

We anticipated that the reactants could be modified by changing the number and/or position of the hydrogen bond acceptors sites, as well as by adding, or removing, functional groups. In this context, we have shown that the self-assembly process is tolerant to changes to both the number and position of the acceptor sites. A study involving the addition of alkyl groups to a hydrogen bond acceptor substrate has also been conducted.

6.1.2.1
Number of Hydrogen Bond Acceptor Sites

The minimalist olefins that may be organized by a linear template within a complex for reaction possess a single hydrogen bond acceptor site. Thus, a single template may organize two stilbazoles in a head-to-head arrangement for a regiocontrolled photodimerization to produce a head-to-head cyclobutane product.

Similar to 2(resorcinol)·2(4,4′-bpe), co-crystallization of resorcinol with 4-chlorostilbazole (4-Cl-sbz) produced the three-component complex (resorcinol)·2(4-Cl-sbz) held together by three O-H···N hydrogen bonds. Cl···Cl interactions also formed between the complexes such that nearest-neighbor assemblies constituted six-component complexes held together by Cl···Cl and O-H···N forces (Fig. 12) [29]. The complexes assembled to form offset linear stacks. As a consequence of these forces, the C=C bonds of the 'super-complex' were organized in close proximity, the two C=C bonds being separated by 3.98 Å. Unlike 2(resorcinol)·2(4,4′-bpe), however, the olefins of the complex adopted an antiparallel orientation, which was expected to render the C=C bonds photostable.

Although the olefins of (resorcinol)·2(4-Cl-sbz) were misaligned, the stilbazoles were photoactive. UV irradiation of (resorcinol)·2(4-Cl-sbz) using 350 nm radiation produced *rctt*-1,2-bis(4-pyridyl)-3,4-bis(4-chlorophenyl) cyclobutane (4-Cl-dpcb) in 100% yield. The generation of 4-Cl-dpcb was attrib-

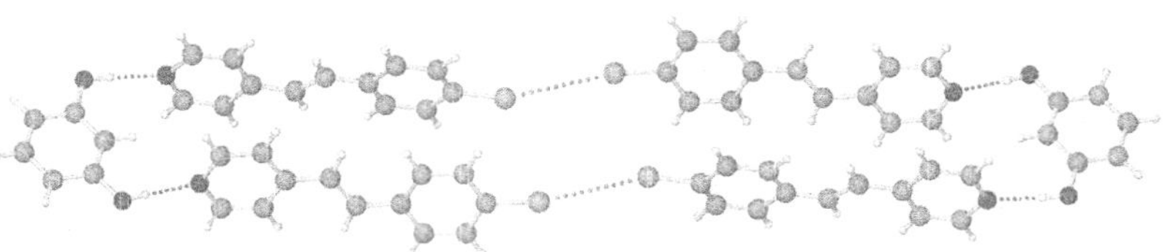

Fig. 12 Perspective of the six-component 'super-complex' 2(resorcinol)·4(4-Cl-sbz)

uted to the reactant undergoing a pedal-like change in conformation in the solid that enabled the C=C bonds to adopt a parallel orientation suitable to form the cyclobutane.

6.1.2.2
Position of the Hydrogen Bond Acceptor Sites

In addition to the number of hydrogen bond acceptor sites, we have demonstrated that the assembly process is tolerant to changes to the positions of the acceptor sites. Such changes were anticipated to force the templates to assemble so as to occupy different positions along the peripheries of the reactants.

Co-crystallization of 1,8-nap with *trans*-1,2-bis(2-pyridyl)ethylene (2,2′-bpe), for example, produced the discrete, four-component complex 2(1,8-nap)·2(2,2′-bpe) held together by four O-H·N hydrogen bonds (Fig. 13a) [28]. The C=C bonds of the complex were organized parallel and separated by 3.91 Å. In contrast to 2(1,8-nap)·2(4,4′-bpe), each template assembled in close proximity to each olefinic bond. UV-irradiation of the solid with broadband UV-radiation produced *rctt*-tetrakis(2-pyridyl)cyclobutane (2,2′-tpcb) in quantitative yield.

The ability of a linear template to orient two identical pyridyl units in a face-to-face stacked arrangement suggested that a linear template could assemble two reactants with two different pyridyl units for a head-to-head photodimerization. Since different combinations of hydrogen bond acceptor sites could be employed for the photoreaction (i.e., *trans*-1-(*n*-pyridyl)-2-(*m*-pyridyl)ethylene (where: *n*, *m*=2, 3, or 4; *n*≠*m*), a general means to establish regiocontrol of the photoreaction could be achieved.

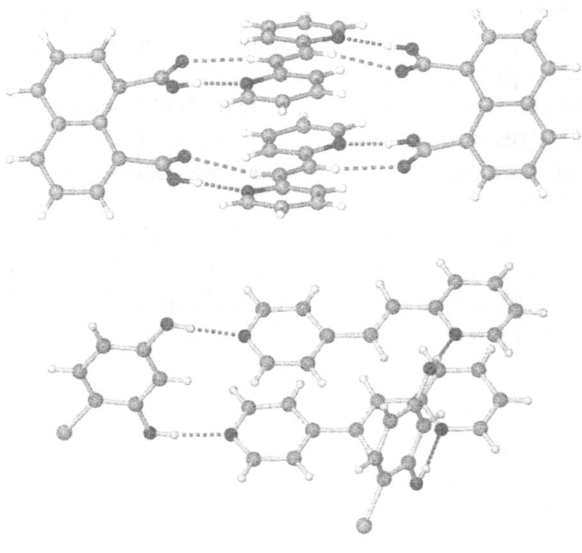

a

b

Fig. 13a, b Perspective of complexes based on different hydrogen bond acceptor sites: **a** 2(1,8-nap)·2(2,2′-bpe); **b** 2(4-Cl-res)·2(2,4-bpe)

Thus, co-crystallization of 4-chlororesorcinol (4-Cl-res) with *trans*-1-(2-pyridyl)-2-(4-pyridyl)ethylene (2,4-bpe) produced the four-component complex 2(4-Cl-res)·2(2,4-bpe) (Fig. 13b) [30]. The bipyridine was organized within the complex in a head-to-head fashion, each diol interacting with either a 2-pyridyl or 4-pyridyl unit. The C=C bonds were aligned parallel and separated by 3.89 Å. UV-irradiation of the solid using broadband UV radiation produced the head-head photoproduct 2,4-tpcb-hh in quantitative yield. Similar to 2(4-Cl-res)·2(2,4-bpe), co-crystallization of 1,8-nap with *trans*-1-(3-pyridyl)-2-(4-pyridyl)-ethylene (3,4-bpe) produced the four-component complex 2(1,8-nap)·2(3,4-bpe) wherein the two olefins were organized head-to-head [31]. The two C=C bonds were aligned parallel and separated by 3.58 Å. UV-irradiation of 2(1,8-nap)·2(3,4-bpe) produced the head-to-head cycloproduct, 3,4-tpcb-hh, in quantitative yield.

6.1.2.3
Functional Groups

Our first study to determine tolerance of the assembly process to functional groups has involved methylene linkages. Specifically, co-crystallization of resorcinol with 1,2-bis(4-pyridyl)ethane (4,4′-bipyeth) produced, in contrast to 2(resorcinol)·2(4,4′-bpe), an infinite 1D polymer, (resorcinol)·(4,4′-bipyeth) [32]. In this arrangement, the hydroxyl groups of the resorcinol unit adopted a divergent conformation. Similar to neat crystalline 4,4′-bipyeth [33], the ethane moiety and pyridyl units of the bipyridine were twisted approximately orthogonal (Fig. 14). Notably, the closest N(bipyridine)···C-H(resorcinol) separations of the 1D array involved the 4- and 6- positions of the diol. Thus, the flexibility of the methylene linkages, as realized by the twisting of the ethane moiety, enabled the diol and bipyridine to form the 1D array, largely prohibiting the components to assemble to form the targeted discrete structure.

6.1.3
Template-Switching

Although the goal to organize two reactants within a discrete hydrogen-bonded complex was formulated as a means to isolate two reactive olefins from structure effects of close packing, it was evident that such a complex was not completely independent of such packing effects. Indeed, in the case of (resorcinol)·

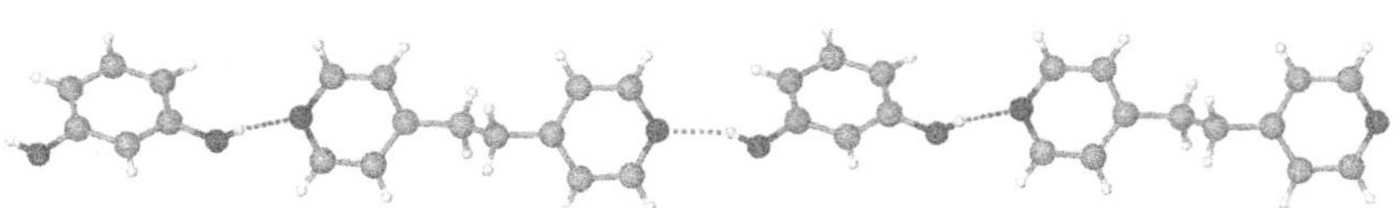

Fig. 14 Perspective of the infinite 1D polymer (resorcinol)·(4,4′-bipyeth)

2(4-Cl-sbz) [29], the C=C bonds of the complex, although photoactive, assembled misaligned while the components of 2(resorcinol)·2(4,4′-bipyeth) [32] assembled to give an infinite, rather than a discrete structure.

To confront further effects of close packing, and thereby establish further control of the [2+2] photodimerization [7], we anticipated that changing the template to derivative, by adding a substituent along the periphery of the template, could be used to promote the formation of a discrete complex that exhibits desired photoreactivity. The addition of a substituent to the template would effectively modify the shape of the template which, de facto, would be expected to lead to a different crystal packing environment [34] that may support the formation of a desired photoactive structure. Although an ability to predict successfully solid-state packing has remained elusive, we anticipated that such 'template-switching' [35] would, in effect, exploit unpredictable structural consequences of crystal packing, enabling the approach to provide a means to increase the likelihood of achieving the topochemical postulates of Schmidt for a [2+2] photoreaction.

6.1.3.1
Olefin Alignment

In line with our template-switching strategy, co-crystallization of 4-ethylresorcinol (4-Et-res) with 4-Cl-sbz produced the three-component complex (4-Et-res)·2(4-Cl-sbz) held together by three O-H···N hydrogen bonds (Fig. 15) [29]. Unlike (resorcinol)·2(4-Cl-sbz), the C=C bonds of the complex, which were separated by 3.92 Å, were aligned parallel. That the C=C bonds of 4-Cl-sbz assembled parallel was consistent with the ability of stacked C=C bonds to adopt either a parallel or nonparallel orientation in a solid [29]. UV irradiation of the solid produced the targeted product, 4-Cl-dpcb, in near quantitative yield.

6.1.3.2
Assembly Process

That 4,4′-bipyeth and resorcinol formed a 1D array also prompted us to apply template-switching to generate a discrete complex involving the bipyridine.

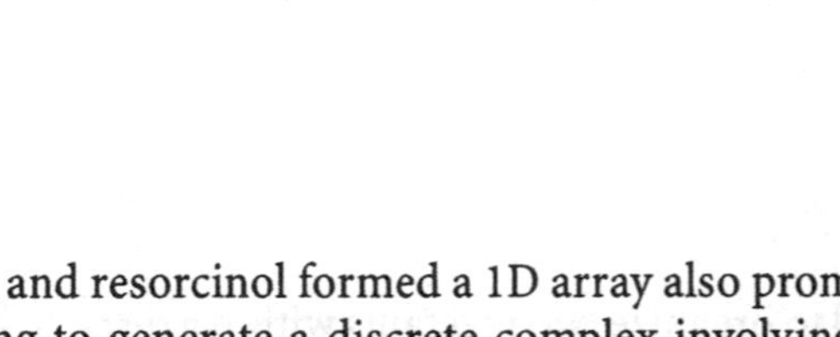

Fig. 15 Achieving olefin alignment: Perspective of the three-component complex (4-Et-res)·2(4-Cl-sbz)

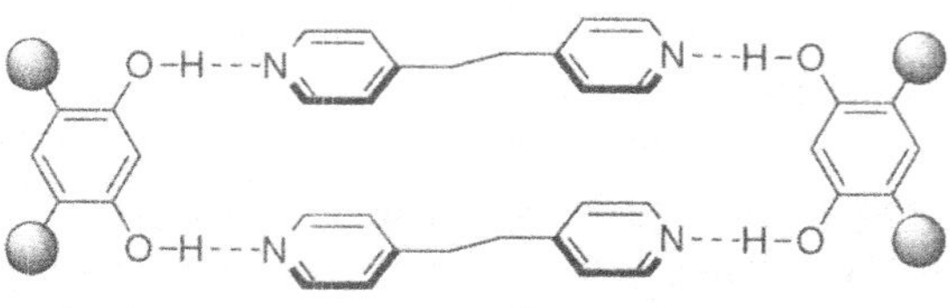

Fig. 16 Scheme of a discrete assembly achieved using a resorcinol derivative

Stacking of methylene groups in the solid state was established [36] while the known flexibility of 4,4′-bipyeth [37] suggested that the bipyridine could be forced to adopt a planar conformation suitable for a discrete structure. Moreover, it was reasoned that substituents placed in the 4- and/or 6- positions of a resorcinol may enable the diol to direct stacking of 4,4′-bipeth within a discrete complex (Fig. 16). In addition to making a 1D assembly energetically unfavorable by sterically protecting each 'side' of the diol [38], the positioning of the substituents adjacent to the hydroxyl groups would promote the diol to adopt a convergent conformation. In effect, the substituents would largely preorganize [39] a resorcinol to form a targeted discrete structure.

As anticipated, co-crystallization of either 4-Cl-res or 4,6-dichlororesorcinol (4,6-di-Cl-res) with 4,4′-bipyeth produced the four-component complexes 2(4-Cl-res)·2(4,4′-bipyeth) and 2(4,6-di-Cl-res)·2(4,4′-bipyeth), respectively, held together by four O-H···N hydrogen bonds. The bipyridines stacked alongside each resorcinol unit such that the methylene linkages were aligned and separated by 4.13 Å [32]. In contrast to (resorcinol)·(4,4′-bipyeth), the bipyridines adopted an approximate planar conformation wherein the ethane moiety and pyridyl units were twisted toward co-planarity (Fig. 17). Thus, in addition to promoting the formation of the discrete complex, the presence of the substituents along the periphery of each diol affected the conformation of the bipyridine.

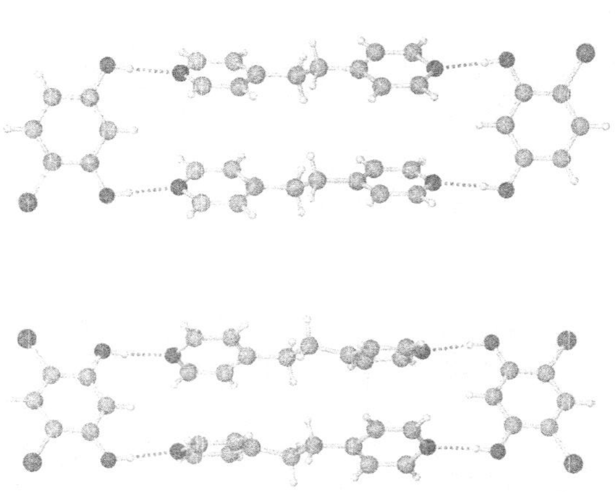

a
b

Fig. 17a, b Achieving a discrete complex: Perspectives of: **a** 2(4-Cl-res)·2(4,4′-bipyeth); **b** 2(4,6-di-Cl-res)·2(4,4′-bipyeth)

6.2
Target-Oriented Synthesis

The tolerance of the assembly process to structural changes to the templates and reactants demonstrated that linear templates could be used to direct reliably the [2+2] photodimerization in the solid state. Moreover, that the templates assembled along the peripheries of the hydrogen-bonded complexes suggested that the templates could be used to synthesize molecules of varying size (e.g., lengthening) and shape (e.g., bending) since the templates could, in principle, adapt to changes to size and shape of the reactants. An increase in size to the reactants could also be accompanied by attachment of additional reactive sites (i.e., C=C bonds) to the reactants. Indeed, such abilities to dictate the size and shape of a molecule would be reminiscent of synthetic flexibilities of the liquid phase where such synthetic 'freedoms' are routinely employed to synthesize molecular targets [4]. In the case of the solid state, the linear templates would effectively serve as molecular-level tools that enable molecular syntheses by design.

6.2.1
Cyclophane

To employ a linear template as a tool to construct a molecule, we focused upon a [2.2]paracyclophane as a target (Fig. 18). The cyclic molecule is a member of a general class of layered aromatic compounds that has gained widespread interest owing to the ability of such molecules to provide challenging targets in organic chemistry and find applications in areas ranging from chemical syn-

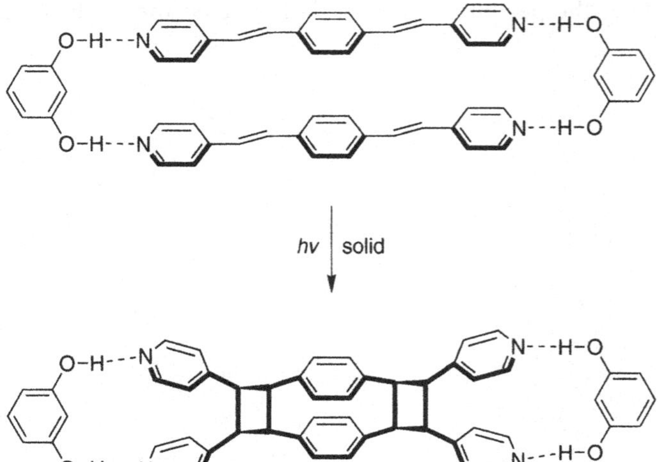

Fig. 18 Scheme illustrating the template-controlled solid-state synthesis of a [2.2]paracyclophane

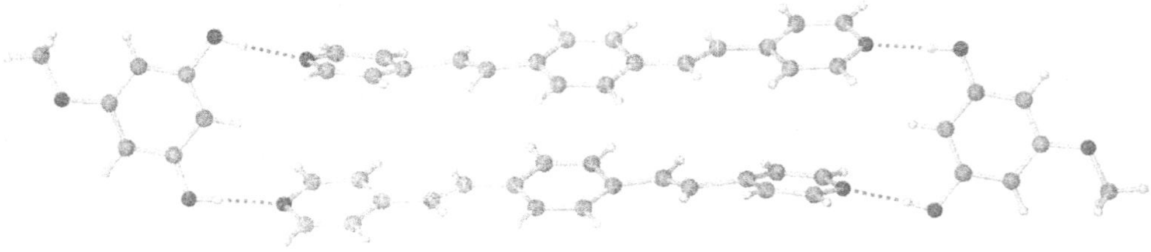

Fig. 19 Perspective of the four-component complex 2(5-OMe-res)·2(1,4-bpeb)

thesis (e.g., catalysis) to materials science (e.g., optical materials) [40]. Specifically, we anticipated that co-crystallization of a resorcinol, or a 1,8-nap, with the diolefin 1,4-bis(2-(4-pyridyl)ethenyl)benzene (1,4-bpeb) would produce the four-component complex 2(resorcinol)·2(1,4-bpeb) with two olefins positioned for a 'double' [2+2] photodimerization. UV-irradiation of the solid would give a tricyclic product with an inner cyclic core reminiscent of a [2.2]paracyclophane. The periphery of the product would, in effect, possess an imprint of the two templates along the exterior of the molecule [8]. The method would also provide a novel entry to cyclic molecules where low yields often requiring high dilution conditions (i.e. large batches of solvent) are common in solution [41].

As anticipated, co-crystallization of 1,4-bpeb with 5-OMe-res produced the discrete, four-component complex 2(5-OMe-res)·2(1,4-bpeb) held together by four O-H···N hydrogen bonds (Fig. 19) [25]. In this arrangement, the olefins of 1,4-bpeb, one of which was disordered across two positions (occupancies: 70:30), were aligned such that the olefinic bond of the major site conformed to the geometry criteria of the topochemical postulates with the ordered olefin, the two C=C bonds being separated by 3.70 Å. UV-irradiation of 2(5-OMe-res)·2(1,4-bpeb) produced the targeted [2.2]paracyclophane, along with a monocyclized dimer and indefinable products, in 60% yield. The formation of the monocyclized product was attributed to face-to-face stacking of the complexes which introduced close contacts (i.e., <4.2 Å) of C=C bonds in the solid. Thus, by adapting to the size of a lengthened reactant, the template served as a tool, providing regio- and stereocontrolled access to a lengthened cyclophane target.

6.2.1.1
Template-Switching to Influence Yield

With a target-oriented synthesis of a paracyclophane realized, we anticipated that template-switching could be used to generate a packing environment that, in contrast to 2(5-OMe-res)·2(1,4-bpeb), could accommodate the cyclophane in quantitative yield. The ability to generate the cyclophane in quantitative yield could also be employed to provide ready access to gram quantities of the product.

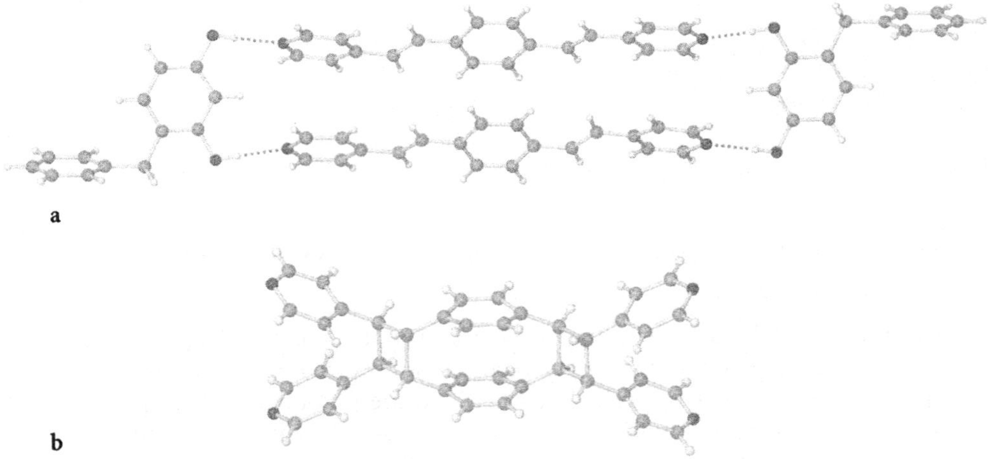

Fig. 20a, b Achieving 100% yield and gram quantity construction of a cyclophane: a perspective of the four-component complex 2(4-benz-res)·2(1,4-bpeb); b the targeted [2.2]paracyclophane

In line with our strategy, co-crystallization of 4-benzylresorcinol (4-benz-res) with 1,4-bpeb produced the discrete, four-component complex 2(4-benz-res)·2(1,4-bpeb) held together by four O-H···N hydrogen bonds (Fig. 20a) [35]. The olefinic groups of the complex assembled parallel and separated by 3.60 Å. In contrast to 2(5-OMe-res)·2(1,4-bpeb), the C=C bonds of the complex were ordered and nearest-neighbor complexes assembled offset such that there were no close C=C contacts involving neighboring complexes in the solid. Moreover, UV-irradiation of 2(4-benz-res)·2(1,4-bpeb) resulted in the quantitative conversion of 1,4-bpeb to the targeted [2.2]paracyclophane, as confirmed by way of single-crystal X-ray analysis (Fig. 20b). Indeed, the quantitative generation of the cyclophane in 2(4-benz-res)·2(1,4-bpeb) provided ready access to gram quantities of the target which enabled us to determine the structure of the product.

7 Solid-State Polymerization of Acetylenes

For a thermal polymerization of a diacetylene to give a polydiacetylene to occur, the diacetylenes, similar to a [2+2] photodimerization, must be organized in a prescribed geometry to react [42]. Specifically, the carbon atoms of a diacetylene must adopt a linear arrangement with an intermolecular repeat spacing of approximately 5.0 Å and tilt angle of 45° (Fig. 21). Most diacetylenes, similar to olefins, however, do not crystallize in an arrangement suitable for solid-state reaction. Moreover, the application of molecules that assemble and, in effect, serve as linear templates has offered a solution to this problem.

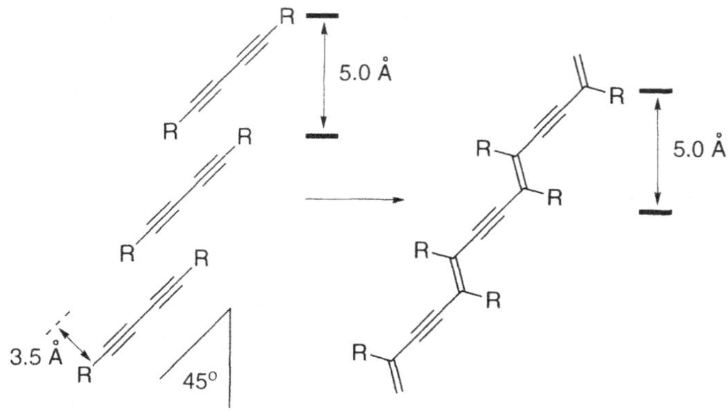

Fig. 21 Topochemical requirement for the polymerization of diacetylenes in the solid state

7.1 Templates to Direct Di- and Triacetylene Polymerization

A strategy based on host molecules that assemble and function as templates that direct the thermal polymerization of diacetylenes has been described by Fowler and Lauher [43]. The method has involved molecules that self-assemble into 1D hydrogen-bonded polymers. The polymers have been shown to impose a characteristic intermolecular repeat distance and tilt angle upon diacetylenes such that the reactants conform to the geometric requirements for solid-state polymerization. Thus, co-crystallization of the oxalamide of glycine with a bis-pyridyl substituted diacetylene was revealed to impose a repeat distance and tilt angle of 4.97 Å and 43°, respectively, upon the diacetylene (Fig. 22) [43]. The resulting solids slowly polymerized at room temperature to give the targeted diacetylene polymer. Notably, thermal annealing at a slightly elevated temperature increased the rate of conversion. In addition to a diacetylene, the

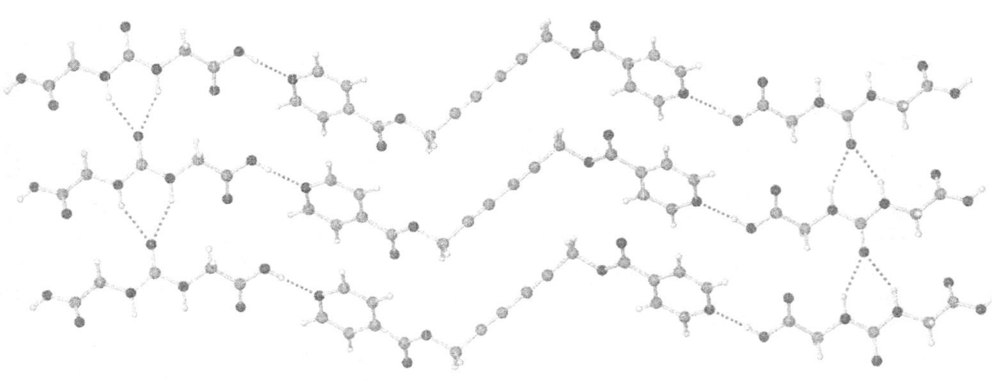

Fig. 22 Perspective of a co-crystal involving an oxalamide of glycine and a bis-pyridyl substituted diacetylene

method has been applied a triacetylene in an unprecedented polymerization that gives a triacetylene polymer [44].

8 Summary and Outlook

Linear templates have served as tools that control reactivity in the solid state. By preorganizing olefins within discrete hydrogen-bonded complexes for [2+2] photodimerization [22, 23, 25, 26], linear templates have been shown to provide a means to isolate reactants from subtle structural effects of close packing such that it is now possible to conduct molecular solid-state synthesis by design [25]. The approach has also been applied to direct polymerizations in the solid state, yielding both di- and triacetylenes [49, 50].

Whereas an early goal of applying linear templates to the solid state has been to control reactivity, it has become clear that the method has also provided an equally intriguing means to employ principles of molecular recognition [20] and self-assembly [21] to direct the syntheses of molecules and polymers. Indeed, the degree of organization provided by the solid state has enabled multiple equilibria involving noncovalent complexes normally encountered in the liquid phase to be avoided such that it is now possible to reliably capture and transform self-assembled complexes [3] into complex covalent structures. Consequently, it now possible to address what kinds of covalent structures can be predesigned and constructed using the approach. Notably, the hydrogen bond acceptor sites attached to the products also present functionality intimately connected to the functions of the templates which makes the products attractive building blocks [21] in coordination- and hydrogen-bond-driven self-assembly [30, 45], while the imprints of the templates, upon removal, may yield 'true' targets. Studies are underway to assess these, and related, possibilities of the linear template, solid-state approach.

References

1. Desiraju GR (1995) Angew Chem Int Ed Engl 34:2311
2. Anastas PT, Warner JC (1998) Green chemistry: theory and practice. Oxford University Press, New York
3. Brunet P, Demers E, Maris T, Enright GD, Wuest JD (2003) Angew Chem Int Ed 42:5253
4. Nicolaou KC, Vourloumis D, Winssinger N, Baran PS (2000) Angew Chem Int Ed 39:44
5. Holman KT, Pivovar AM, Swift JA, Ward MD (2001) Acc Chem Res 34:107
6. Ramamurthy V, Venkatesan K (1987) Chem Rev 87:433
7. Schmidt GMJ (1971) Pure Appl Chem 27:647
8. Anderson S, Anderson HL (2000) Templates in organic synthesis: definitions and roles. In: Diederich F, Stang PJ (eds) Templated organic synthesis. Wily-VCH, New York, pp 1–38
9. Sharma JARP, Desiraju GR (1986) Acc Chem Res 19:222

10. Coates GW, Dunn AR, Henling LM, Ziller JW, Lobkovsky EB, Grubbs RH (1998) J Am Chem Soc 120:3641
11. Sharma CVK, Panneerselvam K, Shimoni L, Katz H, Carrell HL, Desiraju GR (1994) Chem Mater 6:1282
12. Maekawa Y, Kato S, Hasegawa, M (1991) J Am Chem Soc 113:3867
13. Gnanaguru K, Ramasubbu N, Venkatesan K, Ramamurthy V (1985) J Org Chem 50:2337
14. Desiraju GR, Kamala R, Kumari BH, Sarma JARP (1984) J Chem Soc Perkin Trans II 181
15. Jones W, Nakanishi H, Theocaris CR, Thomas JM (1980) J Chem Soc Chem Commun 610
16. Feldman KS, Campbell RF (1995) J Org Chem 60:1924
17. Brett TJ, Alexander JM, Clark JL, Ross CR II, Harbison GS, Stezowski JJ (1999) Chem Commun 1275
18. Takagi K, Usami H, Fukaya H, Sawaki Y (1989) J Chem Soc Chem Commun 1174
19. Toda F (1995) Acc Chem Res 28:480
20. Rebek J Jr (1990) Acc Chem Res 23:399
21. Fyfe MCT, Stoddart JF (1997) Acc Chem Res 30:393
22. Ito Y, Borecka B, Trotter J, Scheffer JR (1995) Tetrahedron Lett 36:6083
23. Ito Y, Borecka B, Olovsson G, Trotter J, Scheffer JR (1995) Tetrahedron Lett 36:6087
24. Aoyama Y, Endo K, Anzai T, Yamaguchi Y, Sawaki T, Kobayashi K, Kanehisa N, Hashimoto H, Kai Y, Masuda Y (1996) J Am Chem Soc 118:5562
25. MacGillivray LR, Reid JL, Ripmeester JA (2000) J Am Chem Soc 122:7817
26. Amirsakis DG, Garcia-Garibay MA, Rowan SJ, Stoddart JF, White AJP, Williams DJ (2001) Angew Chem Int Ed 40:4256
27. Trnka TM, Grubbs RH (2001) Acc Chem Res 24:18
28. Papaefstathiou GS, Kipp AJ, MacGillivray LR (2001) Chem Commun 2462
29. MacGillivray LR, Reid JL, Ripmeester JA, Papaefstathiou GS (2002) Indust Eng Chem Res 41:4494
30. Hamilton TD, Papaefstathiou GS, MacGillivray LR (2002) J Am Chem Soc 124:11606
31. Varshney DB, Papaefstathiou GS, MacGillivray LR (2002) Chem Commun 1964
32. Papaefstathiou GS, MacGillivray LR (2001) Org Lett 3:3835
33. Ide S, Karacan N, Tufan Y (1995) Acta Crystallogr Sect C Cryst Struct Commun C51:2304
34. Ramamurthy V (1986) Tetrahedron 42:5753
35. Friscic T, MacGillivray LR (2003) Chem Commun 1306
36. Menger FM, Lee, JJ, Hagen KS (1991) J Am Chem Soc 113:4017
37. Hennigar TL, MacQuarrie DC, Losier P, Rogers RD, Zaworotko MJ (1997) Angew Chem Int Ed Engl 36:972
38. Whitesides GM, Simanek EE, Mathias JP, Seto CT, Chin DN, Mammen M, Gordon DM (1995) Acc Chem Res 28:37
39. Cram DJ (1988) Science 240:760
40. Cram DJ, Cram JM (1971) Acc Chem Res 4:204
41. Dietrich B, Viout P, Lehn JM (1993) Macrocyclic chemistry. Wiley-VCH, New York
42. Enkelmann V (1984) Adv Polym Sci 63:91
43. Kane JJ, Liao RF, Lauher JW, Fowler FW (1995) J Am Chem Soc 117:12003
44. Xiao J, Yang M, Lauher JW, Fowler FW (2000) Angew Chem Int Ed 39:2132
45. Papaefstathiou GS, MacGillivray LR (2002) Angew Chem Int Ed 41:2070

Topics in Current Chemistry (2004) 248: 223–260
DOI 10.1007/978-3-540-44505-0

Gels as Templates for Nanotubes

Jong Hwa Jung[1] · Seiji Shinkai[2] (✉)

[1] Nano Material Team, Korea Basic Science Institute (KBSI), 52 Yeoeun-dong, Yusung-gu, Daejeon 305-333, Korea
jonghwa@kbsi.re.kr

[2] Department of Chemistry and Biochemistry, Graduate School of Engineering, Kyushu University, Fukuoka 812-8581, Japan
seijitcm@mbox.nc.kyushu-u.ac.jp

Abstract The design of artificial models for "biomineralization" processes that utilize sol-gel polycondensation has led to a merging of inorganic materials science and supramolecular organic chemistry. This chapter introduces self-assembled superstructures of cyclohexane-based, sugar-integrated, and crown-appended cholesterol gelators, which are used as versatile building blocks in organogels and as templates for sol-gel transcription. This review focuses on the use of these self-assembled organic superstructures in the creation of novel inorganic materials with controlled morphologies.

Keywords Cholesterol · Crown ether · Gelator · Nanotube · Organogel · Self-assembly · Sol-gel transcription · Sugar

List of Abbreviations

Tgel	Sol-gel transition temperature
CD	Circular dichroism
SEM	Scanning electron microscope
TEM	Transition electron microscope
TEOS	Tetraethoxy silane
MAS	Magic angle spinning

1 Introduction

Everyone knows what a gel is, but from a scientific point of view the term gel encompasses chemistry with very diverse systems. Well known gel systems include, for instance, dilute solutions of polymers, proteins, and surfactants in water and organic solvents. These gel systems are important in medicine, biology, chemistry, and physics, and find many applications in the photographic, cosmetic, food and petroleum industries [1, 2]. However, as D. Jordan Lloyd wrote in 1926: 'The colloidal condition, the "gel", is one which is easier to recognize than to define...'. Although an exact definition of a gel is still a problem, from a topological point of view gels can be defined as dilute mixtures of least two components, in which both components form a separate continuous phase throughout the system [3]. This definition includes not only gels composed of a solid-like and a liquid phase, but also those composed a solid and a gas phase. For most gels a solid-like phase is the minor component which forms a three dimensional network structure within the fluid or gas phase. For solid-fluid gels, it can be said that the network structure prevents the fluid from flowing, whereas the liquid phase prevents the network from collapsing [4]. The coexistence of a solid network structure together with a liquid phase distinguishes gels from pure solid, liquid crystalline, or fluid materials and gives gels their unique elastic properties.

Often, gels are divided into two groups depending on the type of interactions which hold the network structure together. In chemical gels, both the individual filaments of which the network consists, as well as the connections (junction zones) between the filaments to form the network, are created through the formation of covalent bonds. Cross-linked polymer gels belong to inorganic oxides, like vanadium pentoxide and silica. The properties of chemical gels can be as different as their constituents, but they have a feature in common that formation is irreversible. Opposed to chemical gels are the physical gels. In this type of gels the network structure is build from small molecular subunits, which are held together by non-covalent interactions. The attractive forces between the molecular subunits can be very specific, like hydrogen bonding or π-π stacking, but solvophobic and entropy effects also play an important role. Because the non-covalent interactions stabilizing the network are offers of magnitude weaker than covalent bonds and comparable to the thermal energy, physical gels exhibit a characteristic reversible transition from a gel phase to a solution at moderate temperature. Properties of physical gels, like thermal stability and viscoelastic behavior, are therefore the result of a delicate balance in the interactions between them.

Because the literature on organogelators has been reported in several excellent reviews by Terech and Weiss [5, 6] as well as Esch and Feringa [7], we will give in the following section only a brief overview with some representative examples of various organogelators such as sugar-, cyclohexane- and cholesterol-based gelators, which can be useful as templates in sol-gel reaction to obtain new functional inorganic nanotubes. Finally, we will discuss sol-gel transcription into inorganic materials using organogelators as templates, which is a new area having large potentials and whose development can be achieved only through a better understanding or organogels.

2 Examples of Organogelators

2.1 Introduction

Gelators fall into one of two categories according to the driving force causing their molecular aggregation: hydrogen-bond-based gelators [8–15] and non-hydrogen-bond-based gelators [16–22]. Typical examples of the former are aliphatic amide- or urea-based cyclohexane-, peptide- and sugar-based derivatives that exhibit distinctive helical structures in the fibrous aggregates formed in certain solvents [8–15]. The superstructures observed as fibrous aggregates in the organic gels of aliphatic amide derivatives provide the complementarity needed for consecutive intermolecular hydrogen-bonding interaction.

Typical examples of the latter are cholesterol derivatives [16–22]. The aggregation of these compounds bearing a crown moiety is based on their π-π stacking, van der Waals, and solvophobic properties. Particularly, our research group synthesized a variety of crown-appended cholesterol gelators, which turned out to be very useful to crate inorganic nano-materials. We will discuss these crown-appended cholesterol gelators in detail.

2.2 Carbohydrate Derivatives as Gelators

Methyl 4,6-*O*-benzylidene derivatives of monosaccharides, a relatively new member among the family of the low molecular weight gelators **1–11** [10, 23, 24], in general, are advantageous to study the structural prerequisites for gelation ability, because of their facile molecular design using the rich carbohydrate library. Methyl 4,6-*O*-benzylidene derivatives of monosaccharides have been intensively investigated with regard to their applicability as a low molecular weight gelators. They stabilize organogels through the establishment of rigid, strong and highly directional hydrogen-bonds, which are proved by temperature dependent NMR spectroscopy and FTIR spectroscopes, no v_{OH} peak for a free OH group is detectable for the solid samples (KBr) of all monosaccharides. In the gel state, all signals are more broadened and the v_{OH} peaks appear in two groups between 3220 and 3475 cm^{-1} and between 3573 and 3588 cm^{-1}, which can be assigned to the intermolecular hydrogen bonds and free OH-groups, respectively.

In addition, the gels provided by sugar-integrated gelators offer the unique possibility of probing their thermal stability by monitoring a change in their intermolecular aggregation near the T_{gel} region by ^1H NMR spectroscopy (compounds **1–11**). Due to the formation of strong intermolecular bonds at T_{gel}, the

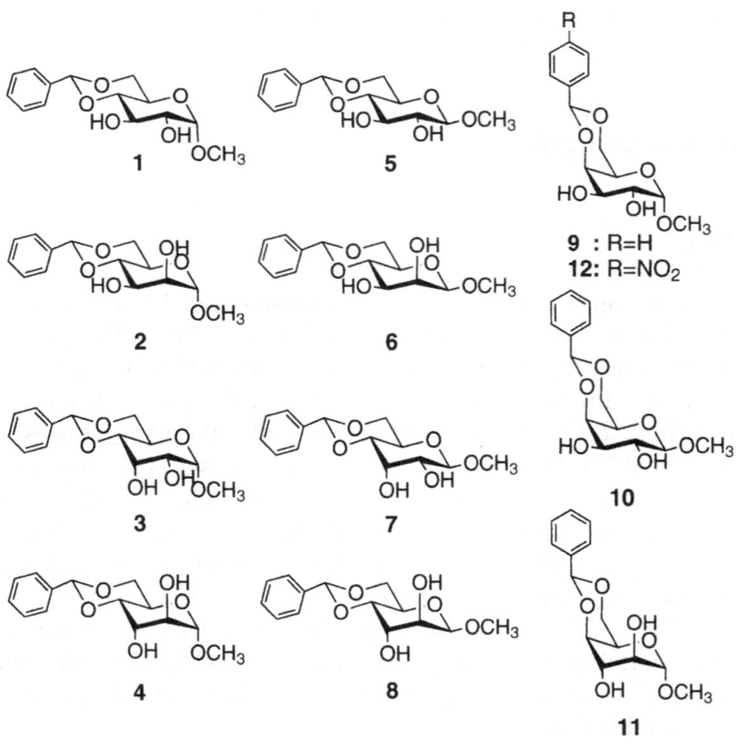

chemical shifts of δ_{OH} have their maximum downfield values at T_{gel}. This phenomenon is reflected by the line broadening effect in the ^{1}H NMR spectra. In the gel phase, the mobility of gelator molecules is significantly suppressed, whereas in the sol phase it is comparable to that of a homogenous solution. Therefore, the half-height width of the peak ($\delta_{1/2}$) of the PhCH methine proton is nearly constant above T_{gel}, while it increases with falling temperature below T_{gel}.

The absolute configuration of the monomer correlates significantly with the ability to gelate solvents. The optimal requirements for these compounds to be classified as a gelator are fulfilled only by derivatives of D-mannose and D-galactose. Methyl 4,6-*O*-benzylidene-α-D-galactopyranoside (**9**) and methyl-4,6-benzylidene-α-D-mannopyranoside (**11**) act as "supergelators" for apolar solvents. Both saccharide derivatives can gelate hydrocarbons (*n*-hexane, *n*-heptane, *n*-octane, *n*-decane and cyclohexane) in concentrations of approximately 0.03–0.05 wt% [25]. Compared to the typical gelator concentration range of 0.9–2.48 mmol/l these values represent the lowest concentrations reported for organic solvents so far.

In addition, four sugar-integrated gelators bearing a *p*-aminophenyl group which are expected to exert an additional hydrogen-bonding effect and a metal coordination effect on the gelation ability were synthesized. α-D-Galactose-based **14b** was only soluble or precipitated and β-D-glucose-based **16b** gelated only two of **15** solvents whereas α-D-glucose-based **13b** acted as an excellent gelator which could gelate **8** solvents [25]. The T_{gel} values for **13b** were higher by 41–78 °C than those for α-D-glucose-based **13a** bearing a *p*-nitrophenyl

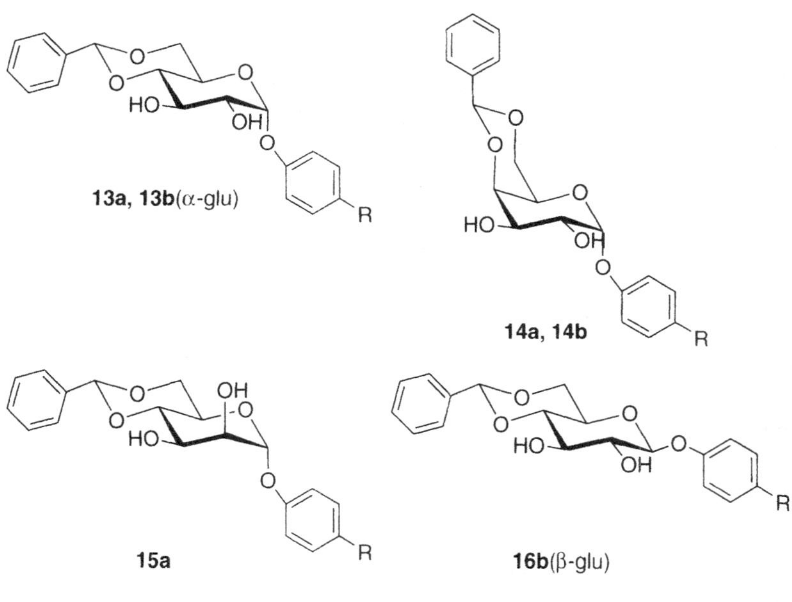

13a, 13b(α-glu)

14a, 14b

15a

16b(β-glu)

a: R= $-NO_2$
b: R= $-NH_2$

group, due to the hydrogen-bonding interaction including the amino group. Also, the T_{gel} values for ethanol gel of **13b** were markedly improved by addition of $AgNO_3$, $CoCl_2$ or $CdCl_2$ by coordination bonds between metal ion and gelator. These results indicate that the bridging effect between gelators with metal ions acts as the origin of the gel reinforcement.

2.3 Amphiphilic Gelators

Hydrogels have been extensively studied because of their applicability to tissue engineering [26] and the development of new materials that reversibly respond to external stimuli [27–30]. They usually consist of covalently or non-covalently cross-linked polymers and contain a large amount of water that fills the interstitial spaces in the network. These hydrogels have complicated intermolecular association modes that are difficult to define. In contrast, organogels are one-dimensional aggregates of low-molecular-weight compounds, for which the definition of the association models is relatively easy. However, only a limited number of hydrogels are formed by low-molecular-weight compounds [31–40].

Very recently, the Shimizu group reported on sugar-derivatives as amphiphilic gelators, which are some of the best aldopyranoside-based organogelators (**17** to **20**) [36,41,42]. Compounds **17** and **18** can gelate seven out of ten organic solvents as well as water in the presence of a small amount of alcohol-based solvent (ca. 1.0 wt%) due to their extremely low solubility in water as shown in Table 1. This indicates that they can act as versatile gelators of organic

Table 1 Gelation ability of **17–20** in organic solvents and water[a]

Solvent	Mono-head amphiphile		Bolaamphiphile	
	17	**18**	**19**	**20**
Methanol	S	S	S	S
Ethanol	S	S	G	G
1-Butanol	G	G	G	G
tert-Butanol	G	G	G	G
Tetrahydrofuran	G	G	G	G
Acetic acid	G	G	G	G
n-Hexane	I	I	I	I
Ethylacetate	G	G	I	I
Acetonitrile	G	S	G	G
Acetone	G	S	G	G
Water	G[b]	G[b]	G	G

[a] Gelator=0.05 to 5.0 wt%.
[b] Stable gel formed in the presence of a trace amount of methanol or ethanol.
G=stable gel formed at room temperature; S=soluble; I=insoluble.

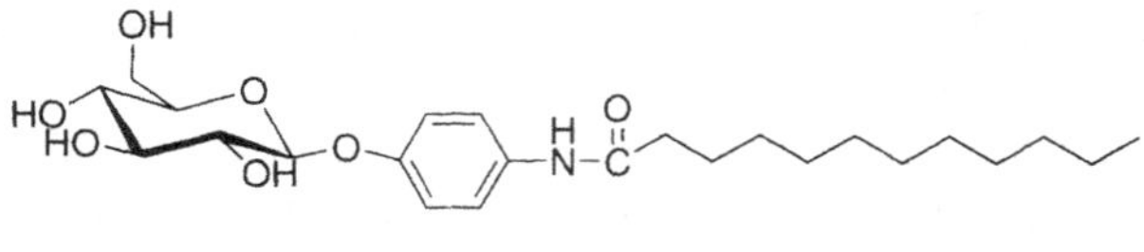

17

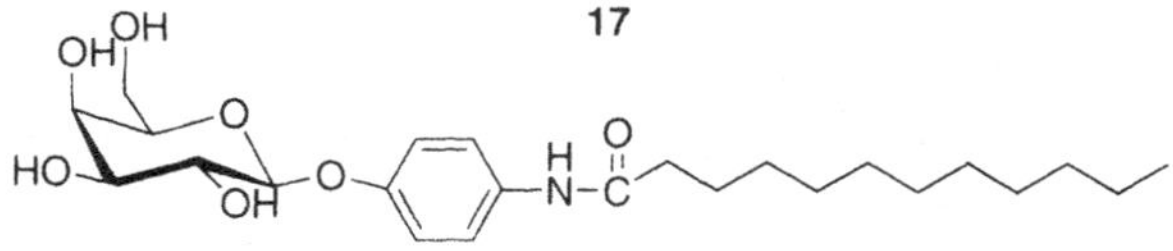

18

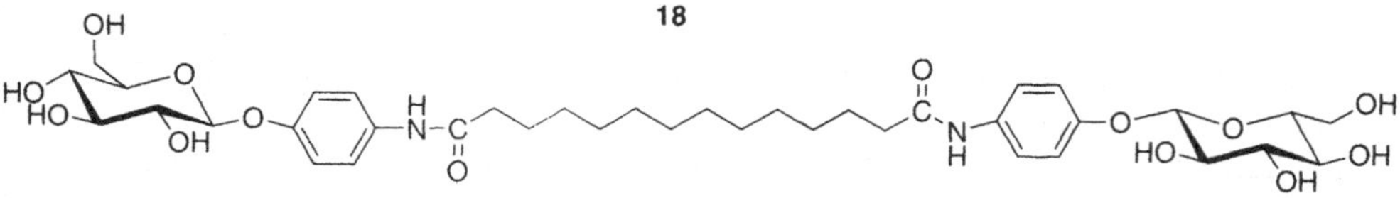

19

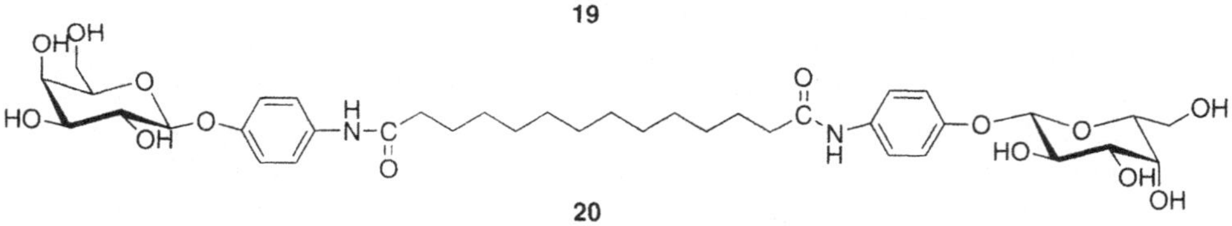

20

solvents and water. In addition, bolaamphiphiles **19** and **20** with two aminophenyl aldopyranoside moieties can also gelate seven out of ten organic solvents, such as alcoholic and aprotic polar solvents. Particularly, they can gelate water in the absence of any organic solvents. This feature is caused by the increased hydrophilicity of **19** and **20** with two aminophenyl aldopyranoside moieties in water.

Hydrogels **17** and **18** display a linear fiber structure with a diameter of 140 to 200 nm and length of several μm (Fig. 1a,b). In contrast, *tert*-butanol gels **17** and **18** exhibit typical structures with three-dimensional networks of fiber bun-

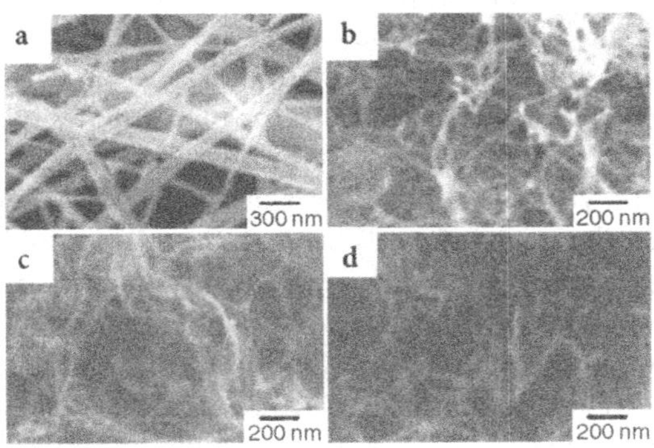

Fig. 1a–d SEM pictures of: **a** aqueous gel **17**; **b** aqueous gel **18** obtained from water; **c** organogel **17**; **d** organogel **18** prepared from *tert*-butanol. Reproduced with permission from [41]

dles. The approximate diameters of the smallest aggregate shown in Fig. 1c,d is 65 to 100 nm, respectively.

Hydrogels **19** and **20**, both of which are transparent, have film-like lamellar structures that are 50 to 100 nm thick at extremely low concentrations (0.05 wt%). In contrast, of particular interest is the xerogel **3** prepared from acetic acid. The 20- to 50-nm diameter spheres are connected to one another like a pearl necklace at the initial stage of gel formation and grow into typical fiber structures with a diameter of 20 to 40 nm. Acetic acid gel **20** displayed a three-dimensional fiber structure with a diameter of 20 to 40 nm and several micrometers. These findings imply that bolaamphiphilic gelators, unlike mono-head amphiphiles, satisfy both the solubility and the maintenance of inter-molecular hydrogen-bond networks in water.

According to powder-XRD experiments, hydrogels **17** and **18** maintained an interdigitated bilayer structure with a thickness of 2.90 nm. In contrast, the organogels **17** and **18** obtained from *tert*-butanol differ considerably from those obtained from water, with a single sharp peak appearing at d=3.48 nm in the small-angle region and one broad reflection in the wide-angle region. These results indicate that the hydrophobic interaction of the alkyl chains with organo-gels **17** and **18** is relatively weak compared to that with hydrogels **17** and **18**.

The powder XRD for bolaamphiphiles **19** and **20** exhibited a single reflection peak with a long period of 3.58 nm, which is compatible with the fully extended molecular length of **19** and **20**, and suggests a layered structure caused by π-π stacking between the aromatic groups, intermolecular hydrogen bonds be-tween the amide groups, and the intermolecular hydrogen-bonding interac-tions of the sugar moieties in the aqueous phase. The similar layered structures were also observed for organogel systems.

2.4 Sugar Lipid Nanotubes

Nanometer-scale hollow cylindrical structures of carbon graphite sheets, so-called carbon nanotubes [43–45], have recently been attracting considerable attention as intriguing new nanostructured materials with many potential elec-trical and photonic applications [46]. Similar morphologies with identical dimensions can be obtained by combining the philosophies of green chemistry and supramolecular chemistry. Lipid nanotubes, characterized by their inner diameter of around 70 nm, wall thickness of 20 to 30 nm, and axial ratios of more than 1000, can be prepared easily through the self-assembly of amphiphilic glucopyranoside derivatives [47, 48]. Synthetic lipid nanotubes may owe their origin to the pioneering work of Yager [49]and Schnur [50, 51] who used di-acetylenic phospholipids, as well as to Nakashima and Kunitake [52] who used glutamate derivatives. Although nearly two decades have passed since the ap-pearance of the first report on self-assembled lipid nanotubes, the number of identified tube-forming amphiphiles is still relatively small. Typical molecular backbones include diacetylenic phospholipids [49, 53], glutamates [52] long-chain diamides [54] anionic glucophospholipids [55], block copolymers [56], a lipid/biotin conjugate [57] and porphyrin derivatives [58]. It is well-known that most lipids are transformed from vesicular assemblies into tubular structures using a helically wound fiber as an intermediate.

Very recently, Jung et al. [47] reported on self-assembling structures of long-chain phenyl glucoside influenced by the introduction of double bonds. For examples, the self-assembly of **22–24** was prepared in aqueous solution. However, **21** proved to be insoluble in water and only formed a typical nanofiber structure (with a diameter of 100 to 350 nm and a length of several micrometers) in a mixture of water and methanol (1:1 v/v). Figure 2 shows TEM images of the self-assembled **22** to **24** in aqueous solutions. Note that **22** exhibits a twisted fiber structure with a width of 50 to 200 nm and a length of several micrometers whereas **23** has a helical ribbon structure, showing the influence of double bonds on the final morphology of the self-assembled structures.

On the other hand, **24** and its three *cis* double bonds in the lipophilic region exhibit a helical-ribbon morphology with an outer diameter of 80 to 100 nm and a nanotubular structure that has an inner diameter of around 70 nm and a wall thickness of 20 to 30 nm (Fig. 2e). As far as can be determined, all the chiral structures have a left-handed helical motif. In addition, the helical ribbon gradually becomes a nanotubular structure that is several micrometers long and with an internal diameter of around 70 nm. Both ends of the nanotubes are

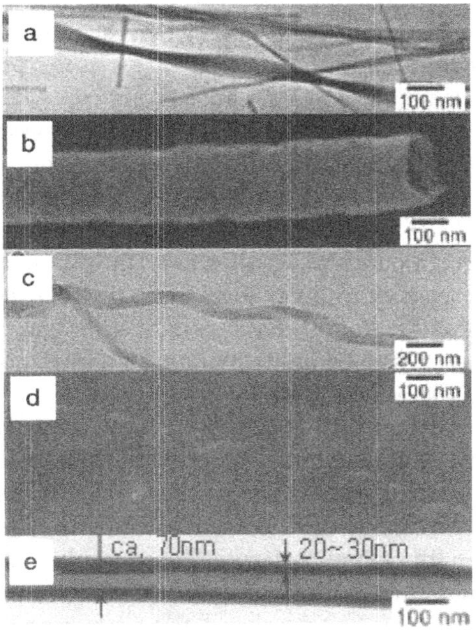

Fig. 2a–e EF-TEM pictures of self-assembled: **a 22**; **b,c 23**; **d,e 24**. Reproduced with permission from [47]

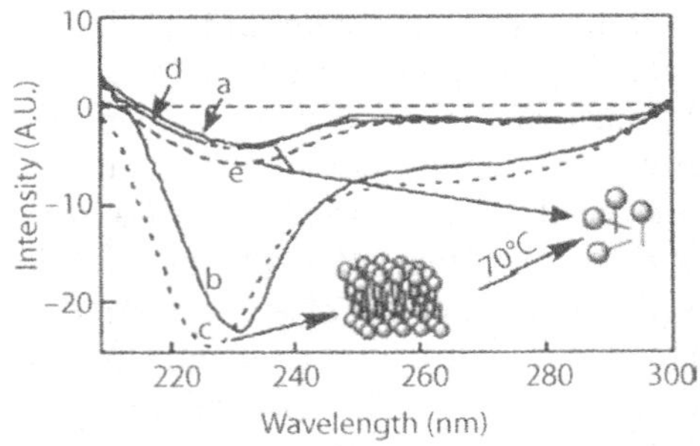

Fig. 3 CD spectra of self-assembled (a) **22**, (b) **23**, (c) **24** at 25 °C, (d) **23** and (e) **24** at 70 °C, respectively. Reproduced with permission from [47]

open and have a hollow inner core and a wall. These observations also support the view that unsaturated units are important to the self-assembly of nanotubes.

To provide further evidence for the nanotube formation of **23** and **24** in the microscopic structural view, the CD spectra of the self-assembled **22** to **24** were observed, as shown in Fig. 3. The CD spectra of the self-assembled **23** and **24** in aqueous solutions exhibit a strong negative band at 225 nm and 237 nm, respectively, and form nanotube structures by chiral assembling. They emit only a weak CD signal at temperatures above the phase transition temperature (Fig. 3) and the maximum shift to longer wavelength, which is probably composed of monomers or small lipid aggregates, such as micelles or vesicles. The CD signal, however, becomes strong again when the lipids self-assemble after several hours to form a nanotubular structure. On the other hand, the CD spectrum of self-assembled **22** in an aqueous solution exhibits a much weaker negative band relative to **23** and **24**, suggesting that self-assembled **22** forms a disordered chiral packing structure. The CD results provide direct evidence for a chiral molecular architecture in the nanotubes and of the proposal that the molecular packing of **23** and **24** shows a loose chirality at temperatures above the phase transition temperature. Furthermore, the phase transition temperatures of **23** and **24** are much higher than those of glucoside amphiphiles **25** to **28** that lack the amide group [48], thus indicating that the stabilization of self-assembled tubes **23** and **24** arises mainly from intermolecular hydrogen-bonding interaction between the amide groups.

According to the powder XRD patterns, the small-angle diffraction patterns of nanotubes **23** and **24** gave at least four ordered reflection peaks with a long period of 4.62 nm, which is less than twice of the extended molecular length of **23** (3.15 nm as determined by CPK molecular modeling) and **24** (3.02 nm) but greater than the length of one molecule. These results strongly suggest that self-assembled **23** and **24** form a bilayer structure as a result of interdigitated hydrophobic interaction. On the other hand, the diffraction diagram for microcrystalline solid **22** (molecular length=3.03 nm) formed a bilayer structure with a molecular length of 3.59 nm, supporting the assumption that **22** maintains a

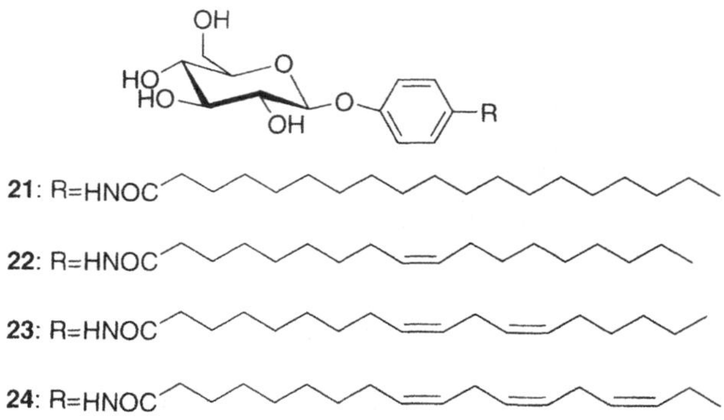

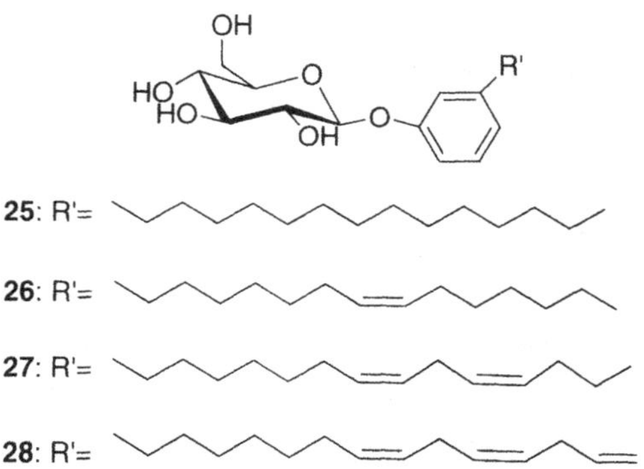

much stronger interdigitated bilayer structure between the lipophilic regions than those of **23** and **24**. According to the CD and powder X-ray diffraction results, the intermolecular hydrogen-bonding interaction of the glucopyranoside moiety of **23** and **24** provides a highly ordered chiral parking structure even though there is only a weak hydrophobic interaction between the lipophilic groups, which leads to the formation of the nanotubular structure.

2.5 Cyclohexane-Based Gelators

The cyclohexane-based gelling agents such as **29** have recently been introduced, first by the Hanabusa group [13] and later also by the Feringa group [11]. Sample diamide compounds **29** and **30** were prepared [13], in the expectation that it may form interlocked macromolecule-like aggregates as a result of two cooperating noncovalent forces, the intermolecular hydrogen bonding between the amide groups of adjacent molecules and the hydrophobic interaction of the long hydrocarbon tails.

Surprisingly, **29** can forms physical gels and harden a wide variety of organic fluids including hydrocarbons, such as alcohols, ketone, ester, ethers, aprotic polar compounds, aromatic compounds, mineral oils and edible oils. In contrast, the racemate consisting of **29** and the S,S-enantiomer **30** only formed an

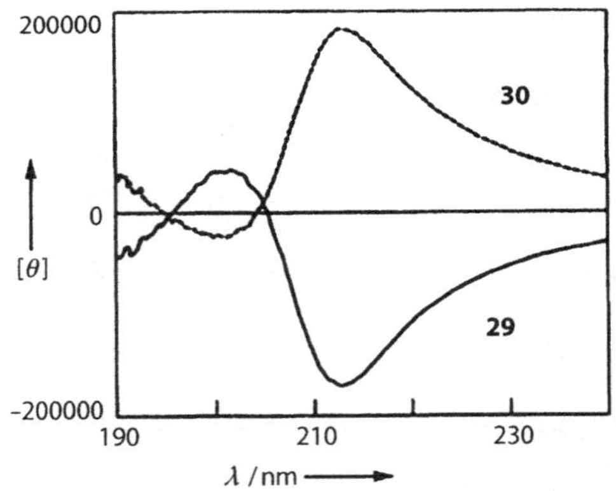

Fig. 4 CD spectra of **29** and **30** in acetonitrile. Reproduced with permission from [13]

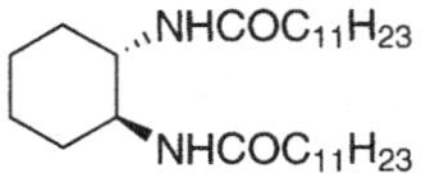

29 *trans(1R, 2R)*
30 *trans(1S, 2S)*

unstable gel, which was converted to cocrystals after several hours, through **30** itself revealed the same gelation behavior as **29**.

To study the chiral structure of the aggregate in the gel, circular dichroism (CD) spectra were measured. The CD spectra of loose gels (1 mmol/l; 0.0677 wt%) for **29** and **30** in acetonitrile are shown in Fig. 4. The CD spectrum of **29** exhibited a negative sign for the first Cotton effect, indicating that dipole moments orientate in an anticlockwise direction in the aggregate of **29**. On the other hand, **30** exhibited a positive sign for the first Cotton effect, indicating that they are oriented in a clockwise direction. These CD spectral data support the view that the aggregates formed from **29** and **30** are cooperatively organized into the aggregates with left- and right-handed helical structures, respectively.

A TEM image of a **29**+acetonitrile gel negatively stained by osmic acid is shown in Fig. 5. Numerous intertwined aggregates form helical fibers with 40–70 nm diameters. The helicity of fibers was always left-handed for **29**, and right-handed for the S,S-enantiomer **30**.

2.6 Cholesterol Gelators

Molecules bearing a highly polar azobenzene group linked to C_3 of a steroidal moiety (**31**–**34**) have been found to gel liquids such as ethanol, acetic acid, amines, dichloromethane, ethers and esters [5, 16, 17]. The α- and β-anomers preferentially gel apolar and polar liquids, respectively. As frequently observed with fiber-like aggregates of optically active molecules [5, 59, 60], CD can be used to follow aggregation properties at the microscopic scale, which is com-

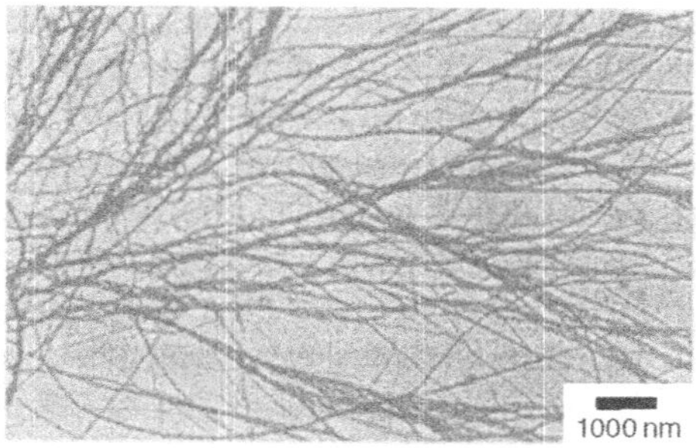

Fig. 5 TEM images of loose **29**-acetonitrile gel (1 mmol/l **29**) stained by osmic acid. Reproduced with permission from [13]

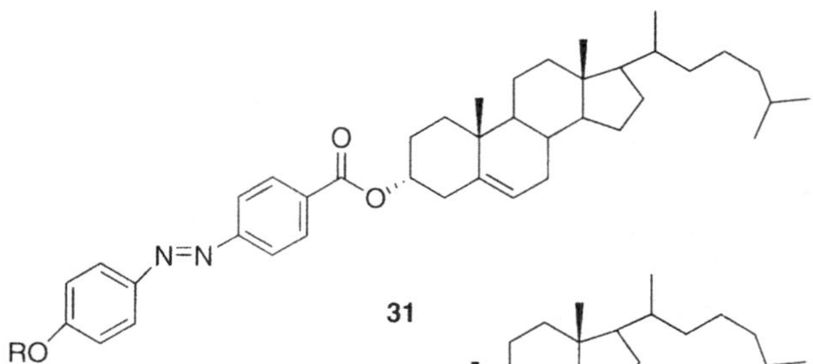

31

R : CH_3, $C2H_5$

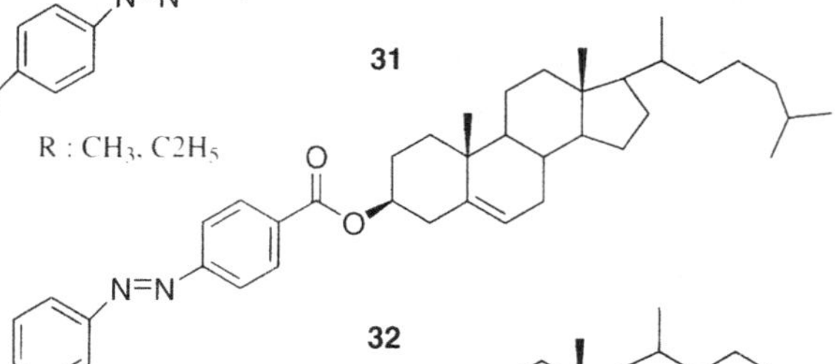

32

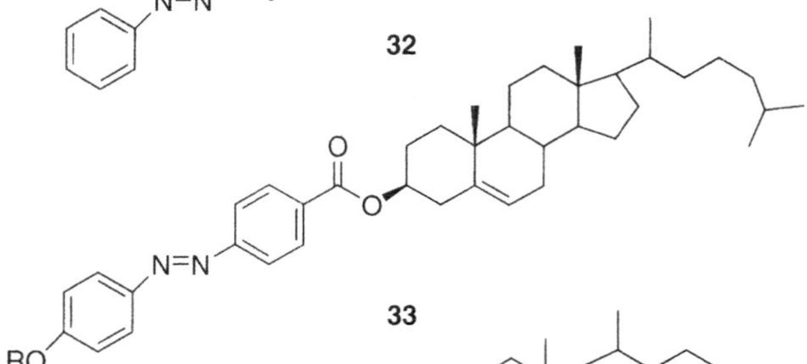

33

R : CH_3, $C2H_5$

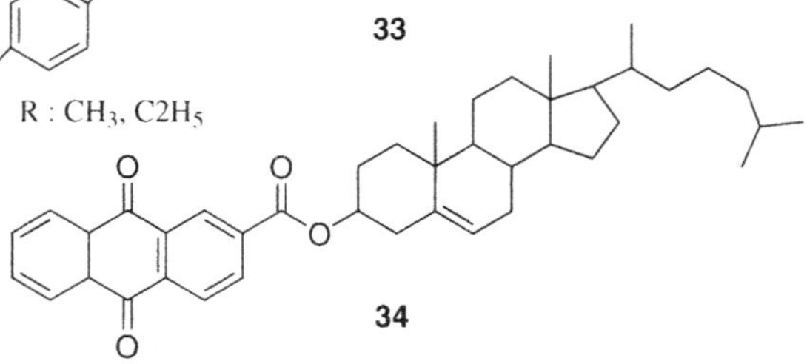

34

pared with those of macroscopic gelation. For instance, from the positive exciton coupling band observed for the α-anomer (**31**), it can be deduced that the dipoles of the azobenzene parts in the aggregate are associated in a clockwise direction (R-chirality). The cooling rate of samples initially above T_{gel} can affect the sign of the gel CD spectra; metastable structures are formed as in the case of **34** gels in hexadecane/1-octanol mixtures [5, 61]. Attempts to correlate the strand helicity with the cooling rate and the α-/β-anomeric nature of the cholesterol linkage have been made on the basis of SEM observations. The absolute configuration at C_3 has also been shown to be an important factor to determine the aggregation modes of monolayers developed at the air-water interface. As witnessed by greater efficiency of gelators with an odd number of carbon atoms on a group attached to the azobenzene moiety, there is a delicate lipophobic/lipophilic structural balance governing gelation properties.

The sensitivity of T_{gel} of gels containing one of **31–33** to the nature of the liquid component suggests that solvent molecules do not participate in the structures of the gelator strands. However, this conclusion cannot be easily generalized and is still suspicious when lyotropic organizations participate in the gel network.

For instance, one dimensional stacking of cholesterol moiety in **35** and **36** causes gelation, which is affected by the ordered azobenzene and the absolute configuration of C_3 in the cholesterol moiety [62, 63]. A TEM image of the acetic acid gels **35** and **36** showed the fiber structure with 50–200 nm outer diameters. However, very interestingly, a TEM image of a mixed organogel of **35** and **36** shows the helical fiber structure as shown in Fig. 6. One can clearly recognize the fiber structure with a diameter of 10–20 nm and a helical pitch of 100–150 nm. In addition, the CD spectrum of the mixed organogels **35** and **36** at R=10 mol%

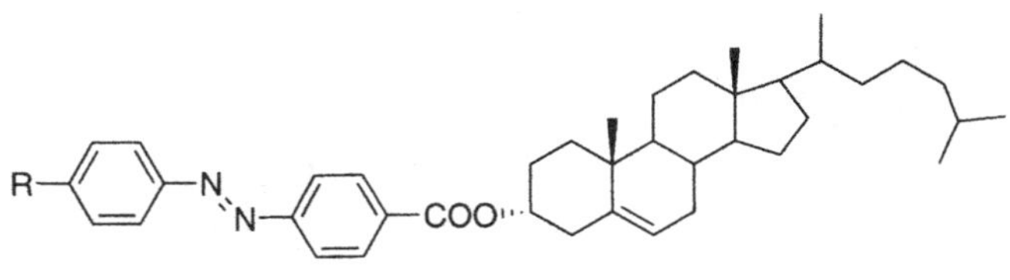

35: R= $O(CH_2)_2Me$
36: R= $O(CH_2)_2NMe_3^+Br^-$

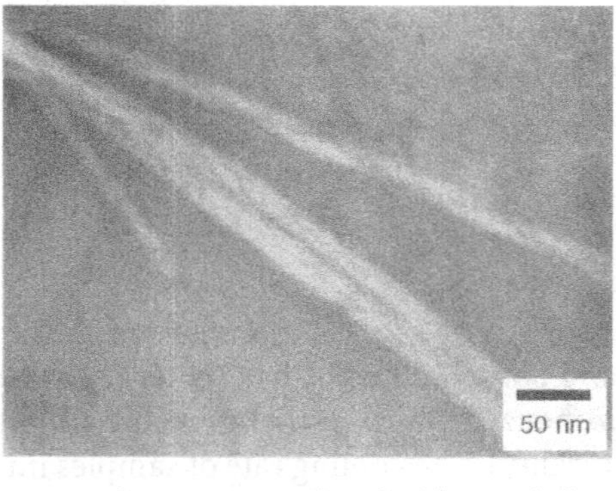

Fig. 6 TEM image of organogel **35+36**. Reproduced with permission from [62, 63]

clearly exhibited a positive sign for the first Cotton effect. These results support the view that the dipole moments of the gelator molecules tend to arrange in the right-handed helical orientation.

2.7
Crown-Appended Cholesterol Gelators

Crown-appended cholesterol gelators were mainly synthesized and characterized on the gelation ability and their physical properties by the Shinkai group [64–70]. In these gel-forming aggregates, the crown ether groups play three essential roles. They are, (i) because of their chain flexibility, introduction of a crown ether group suppresses the formation of a crystalline precipitate and enhances their gelation ability, (ii) oxygen (and nitrogen) atoms in the (aza)crown ring act as a silica binding site with the hydrogen-bonding interaction; moreover, when the crown ring binds metal cations or when the nitrogen is protonated, they can bind anionic silica particles through electrostatic interaction and (iii) they can act as a guest binding site that affects the gelation properties. For instance, the superstructures constructed in organogel **37**, which has a benzo-18-crown-6 moiety, were observed by SEM. The xerogel featured a fibrous network structure with 25–62 nm diameter and the fibers were partially twisted in a helical fashion. In contrast, the SEM image of xerogel **38** did not show the fibrous structure characteristic of organogel systems but rather featured the film-like aggregates with 30~40 nm thickness. One may regard that these films consist of the lamellar structure of **38**. It is noteworthy that some films are curved to form a pseudo-cylindrical structure. The most interesting is the xerogel obtained from a **39**+cyclohexane system. The xerogel featured a tubular structure with 45~75 nm wall thickness and 170~320 nm inside tube diameter. Careful examination of this picture reveals that the wall consists of the multi-layer structure. Presumably, the curved lamellae as seen for **38** have developed to roll up to a tubular structure. It is not clear, however, why the growth of the curved lamellae stops in **38** whereas it continues to grow up to the "roll-paper-like" structure in **39**.

2.8
Multi-Layered Vesicular Morphology of Organogels

Gokel et al. found that certain azacrown-appended cholesterol derivatives form unique vesicular or lamellar structures both in the absence and the presence of metal salts in aqueous solution [71–73]. They possess a polar azacrown head group and a hydrophobic group that lead to the formation of stable aggregates, the morphology of which can be adjusted by changing the ring size of the azacrown head group or the length of the hydrophobic group.

On the other hand, Jung et al. successfully obtained an azacrown-appended cholesterol gelator which possesses the multilayered vesicular structure [68]. The **40**+acetic acid gel produces two different spherical structures, one with a diameter of around 200 nm and the other of around 2500 nm (Fig. 7). The spher-

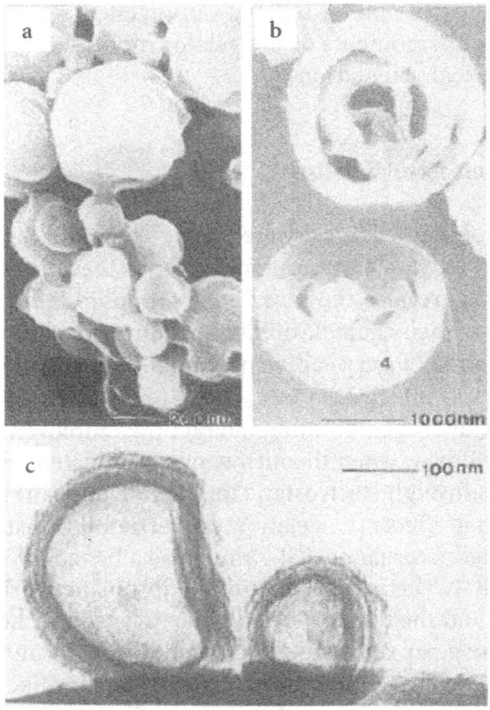

Fig. 7a–c **a, b** SEM images of the acetic acid gel **40**. **c** TEM image of the acetic acid gel **40**. Reproduced with permission from [68]

Fig. 8a, b **a** TEM image of the acetonitrile gel **41**. **b** SEM image of the acetonitrile gel **41**

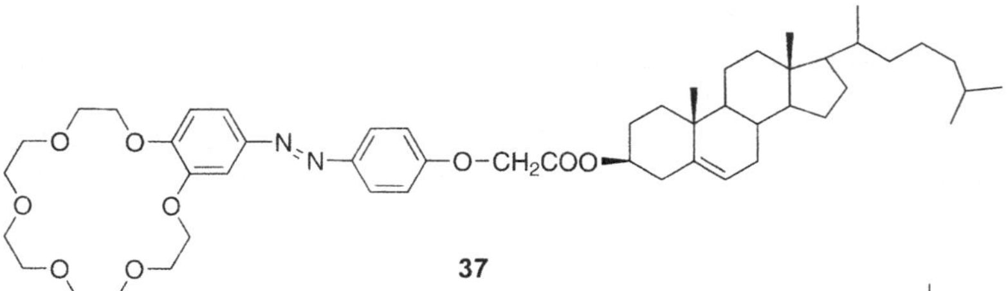

37

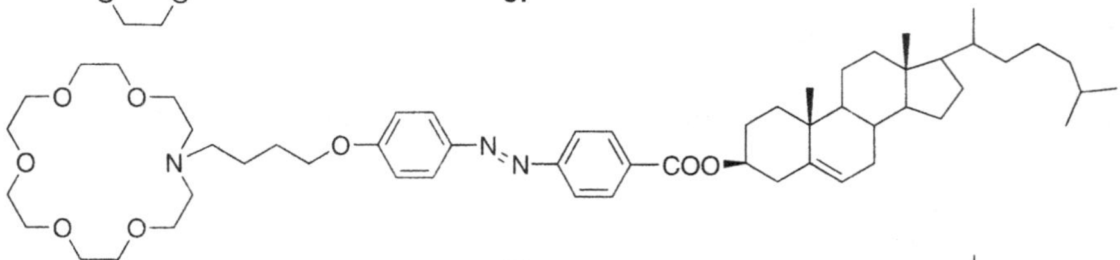

38

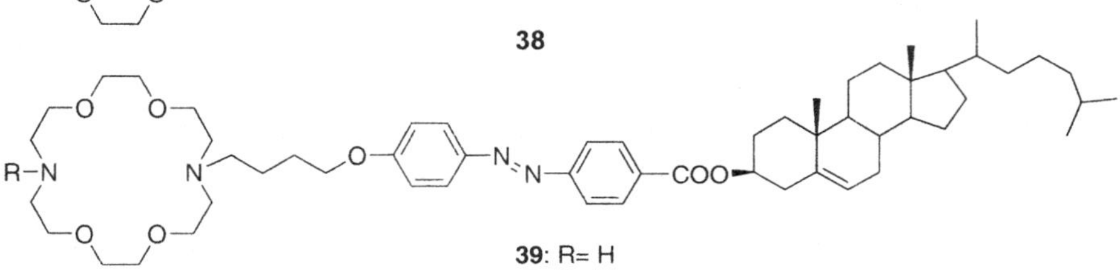

39: R= H
40: R= X

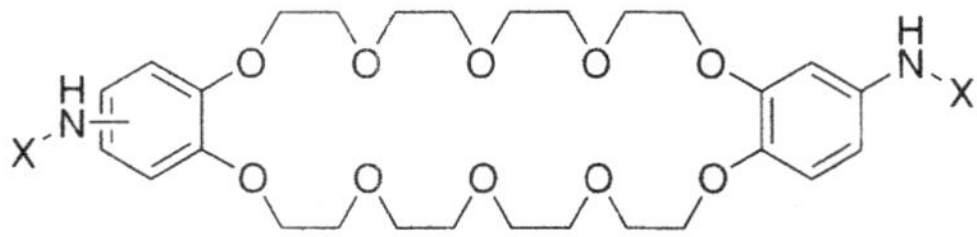

41

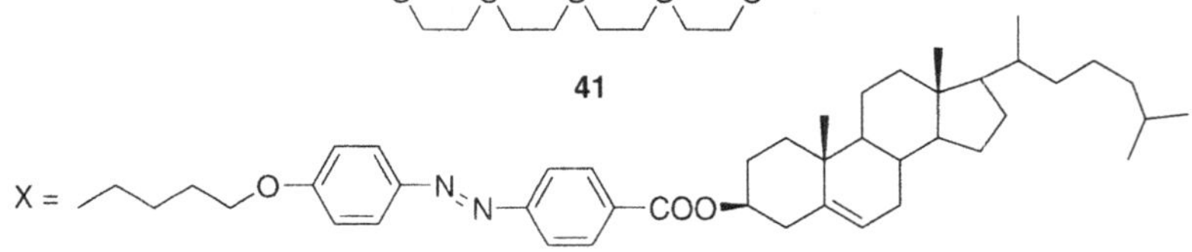

ical morphology is in sharp contrast to the more common types of organogel, which tend to be fibrous or plate-like. Another interesting finding is the fact that the smaller vesicles are connected together like a pearl necklace. When the **40**+acetic acid gel was dissolved by heating and sonication with a 35-W tip sonicator for 15 min, the resulting SEM picture revealed an increase in the concentration of the larger vesicles. One particle was cut by an ultramicrotome. Fig. 7b clearly shows that the particle has a shell wall with a thickness of around 200 nm and an inner sphere with a diameter of around 300 nm. The clearer picture of the larger vesicle was obtained by TEM. The **40**+acetic acid gel was translucent even in the presence of uranyl acetate, and was subjected to TEM observation. The TEM image (Fig. 7c) clearly shows that the shell wall consists of a multi-layered structure with a layer thickness of around 5 nm and a diameter of around 200 to 350 nm.

Fig. 8, on the other hand, shows TEM images of organogel **41** obtained from acetic acid. Interestingly, organogel **41** with a 30-crown-10 moiety consists mainly of tubular structures with an outer diameter of around 520 nm but also exhibits a linear ribbon and a helical ribbon both with a total width of 1300 nm. In addition, the SEM picture of the xerogel **41** reveals a characteristic tubular structure (Fig. 8b). These results indicate that the structure of this organogel involves several metastable intermediate processes, viz. helical ribbon→tubule.

2.9 Charge-Transferred Gelators

Raju et al. reported the relation between the gelation ability and the charge transfer interaction between bile-acids **42–44** and trinitrofluorenone **45** [74]. Interestingly, the location of the pyrene unit on the bile acid appeared to be critical, however, **43** and **44** which did not form gels in the presence of **45**. Pyrene derivative **42** in general formed highly stable gels even with less than 1 wt% of the gelator. Compound **45** in the presence of the methyl pyrene-1-carboxylated did not form a gel in BuOH and $CHCl_3$. A steady increase of T_{gel} was observed with increasing amounts of **45** but when 1.25 equiv. were added there was no measurable increase, suggesting the requirement of a 1:1 stoichiometry for effective gelation. The gels formed from colorless **42** and **43** are colored due to a charge-transfer band. The intensity of this band changes substantially during gelation.

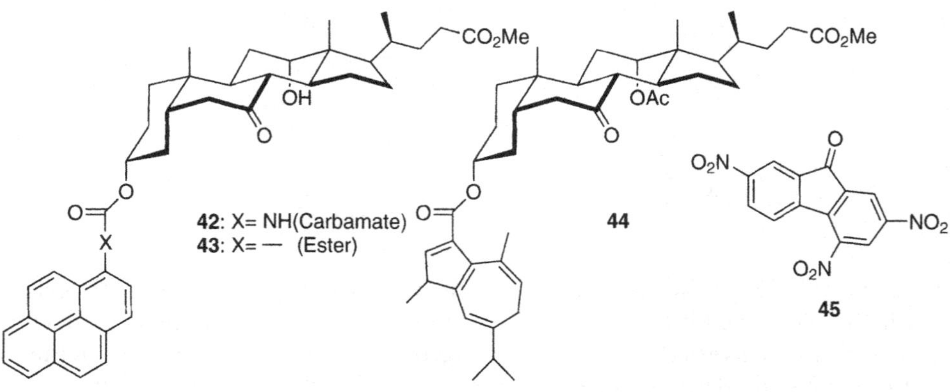

42: X= NH(Carbamate)
43: X= — (Ester)
44
45

3 Sol-Gel Transcription to Create Inorganic Nanotubes

The use of organic compounds as templates for the generation of inorganic structures and materials has been receiving considerable attention over the last decade [75–82]. Inorganic materials, although diverse in their compositions, generally lack the structural variety as seen for supramolecular and other organic structures [5–7]. Although inorganic materials are easily molded at the macroscopic level, our ability to shape them at the microscopic level is, at best, limited. The formation of these inorganic structures is challenging not only from a purely scientific point of view, but also from a fact that these materials have the potential to be applied to a wide range of fields. We can envision the use of hollow, inorganic spheres for the controlled release of a variety of substances, as well as their use as light but strong filler materials or as microreactors. Since these higher-order aggregates can be created with a variety of different architectures, using them as templates for sol-gel polycondensation gives us the means to create a range of novel inorganic architectures that could not be created directly from inorganic materials. As possible templates, proteins, multicellular superstructures, surfactants and DNA have all been utilized to create novel structures from inorganic materials [75–82].

More recently, it has been found that certain cholesterol derivatives can gelate even tetraethoxysilane (TEOS), to produce silica gel by means of sol-gel polycondensation [16]. Interestingly, it has been shown that the sol-gel polycondensation of gelated TEOS solutions followed by calcination produces a novel silica with a hollow fiber structure. This is because, in the polycondensation process, the organogel fibers act as a template to create an internal tube [62]. We have found that self-assembled organogels form not only superstructures such as fibers, tubes, vesicles, helical ribbons and double helixes, but also act as good templates for sol-gel polycondensation. In this section, we report transcription of the superstructures constructed in the organogels into inorganic nanotubes.

3.1
A Lotus-Shaped Silica Nanotube

Sol-gel polycondensation of TEOS was carried out using **13a**, **13b**, **14b**, **15b** and **16b** in the ethanol or water gel phase [83, 84]. The amino group was used expecting the stronger hydrogen-bonding interaction between NH_2 and silica particles than that between NO_2 group and silica particles. The silica obtained from **13b** showed the tubular structure with 20–30 nm outer diameters and 350–700 nm lengths (Fig. 9a,b) whereas the silica obtained from **13a** showed the conventional granular structure. It is hardly conceivable that the aromatic amino group is protonated in the presence of benzylamine used as a catalyst. These results indicate, therefore, that the tubular structure of the silica was successfully transcribed by the hydrogen-bonding interaction between the amino group of **13b** and TEOS (or oligomeric silica particles).

In contrast, β-glucose-type organogel **16b** resulted in the tubular silica with larger outer diameters of 150–200 nm (Fig. 9c,d). Very surprisingly, the TEM images of the silica obtained from **16b** reveal that the silica consists of 50–100 nm inner diameters and 150–200 nm outer diameters. Furthermore, the silica in the inner tube is composed of micro-tubes with 5–10 nm diameters, in total giving rise to a lotus-like structure. The silica obtained from ethanol gels **13b** and **16b** have a Brunauer-Emmett-Teller (BET) surface area of 450 $m^2\,g^{-1}$ and 475 $m^2\,g^{-1}$, respectively. These values are smaller than those of mesoporous silica obtained surfactants.

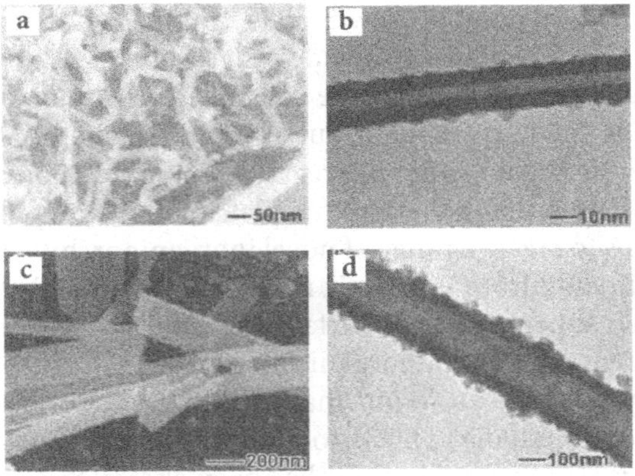

Fig. 9a–d SEM and TEM images of the silica tubes obtained from: **a,b** organogels **13b**; **c,d** organogels **16b**. Reproduced with permission from [83]

3.2 Double-Helical Silica Nanotube

Reports on helical inorganic materials have documented their use as asymmetric reaction catalysts [85, 86], helical sensors [87, 88], inorganic optical materials [89] etc. Although *N*-alkylaldonamide amphiphiles [90, 91] and synthetic nucleotide analogues [92] in artificial systems, as well as DNA, RNA and polypeptides in natural systems, self-assemble into highly ordered superstructures such as single-, double- and triple-helical structures, their transcription into inorganic materials has not yet been achieved [93, 94]. The reasons for this are twofold: (i) the presence of either cationic charges or efficient hydrogen-bond donating moieties is indispensable to adsorb the "anionic" inorganic precursor moieties onto the organic molecular assemblies and (ii) the self-assembled morphology of the organic molecules should be maintained during the sol-gel reaction. Obtaining helical structures that satisfy the above requirements, however, can be accomplished using stable, organic supramolecular self-assembled structures. Jung et al. prepared sugar-based gelator **17** as a template, which, together with aminophenyl glucopyranoside **46**, can form a gel consisting of double-helical fibers that can be transcribed into double-helical silica nanotubes [95].

SEM images of the **17**+**46** mixed gel (1:1 wt%/wt%) exhibited a well-defined double-helical fiber structure with diameters of 3 to 25 nm, lengths of several micrometers [95], and dimensions similar to those of double-helical DNA

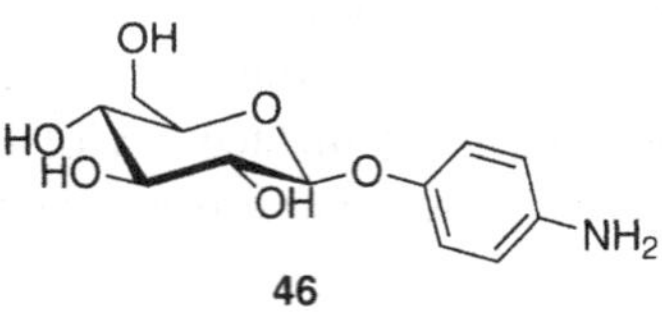

46

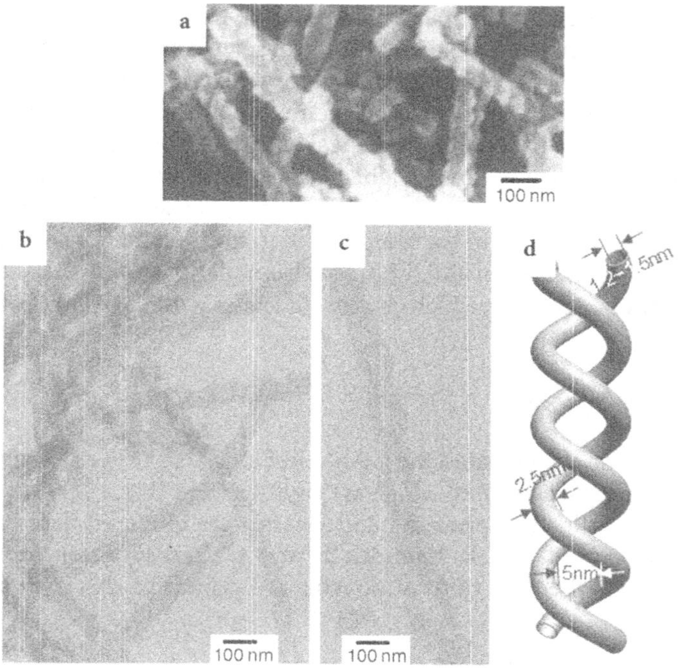

Fig. 10 a FE-SEM image. **b, c** TEM images of double-helical silica nanotube obtained from mixed gel of **17** and **46** (1:1 w/w) after calcination. **d** Schematic representation of double-helical structure of the silica nanotubes as observed by SEM and TEM. Reproduced with permission from [95]

and RNA structures. Sol-gel polycondensation of TEOS was carried out in a water-methanol (10:1 v/v) gel of **17**+**46**. After calcination, the morphology of the silica product turned out to be a well-defined double-helical nanotube with a diameter of 50 to 80 nm and a pitch of 50 to 60 nm (Fig. 10a). These results prove that the double helical structure of the self-assembled gel fibers of **17** and **46** is successfully transcribed into the silica nanotube structure, most likely through hydrogen-bonding interactions between the amine moiety of **46** and negatively charged oligomeric silica particles.The results clearly support the view that (i) the presence of a minimal number of hydrogen-binding sites is indispensable to the transcription of the self-assembled structure into a silica structure, (ii) the helicity of the silica can be accurately transcribed from that of the template and (iii) this method can be applied to the efficient transcription of self-assembled superstructures into inorganic materials.

High-magnification TEM images of the double helical silica tubes reveal a 5.0-nm nanospace between the helical strands (Fig. 10b,c), the structure of which is quite similar to that of double-helical DNA. Furthermore, there is essentially no difference between the silica before and after calcinations, indicating that the double-helical structure of the silica indeed originates from the self-assembled organogel superstructure. Also, the inner cavity of the silica nanotube was proved by the N_2 adsorption-desorption isotherm. The silica had a surface area of 450 to 500 $m^2 g^{-1}$ and pores with two distinct diameters. The TEM and DFT observations revealed the first to be 1.2 to 1.5 nm in diameter. The inner diameter of the double-helical silica nanotube is similar to that of single-wall carbon nanotubes. A slight constriction of this inner diameter of the silica probably occurs during the calcination, because the inner diameter is slightly smaller than that of the template fiber. The second type of pore has an average diameter of 5.0 nm and originates from the space between the two separate helical silica strands, which closely correlates with the TEM observations.

3.3 Chiral Silica Nanotubes

Transcription of cyclohexandiamine-based helical organogels resulted in the helical silica nanotubes, all of which are right-handed (Fig. 11) [96, 97]. Helical silica nanotubes of the opposite handedness were obtained using the enantiomeric mixture of 1R, 2R-diamides **29** and **47**. Transcription proved to be successful only when the ratio of noncharged gelator to charged gelator was

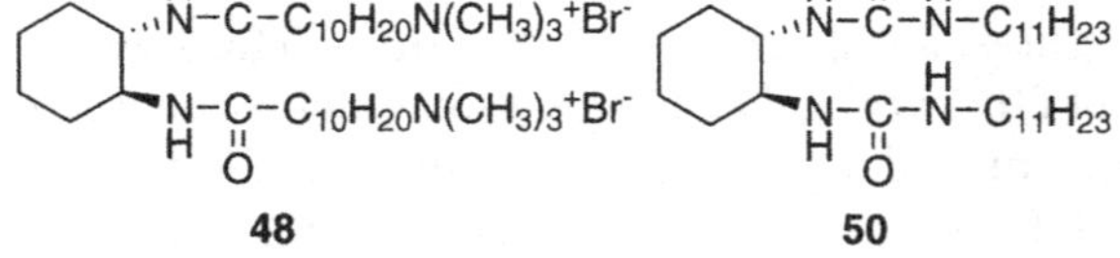

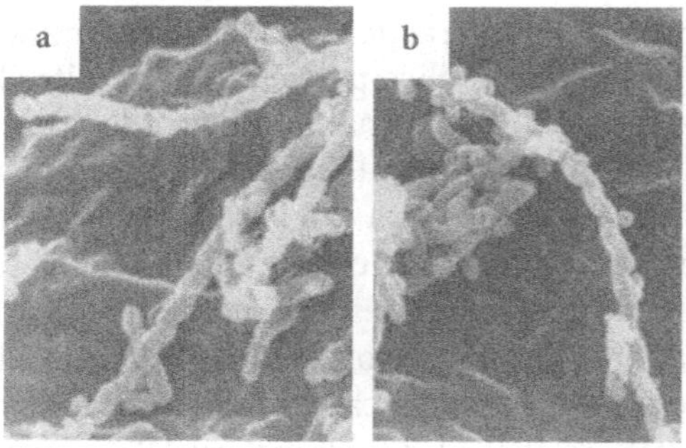

Fig. 11a, b FE-SEM images of single-helical silica nanotube obtained from: **a 47+49**; **b 48+50**. Reproduced with permission from [97]

lower than 70:30, indicating that a moderate amount of cationic charges is indispensable to in the transcription process of organogels into silica nanotubes. Similar results were obtained using the combinations **47/49** and **48/50** [97]. After calcination of the silica fibers, inner channels possessing the same helicity as the fiber itself could be observed, thus proving the view that the organogel fibers had indeed been the template around which the silica structures had formed. Furthermore, it showed that the chirality present in the gelator molecules is transcribed consistently into the macroscopic silica materials. This system is the most striking example for the potential of the organogel approach for the formation of inorganic materials.

Hanabusa et al. reported preparation of a novel TiO_2 materials with a hollow-fiber structure like "macaroni" by using cyclohexandiamine-base gelator **51** [98, 99]. The self-assembled gelator **51** was electrostatically interacted with titania species in the sol-gel polycondensation process of $Ti[OCH(CH_3)_2]_4$. The results of gelation tests for **51** are summarized in Table 2. Compound **51** gelated *n*-butanol and the minimum gel concentration necessary for the gelation was 2 wt%.

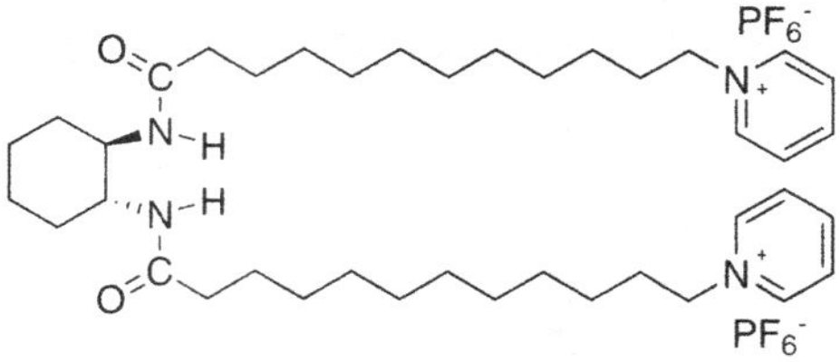

51

SEM images of dried samples prepared under acetic acid and basic conditions using **51** showed an extensive network structure with 150–600 nm diameters under both conditions. The surface of the fiber prepared under basic conditions was rough, whereas the sample under acidic conditions had the very

Table 2 Influence of sol-gel polycondensation conditions on the resultant silica structure[a]

Run	Catalyst	Solvent	Additive	Silica structure
1	$MeCO_2H$	$MeCO_2H$ (306 µL)	–	Hollow
2	$MeCO_2H$	$MeCO_2H$ (306 µL)	Me_4NCl (2.2 mg)	Hollow
3	$MeCO_3H$	$MeCO_2H$ (306 µL)	Me_4NCl (109.7 mg)	Granular
4	1.0 mol/l HCl (15.6 µL)	$MeCO_2H$ (306 µL)	–	Granular
5	Benzylamine (5.7 µL)	Butan-1-ol (395 µL)	–	Hollow
6	Benzylamine (5.7 µL)	Octan-1-ol (346 µL)	–	Hollow

[a] **36** 5.0 mg (6.2×10^{-6} mol), TEOS 48.4 µL, and water 15.6 µL. **Gelator 36** was added as a CH_2Cl_2 solution (0.3 g)

smooth surface. The rough surface under the basic conditions may be the result of the adhesion of the TiO_2 particles to fibrous aggregates. No fibrous structure was observed for the samples prepared under acidic conditions after calcination. In contrast, the fibrous structure under basic conditions was completely preserved after calcination. When sol-gel polycondensation is carried out under basic conditions, the propagation species is considered to be anionic. Hence, the oligomeric titania species are adsorbed onto cationic aggregates of **51** and polycondensation further proceeds along these aggregates. The titania fibers of the calcined sampled possessed hollow structure with outer diameters of 150–600 nm, and the maximum length of the titania fiber was ~20 μm.

The crystal structures of the resulting TiO_2 fibers were analyzed by X-ray diffraction. The X-ray diffraction patterns showed that the TiO_2 fibers crystallize into anatase structures after calcination at 450 °C for 2 h. Nanostructured anatase TiO_2 has been employed in a variety of photoelectric devices. Therefore, this hollow-fiber TiO_2 material has an excellent potential for applications to photovoltaic solar cells, photocatalytic devices and rechargeable lithium ion battery electrodes.

Recently, Moreau et al. reported the solids by means of organogelators **52/53** [100]. These compounds, although similar in structure to those discussed before, differ in that the silica precursor had been covalently attached to the

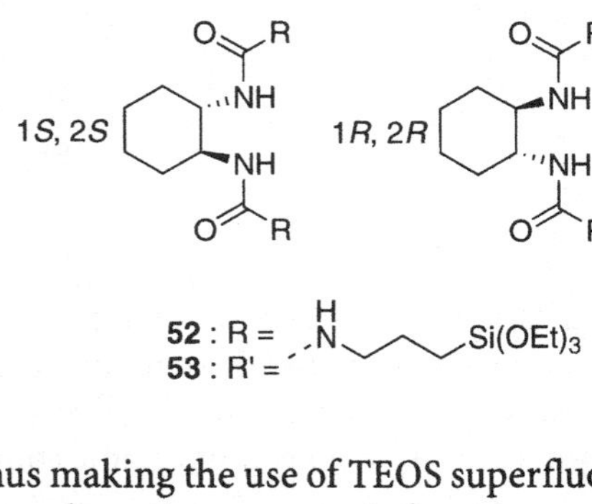

gelator structure, thus making the use of TEOS superfluous. Particularly, however, the silica polycondensation was carried out in pure water, a solvent that cannot be gelatinized by these cyclohexane derivatives. Hence, the obtained materials did not display the well-defined morphologies observed for true organogel systems. Organogelator **53** was also used in the nongelating solvent water. The compound is thought to self-organize in a lamellar fashion, as a result of hydrogen-bonding interactions between the urea moieties and chain-chain interactions. A lamellar organic-inorganic hybrid material with plate thickness of about 50–100 nm was obtained after this reaction.

3.4 Vesicular Structure of the Silica Particle

Mesoporous inorganic materials with vesicular structures have been subjected to special attention because of their potential applications as sorption media,

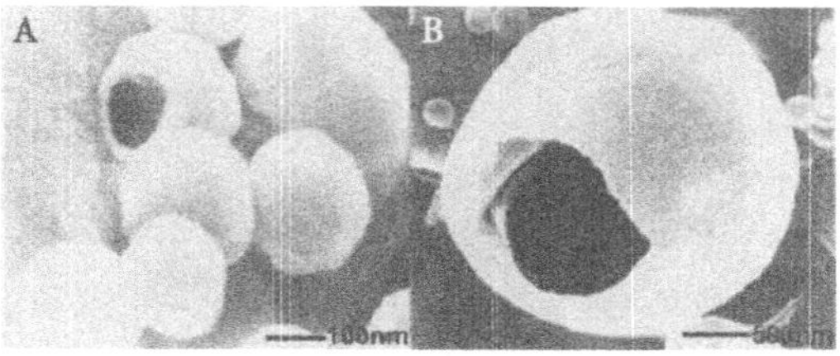

Fig. 12A, B TEM images of the silica microspheres obtained from the **40**+acetic acid gel after calcinations. Reproduced with permission from [68]

molecular sieves and catalysts [101–105]. However, all the vesicular (multi-layered) mesostructures reported to date have shells with an undesirable thickness and shape. The earliest approach to creating vesicular structures of inorganic materials was reported by Ozin et al. [104, 105], who synthesized a lamellar aluminophosphate phase in a solution containing vesicles formed from tetraethylene glycol, amphiphilic alkylamines, and water. The resulting material was composed of millimeter-sized spheroids with patterned surfaces. In addition, recently, Pinnavaia et al. [101–103] reported that arrangements of mesoporous inorganic materials with vesicular structures with shells with a thickness of 3 to 70 nm have been accomplished using "gemini"-type surfactants as templates in aqueous solutions. However, these particles have shells consisting of uni- and multi-lamellae.

Shinkai et al. [68] successfully obtained the multilayered vesicular structure of the silica particles obtained from organogels. For examples, Fig. 12 shows SEM micrographs obtained by sol-gel polycondensation using organogel **40** as a template, taken after calcination. The resultant silica shows the spherical structures with a diameter of around 200 nm and a diameter of around 2500 nm. After the small particle had been sectioned with an ultramicrotome, TEM was used to observe its edges. The resulting TEM images revealed a shell wall to be a multi-layered lamella having a spacing of 5 nm. These results clearly support the view that the multi-layered structure of the organogels is precisely transcribed into the silica structure. This vesicular silica with the unique higher-order morphology is very useful for drug delivery, nanosized microcapsulation, catalysis, compartmentalization, etc.

3.5 Novel Metal Deposition in the Inner Side of the Silica Nanotube

Confirmation of the importance of cationic charges in the transcription process of organogels into silica came from Ono et al. [62, 64], who employed a cholesterol-based gelator functionalized with a crown ether moiety (**37**–**41**) capable of binding potassium cations. It was found that the organogel of **37** could only be successfully transcribed into hollow silica fibers in the presence of substantial amounts of K^+ ions. The addition of the other cations (Li^+, Na^+, Rb^+,

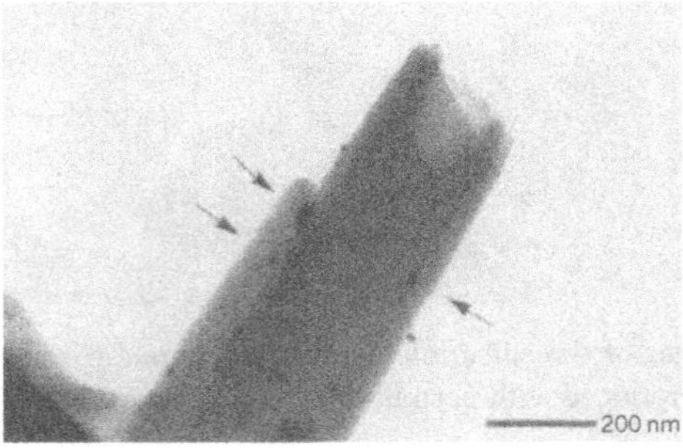

Fig. 13 TEM images of the paper-roll-like silica obtained from the **38**+1-butanol gel in the presence of $AgNO_3$ after calcinations. Reproduced with permission from [65]

Cs^+) did not result in successful transcription and only granular silica was formed. The striking effect of K^+ ion could be ascribed to the high affinity of the benzo[18]crown-6 moiety for the K^+ ion.

The organogel of the diaza-analogue **39**, on the other hand, consists of the fully formed tubular structures [65, 66]. Transcription of these organogels in the presence of K^+ or Cs^+ ions resulted in the formation of "rolled paper-like" silica. Although, like gelator **37**, compounds **38** and **39** do not possess any positive charges, their organogels can be successfully transcribed into tubular silica even in the absence of added metal salts. The driving force for this transcription is the presence of cationic charges arising from protonation of the azacrown moieties. Transcription was also successfully carried out in the presence of added metal salts; however, in all cases the metal was never directly observed in the resultant silica. Silica containing nanosized metal particles (1–5 nm) was obtained for the first time by transcribing the organogel of **38** into silica after the addition of $AgNO_3$. As can be clearly seen in Fig. 13, small silver particles are present in the interlayer space of the rolled paper-like silica. Such metal-containing silica is thought to have interesting applications to catalysts.

3.6 Meso-Sized Silica Nanotube

Phenanthroline-appended cholesterol gelator **54** also induced the fibrous structure with 30–70 nm outer diameters and a few micrometers length [106]. Fig. 14 illustrates N_2 adsorption-desorption isotherms and a BJH pore size distribution plot (insert) calculated from the adsorption branch of the N_2isotherms for calcined silica obtained from the **54**+acetic acid gel. The N_2 adsorption-desorption isotherms of this material exhibited a typical type IV, which is characteristic of mesoporosity. A narrow pore size distribution with a mean value of 4.5 nm is also obtained from both the adsorption and desorption processes. This material has a Brunauer-Emmett-Teller (BET) surface area of 460 $m^2\ g^{-1}$.

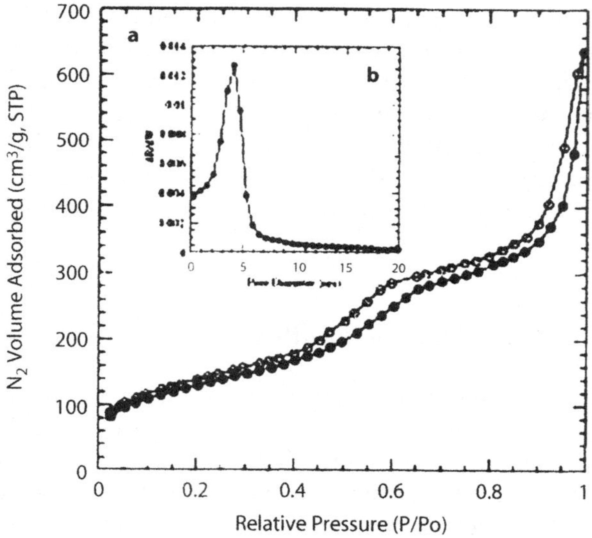

Fig. 14 (**a**) Nitrogen adsorption-desorption isotherms and (**b**) pore size of the silica obtained from the **54**+acetic acid gel after calcinations. Reproduced with permission from [106]

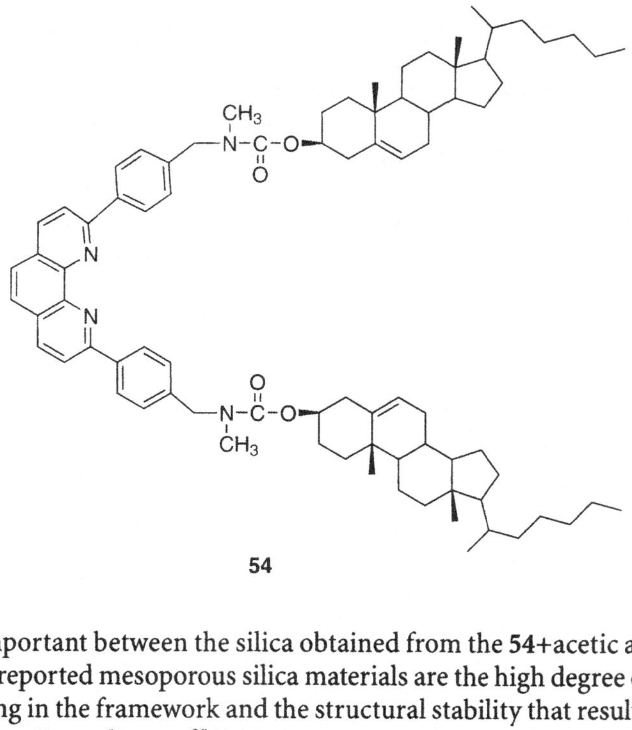

54

More important between the silica obtained from the **54**+acetic acid gel and previously reported mesoporous silica materials are the high degree of SiO_4 unit cross-linking in the framework and the structural stability that results from this cross-linking. According to ^{29}Si MAS NMR experiments, the silica framework consisted of a major fraction of the fully cross-linked Q^4 unit; $Si(OSi)_4$ and a minor fraction of the incompletely cross-linked Q^3 unit; $(OH)Si(OSi)_3$. Generally, surfactant-containing silica mesostructures exhibit Q^4/Q^3 ratios less than 2.0, and their calcined derivatives exhibit the values near 3.0 [81]. However, the Q^4/Q^3 ratio for the silica obtained from the **54**+acetic acid gel gives 7.08, suggesting that the SiOH groups may be site-isolated and buried in the framework. This very high Q^4/Q^3 value is comparable with those of the vesicular silica structures obtained from gemini-type surfactants reported by Pinnavaia et al. (6.2–7.4) [81].

3.7 Double-Walled and Helical Ribbon Silica Nanotubes

Kunitake et al.found that certain amphiphiles can form a tubular structure through the helical ribbon structure in an aqueous solution [52]. They have a polar head and a suitable chiral hydrophobic group for forming stable aggregates. Very interestingly, our group demonstrated that 30-crown-10-appended cholesterol gelator **41** forms the tubular structure in the gel system [69, 70]. Compound **41** has two cholesterol skeletons that act as a chiral aggregate-forming site, two amino groups that act as an acidic proton binding site, and one crown moiety that acts as a cation binding site.

Fig. 15 shows SEM images obtained after calcinations. The silica obtained from **41**+acetic acid possesses a helical ribbon structure with widths of 300 to 800 nm and a tubular silica structure with an outside diameter of around 560 nm. As far as we can observe, all the helicity exhibits a right-handed helical motif. Since the exciton-coupling band of the organogel also exhibits R (right) helicity when observed by CD, we believe that any microscopic helicity is reflected by a macroscopic helicity. These results indicate that the novel helical ribbon structure and the tubular structure of the organogel are successfully transcribed into the silica structures. TEM pictures were taken after

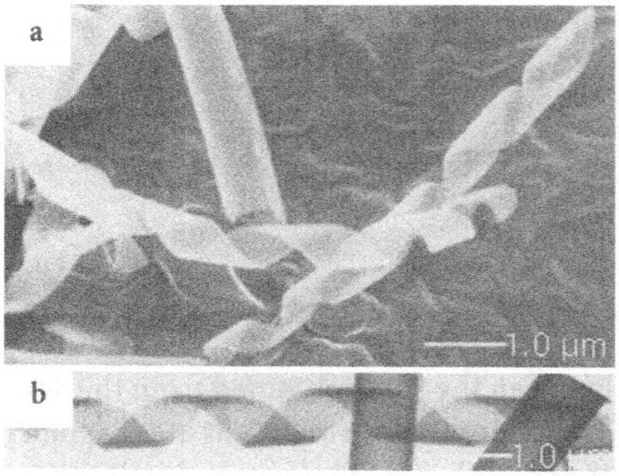

Fig. 15a, b SEM images of double-walled silica nanotubes obtained from the **41**+acetic acid gel after calcinations. Reproduced with permission from [69]

the removal of **41** by calcination (Fig. 15). Interestingly, the silica possesses a helical ribbon structure in which the ribbon width varies (450 to 1500 nm) but in which the tubular structure of the silica has a constant outer diameter of around 560 nm. In addition, the silica has double layers with an interlayer distance of 8 to 9 nm. These results indicate that TEOS (or oligomeric silica particles) is adsorbed onto both surfaces of the double-walled tubules that are 8 to 9 nm thick. Therefore, the tubular silica has two hollow cavities. The smaller cavities, with layers of 8 to 9 nm, are created by the organogel template whereas the larger cavities with inner diameters of 460 to 430 nm are created by the growth of the helical ribbon. This silica with the helical higher-order morphology is useful as unique catalysts for asymmetric syntheses without chiral organic ligand.

3.8 Even-Odd Effect for Silica Tubular Structures

In all of the organogel examples discussed so far, positive charges present in the template enabled transcription of the organogel into the inorganic materials. In a search to find an organogel system that would be able to act as a neutral template, Jung et al. synthesized compound **55*n*** [84, 107]. They were designed to be gelator analogues of the diamine surfactants used by Pinnavaia and co-workers to produce mesoporous silica, with transcription occurring as a result of hydrogen bonding, rather than ionic interactions, between the template and the silica precursor.

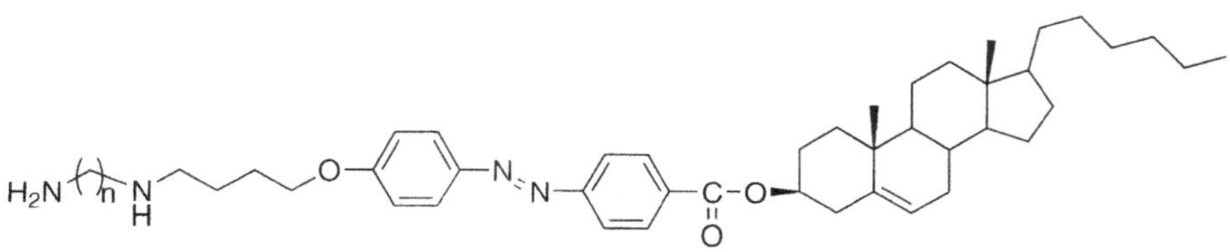

55

Similarly, transcription of organogels of compounds **55*n*** was achieved under basic conditions, that is, without positive charges being present in the template. The attractive forces between the negatively charged silica precursor species and the amine moieties of the gelator are large enough to result in a successful transcription of the template. Interestingly, different morphologies were observed for the organogels and resultant silica products of gelators **55*n*** for different values of n. In the cases of n=1 or 3, the organogels possessed curved lamellar morphology, which resulted in silica with a structure like a paper roll. For even numbers of n, the morphology of the organogel was unmistakably fibrous in nature and resulted in hollow silica fibers (after transcription and calcination). These results clearly demonstrate that minor changes in the gelator structure can have a dramatic effect on the overall morphology of the inorganic materials.

3.9 Narrow Dispersed Silica Nanotube Using a Sugar-Appended Porphyrin Gelator

One-dimensional alignment of porphyrins and phthalocyanines is of much concern in relation to the creation of novel supramolecular architectures such as nanowires, discotic liquid crystals, helical ribbon structures, etc. The major driving force in these architectures is considered to be a π-π stacking interaction.

For sugar-appended porphyrin gelator, Shinkai group reported the gelation ability and its sol-gel transcription into narrow dispersed silica nanotubes [108]. Compound **56** was insoluble in most solvents (water, *i*-PrOH, *n*-BuOH, chloroform, dichloromethane, *n*-hexane, cyclohexane, benzene, toluene, THF, MeCN, anisole, ethyl acetate, etc.). It swelled in MeOH, EtOH, and benzyl alcohol (BzOH) but was not dissolved into these solvents even at reflux temperature. In contrast, **56** gelated DMF/alcohol mixtures such as DMF/BzOH (1:3 v/v). It is seen that the critical gelation concentration (CGC) for **56** is very low (ca. 2 wt% in 1:2 (v/v) DMF/BzOH and ca. 0.4 wt% in 1:3 (v/v) DMF/BzOH) and T_{gel} values at [**56**] >1.3 wt% are higher than 130 °C. The T_{gel} values can be further enhanced at higher gelator concentrations, e.g., 167 °C at 2.5 wt% in 1:3 (v/v) DMF/BzOH and 175 °C at 3.0 wt% in 1:2 (v/v) DMF/BzOH concentration.

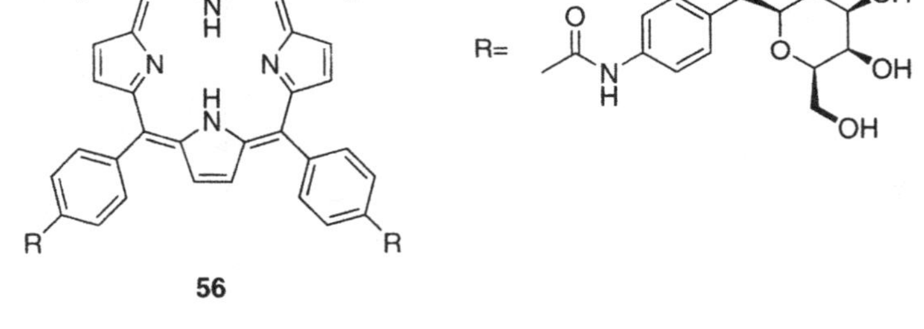

56

Direct evidence for the porphyrin-porphyrin stacking was obtained from absorption and CD spectral examinations. **56** in homogeneous solution gave the Soret band at 422.5 nm and Q-bands at 516.0, 551.0, 592.5 and 648.0 nm. In the gel phase, the Soret band shifted to short wavelength (401.0 nm) with the appearance of a shoulder at 420.0 nm. This blue-shift implies that the porphyrin moieties form a stacked aggregate, but a significant amount of **56** still remains as a monomer in bulk solution. The Q-bands in the gel phase appeared at 522.5, 558.0, 599.0 and 654.0 nm, which were all red shifted compared with those in homogenous solution. The red-shift was also compatible with the formation of porphyrin-porphyrin stacked aggregates.

Sol-gel polycondensation of TEOS was carried out using a gel of **56** by two different methods. In Method A, the mixture was heated at 80 °C again and then

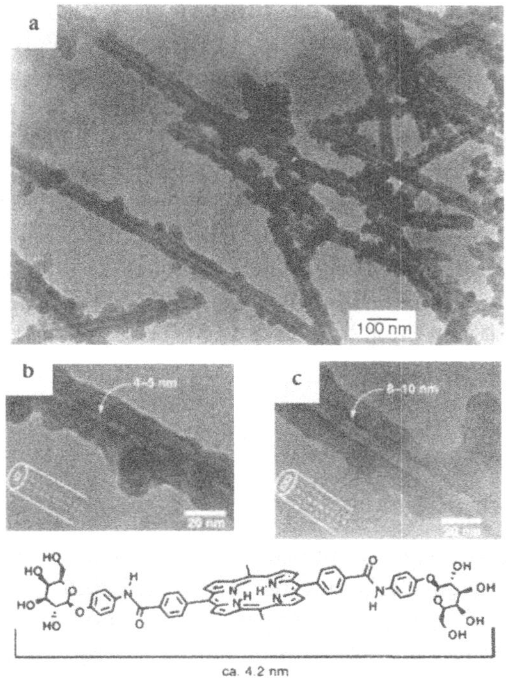

Fig. 16a–c TEM images of the silica nanotubes obtained from the organogel **56**. Reproduced with permission from [108]

left at room temperature for two weeks. However, in Method B the mixture was just let at room temperature without the heating treatment. The silica before calcination shows a fibrous structure with an approximate 50 nm outer diameter (Fig. 16). This morphology scarcely changed, even after calcination. It is clearly seen from the TEM image that these silica fibers have an inner hollow with 4–12 nm diameters. Careful examination reveals that there are two different hollow diameters: one about 5 nm, which is comparable with the length of the long axis in **56** (4.2 nm: Fig. 16), and another about 9 nm, which is comparable to twice of the long axis length, and bears a stripe in the center (Fig. 16). The TEM image scarcely changed, even after calcination, but the central stripe in Fig. 16 disappeared.

In Method A, the solution containing benzylamine was heated so that the formation of bundles of organogel fibers would be suppressed while the rate of sol-gel polycondensation of TEOS would be accelerated. Clearly, this condition facilitates the entrapment of incipient organogel fibers comprised of a unimolecular stack or a bimolecular stack. In the absence of heating after benzylamine addition (Method B), the bundle growth of organogel fibers can compete with sol-gel polycondensation to give a helically bundled silica structure from transcription of well-grown, helically bundled organogel fibers.

More recently, it was demonstrated that single-walled carbon nanotubes (SWNTs) are also useful to design long silica nanotubes. Poly(*N*-vinylpyrrolidone) (PVP) is known not only to disperse SWNTs into water but to adsorb silica particles because of its amphiphilic nature. [109] Thus, sol-gel polymerization carried out in the presence of SWNTs-PVP complex results in triple-layered silica fibers with several μm length which is comparable with that of SWNTs used as a template. At 500 °C, SWNTs are still stable whereas PVP is decomposed. One can thus remove PVP selectively by calcinations at 500 °C, giving rise to novel double-layered silica nanotubes. Since PVP acting as a "glue" to bind between the inside silica wall and SWNTs disappears, one can recognize a gap between them by TEM and SWNTs easily slip out of these silica nanotubes.

3.10 Double Helical Silica Fiber

The French group reported that "cationic" Gemini surfactants having chiral tartrate counter ions assemble into twisted ribbons both in chlorinated and aromatic organic solvents, and in water [110]. The handedness of the ribbons and their helical pitch can be tuned continuously upon varying the ratio between L and D tartrate. This kind of "fine tuning" is nearly impossible in other systems and may make it possible to create the observation of novel double helical silica fibers with a tunable pitch [110].

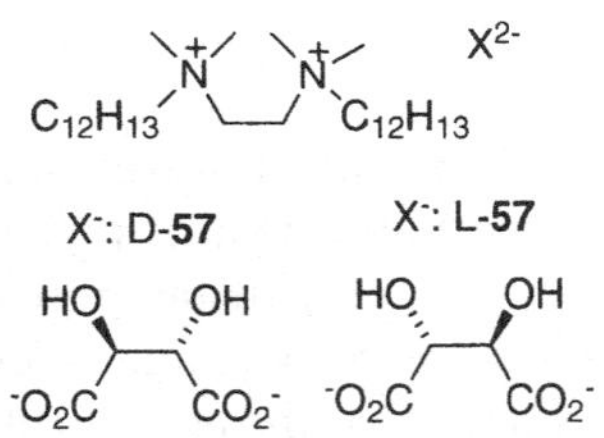

Fig. 17a shows a SEM image of the silica after calcination obtained from a gel of pure L-57 (100% ee). When the ee is decreased from 100 to 20%, double helices with increasing pitch are observed (Fig. 17b). The aspect ratio of these inorganic fibers was found to be T/W=1.6 for 100% ee, 4.3 for 50% ee, 5.1 for 30% ee and 9.8 for 20% ee. These values are somewhat smaller than those of the precursor twisted ribbons, but one may rationalize this discrepancy from a possible shrinkage of silica fibrils during the calcination process. The double stranded helical structure appears particularly clearly for silica fibrils obtained at 50% ee and 33% ee (Fig. 17b,c).

Thus, the helical shape, the handedness and even the tunable chirality of the organic twisted ribbons are efficiently transcribed to the inorganic replicas. However the question remains of how these surprising inorganic double helices arise from flat twisted structures. When a prolonged sol-gel polycondensation

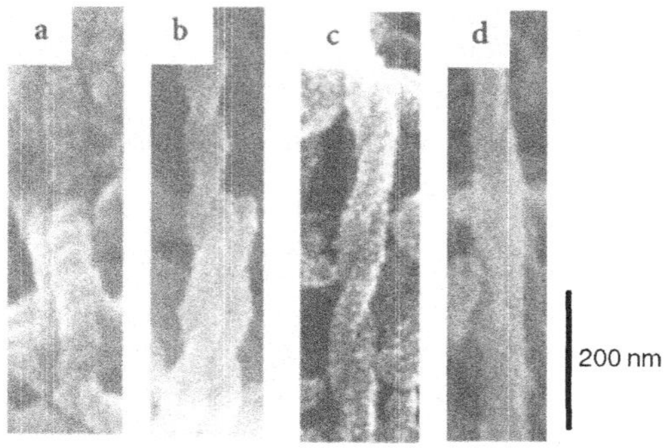

Fig. 17a–d Influence of the ee (L-57 in excess) on the helical pith of double strand silica: **a** 100% ee; **b** 50% ee; **c** 33% ee; **d** 20% ee. Reproduced with permission from [109]

is performed, the structure of the result silica is very similar to the original organic ribbons. This suggests that the organic structures do act as templates, and that the double helical silica fibrils are formed during early stages of the polycondensation. One important observation is that the silica fibers are not hollow: the electron density across the fibers does not decrease in their center. This implies that they are formed at the surface of the organic template but do not completely surround it. Two possible rationales come to our mind to explain double helix formation. TEOS polycondensation may proceed preferentially along the gutter-shape which is more exposed to the solvent. Alternatively, it may proceed preferentially along the edges of the ribbons, where the density of positive charge is lower but the cationic charges are more exposed to the solvent. Since the ribbons have two edges and two faces, two intertwined silica fibrils would be expected to be of the order of a ribbon thickness when condensation occurs on the faces and of the order of a ribbon width when condensation occurs on the edges. It is considered that the edge mechanism is more likely [94].

3.11
A Novel Role for Benzylamine in the Sol-Gel Transcription

The α-glucose-based gelator **58** was shown to gelate a range of organic solvents as well as water. Recently, our research group demonstrated that **58** can form fibers in two different ways [111]. In organic solvents, the polar sugar parts can form a chain of intermolecular hydrogen bonds upon stacking, which exposes the relatively apolar aromatic parts to the solvent. In water, molecules are pre-

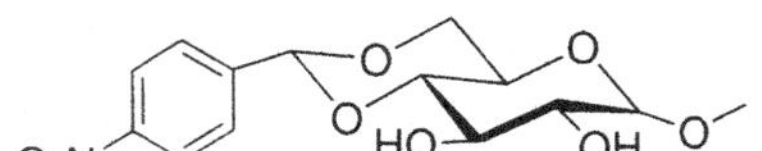

58(α-glu)

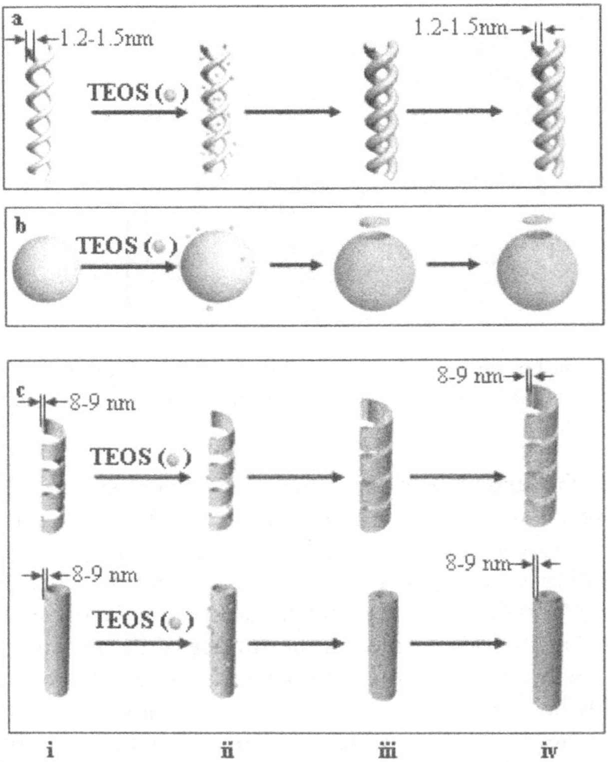

Fig. 18a–c Schematic representation for the creation of various silica structures from the organogel state of **17+46** (a), **40** (b) and **41** (c) by sol-gel polycondensation: (i) gelators; (ii) sol-gel polycondensation of TEOS; (iii) adsorption onto the gelators; (iv) double-helical structure, vesicular structure, double-walled and helical ribbon structure of the silica materials formed after calcinations. Reproduced with permission from [42]

sumed to form fibers through π-π stacking, thus exposing the sugar parts to solvent. The stacking model indicated that other aromatic additives might also be incorporated into the fibers through stacking. The additive of choice was benzylamine, a catalyst frequently used in the silica-transcription process. Benzylamine can interact with the nitrobenzene moieties through π-π stacking or through hydrogen-bonding interactions. Formation of the hydrogel of **58** in the presence of one equivalent of benzylamine, followed by transcription and subsequent calcination, resulted in the isolation of hollow silica fibers, as a template for silica transcription, even though they lacked the cationic charge or amine moieties hitherto deemed indispensable. Additional experiments indeed proved that some part of the added benzylamine (about 20%) is incorporated into the gel fibers through π-π stacking and hydrogen-bonding interactions. This incorporation results in the template structure now possessing enough amine moieties to enable transcription. The major part (80%), which stays in solution, still acts as the conventional catalyst in the transcription process. Control experiments involving other catalysts unable to interact strongly with the gelator molecules did not result such efficient in transcription, thus proving the necessity of having good compatibility between the catalyst and gelator. Since cationic charges and amine moieties are no longer prerequisites for gelator structures, these findings increase the variety of gelators that can be used for transcription into inorganic materials; compounds bearing a moiety that can interact strongly with a catalyst can now also be used.

3.12 On the Mechanistic Aspect of Sol-Gel Transcription

Finally, we now propose the mechanism for the formation of the novel double-helical silica structure obtained from hydrogel **17+46** (Fig. 18a), the vesicular silica structure obtained from acetic acid gel **40** (Fig. 18b) and the double-walled as well as the helical ribbon silica obtained from acetic acid gel **41** (Fig. 18c). Oligomeric silica species are adsorbed onto the surface of the organogel templates by the hydrogen-bonding or electrostatic interactions and the polymerization further proceeds along these fibrils (Fig. 18: (ii)). This propagation mode eventually yields the nano-sized silica with hollow cavity after combustion of gelators by calcination (Fig. 18: (iii), (iv)).

4 Concluding Remarks

Over the last few years, the crossover between organic and inorganic chemistry has yielded several techniques that have enabled the synthesis of inorganic materials with controlled morphologies. Organogels present an even wider variety of morphologies such as linear fibers, single helixes, double helixes, tubes, ribbons, and hollow spheres, that have been transcribed so far. Moreover, the helical forms of some organogel templates have been shown to induce chirality in the corresponding inorganic materials. The structural complexity that can be achieved by using biomaterials as templates for transcription processes is the major advantage of these particular systems. However, suitable biotemplates are often difficult to obtain and modify to desired morphologies.

Prospects for the future include the development of novel transcription methodologies, as well as the creation of new transcription templates which present a greater degree of structural complexity or can give rise to extended, patterned architectures. One of the principle challenges remaining is to find novel methodologies for obtaining transcribed structures with a high degree in the 2D or 3D level.

Acknowledgements We thank Miss Jeong-Ah Rim and Mr. Sujin Lee for their excellent help with drawing chemical structures and figures.

References

1. Guenent JM (1992) Thermoreversible gelation of polymer and biopolymers. Academic Press, London
2. Kajiwar K, Osada A (1991) Polymer gels: fundamentals and biomedical applications
3. Flory PJ (1974) Dtsc Faraday Soc 57:8
4. Tanaka T (1981) Sci Am 244:110
5. Terech P, Weiss RG (1997) Chem Rev 97:3133
6. Abdallah D, Weiss RG (2000) Adv Mater 12:1237
7. Esch JV, Schoonbeek F, Loos M, Veen M, Kellogg R, Feringa BL (1999) Where it is and where it is going: low molecular weight gelatiors for organic solvents. Kluwer Academic Publishers, The Netherland
8. Gronwald O, Shinkai S (2001) Chem Eur J 7:4329
9. Yi T, Sada K, Sugiyasu K, HatanoT, Shinkai S (2003) Chem Commun 344
10. Yoza K, Ono Y, Yoshihara K, Akao T, Shinmori H, Takeuchi M, Shinkai S, Reinhoudt DN (1998) Chem Commun 907
11. Esch JV, Schoonbeek F, Loos MD, Kooijiman H, Speak AL, Kellogg RM, Feria BL (1999) Chem Eur J 5:937
12. Loos MD, Esch JV, Stokroos I, Kellogg RM, Feringa BL (1997) J Am Chem Soc 119:12675
13. Hanabusa K, Yamada M, Kimura M, Shirai H (1996) Angew Chem Intl Ed Engl 35:1949
14. Hanabusa K, Matsumoto M, Kimura M, Kakehi A, Shirai, H (2000) J Colloid Interface Sci 224:231
15. Hanbusa K, Kawakami A, Kimura M, Shirai H (1997) Chem Lett 191
16. Murata K, Aoki M, Suzuki T, Harada T, Kawabata H, Komori T, Ohseto F, Ueda K, Shinkai S (1996) J Am Chem Soc 116: 6664
17. Shinkai S, Murata K (1998) J Mater Chem 485
18. Snip E, Shinkai S, Reinhoudt DN (2001) Tetrahedron Lett 21153
19. Ishi-I T, Iguchi R, Snip E, Ikeda M, Shinkai S (2001) Langmuir 17:5825
20. Wang R, Geiger C, Chen L, Swanson B, Whitten DG (2000) J Am Chem Soc 122:2399
21. Duncan DC, Whitten DG (2000) Langmuir 16:6445
22. Geiger C, Stanescu M, Chen L, Whitten DG (1999) Langmuir 15:2241
23. Yoza K, Amanokura N, Ono Y, Akao T, Shinmori H, Takeuchi M, Shinkai S, Reinhoudt DV (1999) Chem Eur J 5:2722
24. Gronwald O, Snip E, Shinkai S (2002) Curr Opin Colloid Interface Sci 7:148
25. Amanokura N, Kanekiyo Y, Shinkai S, Reinhoudt DN (1999) J. Chem Soc Perkin Trans 2:1995
26. Yong K, Mooney DJ (2001) Chem Rev 101:1869
27. Zhao B, Moore JS (2001) Langmuir 101:1869
28. Luo Y, Dalton PD, Shoichet MS (2001) Chem Mater 13:4087
29. Pardo-Yissar V, Gabai R, Shipway AN, Bourenko T, Willner I (2001) Adv Mater 13:1320
30. Hu Z, Lu X, Gao J (2001) Adv Mater 13:1708
31. Bhattacharya S, Acharya SNG (1999) Chem Mater 11:3504
32. Estroff LA, Hamiton AD (2001) Angew Chem Int Ed 39:3447
33. Bhattacharya S, Acharya SNG (2000) Langmuir 16:87
34. Menger FM, Caran KL (2000) J Am Chem Soc 122:11679
35. Amaike M, Kobayashi H, Shinkai S (2001) Chem Lett 620
36. Jung JH, John G, Masuda M, Yoshida K, Shinkai S, Shimizu T (2001) Langmuir 17:7229
37. Maitra U, Mukhopadhyay S, Sakar A, Rao P, Indi SS (2001) Angew Chem Int Ed 40:2281
38. Simmons BA, Irvin GC, Agarwal V, Rose A, John VT, Mcpherson GL, Balsara NP (2002) Langmuir 18:624

39. Suzuki M, Yumoto M, Kimura M, Shirai H, Hanabusa K (2002) Chem Commun 884
40. Suzuki M, Yumoto M, Kimura M, Shirai H, Hanabusa K (2002) New J Chem 26:817
41. Jung JH, Shinkai S, Shimizu T (2002) Chem Eur J 8:2684
42. Jung JH, Shinkai S, Shimizu T (2003) Chem Rec 3:212
43. Iijima S (1991) Nature 363:603
44. Iijima S, Ichihashi T (1993) Nature 363:603
45. Bethune DS, Ching CH, de Vries MS, Gorman G, Savoy R, Vazquez J, Beyers R (1993) Nature 363:605
46. Tanaka K (ed) (2001) Carbon nanotube. Kagaku Doujin, Kyoto
47. Jung JH, John G, Yoshida K, Shimizu T (2002) J Am Chem Soc 124:10674
48. John G, Masuda M, Okada Y, Yase K, Shimizu T (2001) Adv Mater 13:715
49. Yager P, Schoes P (1984) Mol Cryst Liq Cryst 106:371
50. Schnur JM (1993) Science 262:1669
51. Schnur JM, Ranta BR, Selinger JV, Singh A, Jyothi G, Easwaran KRK (1994) Science 264:945
52. Nakashima N, Asakuma S, Kunikate T (1985) J Am Chem Soc 107:509
53. Thomas BN, Safinya CR, Plano RJ, Clark NA (1995) Science 267:1635
54. Fuhrhop JH, Spiroski D, Boettcher C (1993) J Am Chem Soc 115:1600
55. Giulierik F, Guilod D, Boettcher C (1996) Chem Eur J 2:1335
56. Stewart S, Liu G (2000) Angew Chem Int Ed 39:340
57. Ringler P, Mueller W, Ringsdorf H, Brisson A (1997) Chem Eur J 3:620
58. Fuhrhop JH, Bindig U, Siggel U (1993) J Am Chem Soc 115:11036
59. Lin YC, Kachar B, Weiss RG (1989) J Am Chem Soc 111:5542
60. Terech P (1991) Liquid Cryst 9:59
61. Furman I, Weiss RG (1993) Langmuir 9:2084
62. Ono Y, Nakashima K, Sano M, Kanekiyo K, Inoue K, Hojo J, Shinkai S (1998) Chem Commun 1477
63. Ono Y, Nakashima K, Sano M, Hojo J, Shinkai S (2001) J Mater Chem 11:2412
64. Ono Y, Kanekiyo K, Inoue K, Hojo J, Shinkai S (1999) Chem Lett 23
65. Jung JH, Ono Y, Shinkai S (2000) Langmuir 16:1643
66. Jung JH, Ono Y, Shinkai S (1999) J Chem Soc Perkin Trans 2 1289
67. Jung JH, Ono Y, Shinkai S (2000) Angew Chem Int Ed 39:1862
68. Jung JH, Ono Y, Sakurai K, Sano M, Shinkai S (2000) J Am Chem Soc 122:8648
69. Jung JH, Kobayashi H, Masuda M, Shimizu T, Shinkai S (2001) J Am Chem Soc 123:8785
70. Jung JH, Lee SH, Yoo JS, Yoshida KM, Shimizu T, Shinkai S (2003) Chem Eur J 9:5307
71. Echegoyen LE, Hernandez J, Kaifer AE, Gekel GW (1988) J Chem Soc Chem Commun 836
72. Nakano A, Hernadez JC, Dewall SL, Berger DR, Gokel GW (1997) Supramol Sci 8:21
73. Echegoyen LE, Portugal L, Miller SR, Echegoyen L, Gokel GW (1998) Tetrahedron Lett 29:4065
74. Maritra U, Kumar PV, Chandra ND, Souza LJ, Prasanna MD, Raju RA (1999) Chem Commun 595
75. Mann S (1996) Biomimetic materials chemistry. In: Mann S (ed) VCH, New York
76. Shenton W, Douglas T, Young M, Stubbs G, Mann S (1999) Adv Mater 11:253
77. Douglas T, Young M (1998) Nature 393:152
78. Shenton W, Pum D, Sleytr U, Mann S (1997) Nature 389:585
79. Davis SA, Burkett SL, Mendelson NH, Mann S (1997) Nature 385:420
80. Zhao D, Feng J, Huo Q, Melosh N, Fredrickson GH, Chmeka BF, Stucky GD (1998) Science 279:548
81. Kim SS, Zhang W, Pinnavaia TJ (1999) Science 282:1302
82. Yang H, Coombs N, Ozin GA (1997) Nature 386:692

83. Jung JH, Amaike M, Shinkai S (2000) Chem Commun 2343
84. Jung JH, Amaike M, Nakashima K, Shinkai S (2001) J Chem Soc Perkin Trans 2:1938
85. Soai K, Osanai S, Kadowaki K, Yonekubo S, Shibata T, Sato I (1999) J Am Chem Soc 121:11235
86. Shibata T, Yamamoto J, Matsumoto N, Yonekubo S, Osanai S, Soai K (1998) J Am Chem Soc 120:12157
87. Bodenhöfer K, Hierlemann A, Seemann J, Gauglitz G, Koppenhoefer B, Göpel W (1997) Nature 387:577
88. Cornell BA, Braach-Maksvytis VLB, King LG, Osman PDJ, Raguse B, Wieczorek L, Pace RJ (1997) Nature 387:580
89. Hodgkinson I, Wu QH (2001) Adv Mater 13:889
90. Fuhrhop JH, Boettcher C (1990) J Am Chem Soc 112:1768
91. Feiters MC, Nolte RJM (2000) In: Gokel GW (ed) Advances in supramolecular chemistry: chiral self-assembled structures from biomolecules and synthetic ananlogues. JAI Press, America
92. Shimizu T, Iwaura R, Masuda M, Hanada T, Yase K (2001) J Am Chem Soc 123:5947
93. Ringler P, Mueller W, Ringsdorf H, Brisson A (1997) Chem Eur J 3:620
94. Sugiyasu K, Tamaru S, Takeuchi M, Berthier D, Huc I, Oda R, Shinkai S (2002) Chem Commun 1212
95. Jung JH, Yoshida K, Shimizu T (2002) Langmuir 23:8724
96. Jung JH, Ono Y, Hanabusa K, Shinkai S (2000) J Am Chem Soc 122:5008
97. Jung JH, Ono Y, Shinkai S (2000) Chem Eur J 6:4552
98. Kobayashi S, Hamasaki N, Suzuki M, Kimura M, Shirai H, Hanabusa K (2002) J Am Chem Soc 124:6550
99. Kobayashi S, Hanabusa K, Hamasaki N, Shirai H, Shinkai S (2000) Chem Mater 12:1523
100. Moreau JJE, Vellutini L, Wong Chi Man M, Bied C (2001) J Am Chem Soc 123:1509
101. Tanev PT, Ling Y, Pinnavaia TJ (1997) J Am Chem Soc 119:8616
102. Caruso F (2000) Chem Eur J 6:413
103. Caruso F, Caruso RA, Möhwald H (1998) Science 282:1111
104. Ozin GA (1997) Acc Chem Res 30:313
105. Oliver S, Kuperaman A, Coombs N, Ozin GA (1995) Nature 378:47
106. Jung JH, Nakashima K, Shinkai S (2001) Nano Lett 1:145
107. Jung JH, Ono Y, Shinkai S (2000) Chem Lett 636
108. Tamaru S, Takeuchi M, Sano M, Shinkai S (2002) Angew Chem Int Ed 41:853
109. Asai M, Fujida N, Sano M, Shinkai S (2003) J Mater Chem 13:2145
110. Oda R, Huc I, Candau SJ (1998) Angew Chem Int Ed 37:2689
111. Friggeri A, Gronwald O, van Bommel KJC, Shinkai S, Reinhoudt DN (2001) Chem Commun 2434

Author Index Volumes 201–248

Author Index Vols. 26–50 see Vol. 50
Author Index Vols. 51–100 see Vol. 100
Author Index Vols. 101–150 see Vol. 150
Author Index Vols. 151–200 see Vol. 200

The volume numbers are printed in italics

Subject Index

The manufacturer's authorised representative in the EU is Springer Nature Customer Service Centre GmbH, Europaplatz 3, 69115 Heidelberg, Germany. If you have any concerns regarding our products, please contact ProductSafety@springernature.com

Printed and bound by CPI Group (UK) Ltd, Croydon, CR0 4YY
07/01/2026
02031056-0011